Hydroprocessing Catalysts and Processes

The Challenges for Biofuels Production

CATALYTIC SCIENCE SERIES

ISSN 1793-1398 (Print)
ISSN 2399-4495 (Online)

Series Editor: Graham J. Hutchings *(Cardiff University)*

Catalysis is at the forefront of the chemical industry and is essential to many fields in the chemical sciences. This series explores all aspects of catalysis in authored and edited volumes drawing on expertise from around the globe in a focussed manner. Volumes are accessible by postgraduate students and professionals in academia and industry.

Published

More information on this series can be found at http://www.worldscientific.com/series/css

(Continued at end of book)

CATALYTIC SCIENCE SERIES — VOL. 17

Series Editor: Graham J. Hutchings

Hydroprocessing Catalysts and Processes

The Challenges for Biofuels Production

edited by

Bo Zhang

Wuhan Institute of Technology, China

Duncan Seddon

Duncan Seddon & Associates Pty. Ltd., Australia

World Scientific

NEW JERSEY · LONDON · SINGAPORE · BEIJING · SHANGHAI · HONG KONG · TAIPEI · CHENNAI · TOKYO

Published by

World Scientific Publishing Europe Ltd.

57 Shelton Street, Covent Garden, London WC2H 9HE

Head office: 5 Toh Tuck Link, Singapore 596224

USA office: 27 Warren Street, Suite 401-402, Hackensack, NJ 07601

Library of Congress Cataloging-in-Publication Data

Names: Zhang, Bo (Writer on energy resources), editor. | Seddon, Duncan, editor.
Title: Hydroprocessing catalysts and processes : the challenges for biofuels production /
    edited by Bo Zhang (Wuhan Institute of Technology, China),
    Duncan Seddon (Duncan Seddon & Associates Pty. Ltd., Australia).
Description: New Jersey : World Scientific, 2018. | Series: Catalytic science series ; volume 17 |
    Includes bibliographical references.
Identifiers: LCCN 2017052793 | ISBN 9781786344830 (hc : alk. paper)
Subjects: LCSH: Heterogeneous catalysis. | Catalysis. | Liquid fuels. |
    Petroleum--Refining. | Biomass energy.
Classification: LCC QD505 .H935 2018 | DDC 541/.395--dc23
LC record available at https://lccn.loc.gov/2017052793

British Library Cataloguing-in-Publication Data
A catalogue record for this book is available from the British Library.

For any available supplementary material, please visit
http://www.worldscientific.com/worldscibooks/10.1142/Q0141#t=suppl

Desk Editors: Sundar Madhan/Jennifer Brough/Koe Shi Ying

Typeset by Stallion Press
Email: enquiries@stallionpress.com

Preface

The demand for hydroprocessing catalysts has shown an increasing trend, because of their applications in refining of petroleum and biofuels, in order to comply with the strict environmental regulations controlling the emissions from transportation vehicles. Transportation fuel standards are constantly changing to meet new environmental targets whilst also improving the efficiency of engines. New challenges in petroleum refining and biorefining require improvements of currently used catalysts and development of entirely new catalyst formulations.

This book provides an introduction to the mechanism of hydroprocessing reactions, application of different metals in hydroprocessing, the effect of catalyst supports, applications in refining new feedstock, renewable fuels standards, the management of spent hydroprocessing catalysts, and hydrogen production. The first chapter provides an overview of hydroprocessing chemistry. This includes hydrodesulfurization, hydrodenitrogenation, hydrodeoxygenation, hydrogenation, hydrocracking, and hydrodemetallization. Reactivity and reaction pathways of typical chemicals resulting from biofeedstock and petroleum are discussed. The surface chemistry of hydroprocessing catalysts and hydroprocessing catalyst models are also covered. The second chapter focuses on the challenge of instability of the pyrolytic bio-oil in the hydrotreating technology. The physical and chemical methods for bio-oil stabilization are summarized in detail, with a specific focus on bio-oil catalytic hydrogenation for stabilization. Chapter 3 reviews current developments in the species and application of inexpensive Ni-based nonsulfided hydroprocessing catalysts. Chapter 4 gives an example of the development of cobalt–molybdenum carbide catalysts for hydroprocessing bio-oil. Chapter 5 reviews the state-of-the-art in hydroprocessing of microalgae-based biofuels, as well as the catalyst development, the effect of process parameters on hydrotreated algal fuels, and the standards suitable for characterization of algal biofuels. Chapter 6 emphasizes the importance of the support effects of hydroprocessing catalysts. Chapter 7 introduces the development and status of commercial hydroprocessing processes for biofeedstock as described in the patent literature and commercial brochures. Chapter 8 discusses current and future renewable fuels standards and illustrates how the hurdles experienced by ethanol and biodiesel blends have been overcome. Chapter 9 introduces the management of spent hydroprocessing catalysts. Chapter 10 reviews the current development of hydrogen production technologies that is required to fulfill the need of hydroprocessing processes.

The completion of this book would not have been possible without assistance from a large number of people. Most important are the contributors, who prepared their work in a timely and professional manner. We are also very grateful to all the peer reviewers whose time and efforts in evaluating individual chapters have enhanced the quality of this book. We express our sincere appreciation to the School of Chemical Engineering & Pharmacy at the Wuhan Institute of Technology for the general support of this book project.

Bo Zhang
Duncan Seddon

About the Editors

Dr. Bo Zhang is the Chutian Scholar Distinguished Professor in Chemical Engineering at the Wuhan Institute of Technology (China) and a registered Professional Engineer in North Carolina (USA). He earned his Ph.D. at the prestigious Department of Chemical Engineering and Materials Science of the University of Minnesota (USA). He has authored more than 30 peer-reviewed publications and holds many issued patents. He is the editor-in-chief of *Trends in Renewable Energy*, and the editor of *Biomass Processing, Conversion and Biorefinery* (2013). He is a senior member of American Institute of Chemical Engineers (AIChE). His main research interests include developing new and improving existing biological and thermochemical technologies for converting biomass into useful products for fuels or chemical applications and improving fundamental understanding and developing new technologies for upstream bioprocess engineering.

Dr. Duncan Seddon's industrial career started with ICI Petrochemicals Division in 1974; he moved to ICI Australia in 1980 and worked on the conversion of natural gas to methanol and olefins. In 1983, he moved to BHP and worked on converting gas to liquid transport fuels. Since 1988, he has been practicing as an independent consultant offering a broad range of services to companies and government bodies with an interest in refining, petrochemicals and fuels, including chemical additives for clean fuel production. He is the author of over 120 papers, patents and articles and two books — *Gas Usage & Value: The Technology and Economics of Natural Gas Use in the Process Industries* (PennWell, 2006) and *Petrochemical Economics: Technology Selection in a Carbon Constrained World* (Imperial College Press, 2010). He is a fellow of the Royal Australian Chemical Institute and a Member of the Society of Petroleum Engineers.

Contents

Chapter 1

Hydroprocessing and the Chemistry

Bo Zhang

School of Chemical Engineering and Pharmacy,
Wuhan Institute of Technology, Hubei, China
bzhang_wh@foxmail.com

Abstract

Hydroprocessing is a conventional operation in the petroleum refinery. Recent developments in biofeedstock-derived fuels and stringent environmental legislation brought up new challenges in this traditional area. This chapter reviews hydroprocessing reactions including hydrodesulfurization, hydrodenitrogenation, hydrodeoxygenation, hydrogenation, hydrocracking, and hydrodemetallization, reactivity and reaction pathways of typical chemicals resulted from biofeedstock and petroleum, surface chemistry of hydroprocessing catalysts, and hydroprocessing catalyst models.

Keywords: Hydrodesulfurization, Hydrodenitrogenation, Hydrodeoxygenation, Hydrogenation, Hydrocracking, Hydrodemetallization, Hydroprocessing catalyst model

1.1. Introduction

Hydroprocessing refers to a variety of catalytic hydrogenation (HYD) processes which saturate unsaturated hydrocarbons and remove sulfur (S), nitrogen (N), oxygen (O), and metals from petroleum streams in a refinery [1]. Most recently, the hydroprocessing process has been applied to the biobased feedstock such as vegetable oils and algal oil. Generally, hydroprocessing can be categorized as hydrotreating and hydrocracking. Hydrotreating removes heteroatoms and metals and saturates carbon–carbon bonds with minimal cracking, while hydrocracking that is conducted under more severe conditions breaks carbon–carbon bonds and drastically reduces the molecular weight of the feed [2].

1.1.1. *Hydroprocessing of fossil fuels*

Historically, the hydroprocessing technology was first applied to coal HYD, which was developed by a French chemist, Pierre Eugène Marcellin Berthelot [3]. Later, the industrial development of the coal-HYD process was accomplished by a German scientist, Friedrich Karl Rudolf Bergius, who invented the Bergius process [4]. This HYD process converted coal, tar, or heavy oil into gasoline, and it was found that cobalt (Co), molybdenum (Mo), and tungsten (W) sulfides were especially active catalysts [5]. The patent of the Bergius process was sold to the BASF, and several plants were built with an annual capacity of 4 million tons of synthetic fuel before World War II [6]. After World War II, a "hydroforming process" using the Co–Mo catalysts was investigated by Union Oil (US) for the decomposing all types of sulfur compounds of petroleum fractions [7]. Since then the major use of mixed sulfides has been shifted to hydroprocessing of petroleum feedstocks with Co or nickel (Ni) promoted Mo or W catalysts usually supported on γ-Al_2O_3. The basic components of hydroprocessing catalysts have remained the same for over 50 y.

Environmental regulations are the major driving force for the hydroprocessing operations. For example, fuel quality and it emissions are regulated by US Environmental Protection Agency (EPA) under the authority of the Clean Air Act Amendments of 1990 in the United States. Environmental regulations limit sulfur levels in diesel fuels. A sulfur limit of 500 ppm (0.05 wt.%) was applied for the "low-sulfur" diesel fuel used in heavy-duty highway engines in October 1993. Beginning in June 2006, the maximum sulfur level of 15 ppm was enforced for the ultra-low-sulfur diesel (ULSD) used as the highway diesel fuel. For gasoline, the refinery annual averages required by the Tier 2 and Tier 3 gasoline sulfur standards, which have been effective from Jan 1, 2017 for most refiners, are 30 ppm and 10 ppm, respectively [8]. Table 1.1 summarizes the trends of

Table 1.1. Sulfur standards for diesel in selected regions and countries. (Adapted with permission from Ref. [9]. Copyright © 2010 Elsevier.)

Regions/countries	Sulfur (ppm)	Year of implementation
Argentina	50	2008
Brazil	500	2008
	50	2009
Chile	350	2007
	50	2010
Mexico	500	2005
	15	2009
Peru	50	2010
Uruguay	50	2009
China	20,000	2000
	500	2005
	50	2012
Hong Kong	50	2007
India	350	2005
	350	2010
	50	2010
Singapore	50	2005
Taiwan	50	2007
Australia	500	2002
	10	2009
Kuwait	2,000	2008
	50	2010
Saudi Arabia	800	2008
	10	2013

(Continued)

Table 1.1. (*Continued*)

Regions/countries	Sulfur (ppm)	Year of implementation
Bahrain, Lebnon, Oman, Qatar, UAE	50	2008
Qatar, UAE	10	2010
Bahrain	10	2013
Iran	50	2008
Jordon	10,000	2008
	350	2012
Russia	500	2008
	50	2010
EU	500	1996
	350	2000
	50	2005
	15	2009
USA	500	1993
	15	2006
Japan	500	1997
	50	2003
	15	2005

diesel sulfur fuel specification for highway transportation vehicles in selected regions and countries [9]. To date, most countries require the sulfur content in transportation fuels to be below 50 ppm.

Thus, hydroprocessing is a more important unit operation in petroleum refining industry and synthetic crude upgrading [10]. The increased use of refinery catalysts can help refiners meet fuel standards, manage operational efficiency, and enhance conversion and selectivity. Such demands require that the hydroprocessing catalysts possess not only high activity but also different activity profiles with respect to various functionalities [11]. Accordingly, the worldwide hydrotreating catalyst consumption in 1990 was approximately 35,000 t/y and in the range of 150,000–170,000 t/y in 2007 with an anticipated 4–5% annual increase [12].

1.1.2. *Hydroprocessing of biomass-derived fuels*

The study on conversion of cellulosic biomass into liquid fuels (i.e., bio-oil or biocrude oil) has a history of over 70 y [13, 14]. Over the past 35 y, hydroprocessing including catalytic hydrotreating and hydrocracking of bio-oil in both batch-fed and continuous flow bench-scale reactor systems has been underway, primarily in the US and Europe [15]. This application of hydroprocessing is an extension of petroleum processing.

Typical thermochemical processes for production of bio-oils from biomass include hydrothermal liquefaction and pyrolysis. Both processes convert feedstock to liquid products. An advantage of the thermochemical process is that it is relatively simple, usually requiring only one reactor and thus having a low capital cost. However, this process is nonselective, producing a wide range of products including a large amount of bio-oil [16].

Hydrothermal liquefaction was historically lined to HYD and other high-pressure thermal decomposition processes that employed reactive hydrogen or carbon monoxide carrier gases to produce a liquid fuel from organic matter at moderate temperatures (300–400°C), longer residence times (0.2–1.0 h), and relatively high operating pressure (5–20 MPa) [17]. Drying the feedstock is not needed for the hydrothermal process, which makes it especially suitable for naturally wet biomass. However, a reducing gas and/or a catalyst are often included in the process to increase the oil yield and quality.

Pyrolysis is the thermal decomposition of dry organic matter occurring in the absence of oxygen [18]. Pyrolysis can be further categorized as slow pyrolysis and fast (flash) pyrolysis. Slow pyrolysis is characterized by a relative low heating rate (less than 10°C/s) and long gas and solids residence time, resulting in more tar and char than fast pyrolysis. The primary products are tar and char. Flash pyrolysis describes the rapid, moderate temperature (450–600°C) pyrolysis that produces quantities of liquids. With a fast heating rate of 100–10,000°C/s and a short residence times (less than 2 s), the oil products are maximized the expense of char and gas. Fast pyrolysis of biomass has been demonstrated by multiple institutes and companies [19].

Biomass-derived oils (bio-oils) contain complex mixtures of reactive oxygenate compounds, such as carboxylic acids, aldehydes, ketones, furans, sugars, carbohydrates, and water [20]. The presence of these high oxygen content compounds makes the bio-oil unsuitable to be directly used as the transportation fuel. Generally, there are two routes for catalytic treating of bio-oil: hydrodeoxygenation (HDO) and cracking. Because HDO produces high grade oil product, it is often considered as the most feasible and competitive route for upgrading bio-oil to transportation fuels [21–23].

1.1.3. *Hydroprocessing of vegetable oils*

Due to factors such as capital interests, fluctuations in oil prices, and other geopolitical issues, there is a tremendous growth in biofuel industry globally. Considerable work has been done focusing on biobased fuels (like green diesel) production from hydroprocessing of rapeseed oil, soybean oil, palm oil, karanja oil, and waste cooking oil [24]. Vegetable oils are ideal sources of diesel fuels because of their chemical structures with 16–24 carbon atoms. Choice of feedstock depends on crop growing pattern of local regions. Soybean oil is a potential source in United States, while production of rapeseed oil, palm oil, and sunflower oil dominates in Europe [25].

HDO can be a promising process to remove oxygen content from the feedstock, thus aiding in the production of biobased transportation fuels. The HDO technology for biobased fuels that contain triglycerides is nearly mature. Industrialization of HDO process has been done by multiple companies such as Neste oil and ENI.

1.1.4. *Hydroprocessing of algae-derived fuels*

Algae are recently considered as a promising biofuel feedstock due to their superior productivity, high oil content, and environmentally friendly nature [26, 27]. The algal technology for biofuels production was greatly advanced in the past decade [28]. Two algae conversion technologies, namely the algal lipid upgrading (ALU) process and the hydrothermal liquefaction (HTL) process, are chosen by the US Department of Energy as the two most promising approaches [29–31]. The ALU process yields algal crude lipids that are similar to vegetable oils, and only HDO is needed to upgrade the algal crude lipids. But algal bio-oils produced via HTL contain heterocyclic nitrogenates (such as pyrroles, indole, pyridines, pyrazines, imidazoles, and their derivatives) [32], cyclic oxygenates (like phenols and phenol derivatives with aliphatic side chains), and cyclic nitrogen and oxygenated compounds (e.g., pyrrolidinedione, piperidinedione, and pyrrolizinedione compounds) [33]. Algae-derived bio-oils require further catalytic hydroprocessing to remove both oxygen and nitrogen (i.e., hydrodenitrogenation or HDN) [34].

A range of heterogeneous catalyst materials have been tested, including conventional sulfided catalysts developed for petroleum hydroprocessing and precious metal catalysts. The important processing differences have been identified, which required adjustments to conventional hydroprocessing as applied to petroleum feedstocks. However, the technology is still under development, but it can play a significant role in supplementing increasingly expensive petroleum.

1.2. Hydrodesulfurization (HDS)

For a long time, the most important hydrotreating reaction has been the removal of sulfur from various fuel fractions, i.e., hydrodesulfurization (HDS). Sulfur is the most abundant heteroatom impurity in the crude oil, ranging from 0.1 wt.% to 2–5 wt.%. The nitrogen content of the crude oil is usually between 0.1 wt.% and 1 wt.%, while the oxygen content is often less than 0.1 wt.%. Some typical sulfur containing compounds found in petroleum fractions are summarized in Table 1.2.

HDS of model chemicals has been comprehensively studied. Generally, HYD of the heteroatom ring occurs prior to breaking of C–S, C–N or C–O bond (hydrogenolysis), because the bond between the heteroatom and an aromatic-type carbon atom (sp^2 bonding) is stronger than that with an aliphatic-type carbon (sp^3 bonding). And no fundamental difference in the reaction pathway or the mechanism was found when applying catalysts of Mo/Al_2O_3, $CoMo/Al_2O_3$, $NiMo/Al_2O_3$ and NiW/Al_2O_3 catalyst.

1.2.1. *Reactivity*

HDS reactivity depends critically on the molecular size and the structure of the sulfur-containing compounds. When HDS was catalyzed by sulfided $CoMo/\gamma$-Al_2O_3, the desulfurization rates had following order:

Thiophene > benzothiophene > dibenzothiophene [35].

Table 1.2.   Some typical sulfur-containing compounds found in petroleum fractions.

Thiols (mercaptans)	R–SH
Sulfides	R–S–R′
Disulfides	R–S–S–R′

Thiophenes

Thiophene	2-Ethylthiophene	2-Propylthiophene

Benzothiophenes

Benzothiophene	3-Methyl-1-benzothiophene	3-Ethyl-2-methyl-1-benzothiophene

Dibenzothiophenes

Dibenzothiophenes	4-Methyldibenzo-thiophene	4,6-Dimethyldibenzo-thiophene

The hydrogenated compound (e.g., tetrahydrothiophene) was more reactive than the corresponding aromatic compound (e.g., thiophene). Methyl substituents on benzothiophene had almost no effect on reactivity, whereas methyl substituents on dibenzothiophene (DBT) located at a distance from the S atom slightly increased the reactivity, and those in the 4-position or in the 4 and 6 positions significantly retarded HDS [36].

The increased reactivity attributed to the increased electron density on the S atom (i.e., inductive effect), those adjacent to the S atom decrease reactivity due to steric effects. The typical sample of the steric effect is 4,6-dimethyldibenzothiophene (46DMBT), which was roughly 10 times less active than the compound without methyl groups (i.e., DBT). During hydroprocessing of the real feeds, alkyl- and dialkyl-substituted DBTs were the most difficult desulfurization compounds. The dialkyl-substituted DBTs in a light oil (especially 46DMBT) remained intact until the final stage of the reaction (390°C), while alkyl-benzothiophenes were completely desulfurized at 350°C [37].

1.2.2. *Reaction pathways*

HDS of thiophene forms butene via two alternative pathways (Figure 1.1). The route through butadiene (a, b) was suggested to be the major pathway, due to the presence of butadiene and the absence of tetrahydrothiophene in atmospheric pressure studies [38, 39]. HDS of benzothiophene was suggested to follow a similar route, because of the presence of styrene and ethylbenzene in the products [36]. Accordingly, the HDS of thiophenic compounds, in which the C–S bond (sp^2) is relatively weak, could directly result in sulfur removal without saturation of the heteroatom ring. Molecular orbital calculations showed that the C–S bond cleavage need not occur prior to ring HYD and a concurrent HYD of the γ-carbons may occur [40]. The HDS of thiophene in the gas phase may directly form butene (pathway f) [41]. HDS of tetrahydrothiophene at low pressure via either pathway (e, b) or (c, a, b) was suggested by detection of thiophene and butadiene, while HDS of thiophene under high pressure yielded mostly tetrahydrothiophene (pathway c).

HDS of DBT catalyzed by sulfided CoMo/γ-Al$_2$O$_3$ at 300°C and 10 MPa could follow two parallel pathways (Figure 1.2): the predominant reaction of direct desulfurization (DDS) removes S as H$_2$S, leading to the major product of biphenyl; and HYD to form hexahydrodibenzothiophene, which is followed by hydrogenolysis to yield a minor product of cyclohexylbenzene [42]. In terms of the probability of an electron, the electron density of the sulfur atom in DBT is delocalized over an

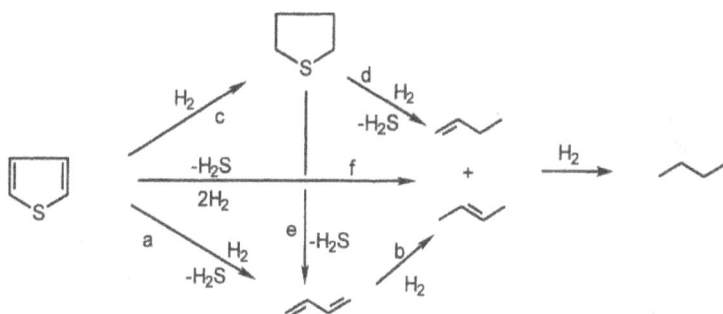

Figure 1.1. Proposed reaction pathways for thiophene HDS. (Reprinted with permission from Ref. [1]. Copyright © 1996 Springer.)

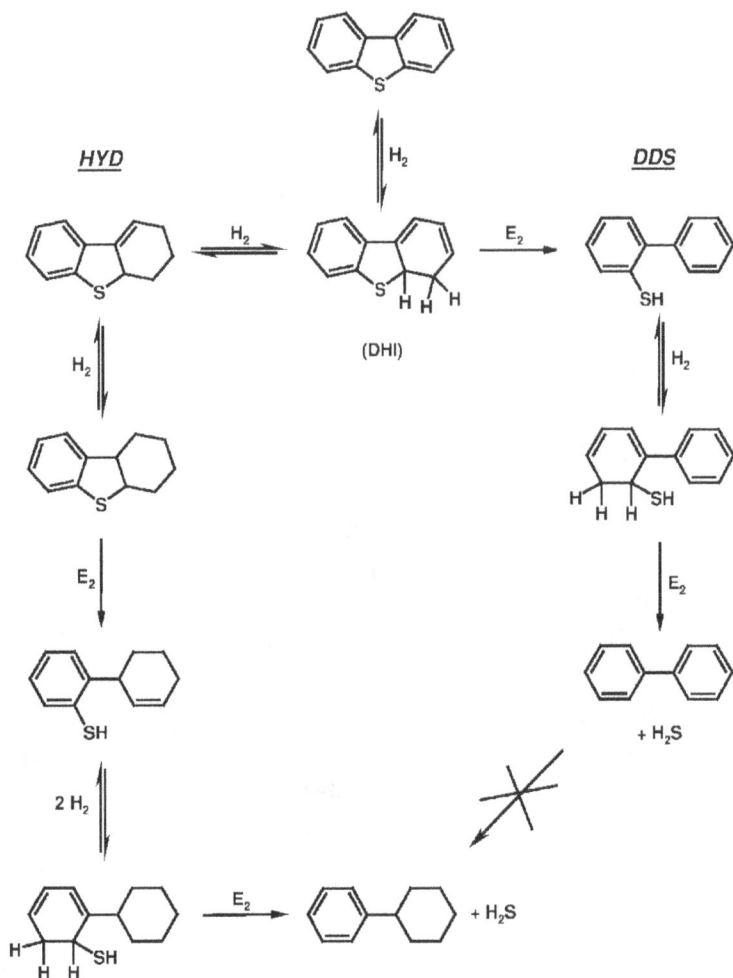

Figure 1.2. Reaction network for the HDS reaction of DBT via two possible pathways (DDS and HYD) with dihydrodibenzothiophene as intermediate. (Reprinted with permission from Ref. [44]. Copyright © 2001 Springer.)

extensive π system and steric crowding around the sulfur atom is present, the first step in the reaction pathway is probably a π-complex formation which is followed by HYD and desulfurization [43].

HDS of the refractory compounds such as 4,6-dimethyldibenzothiophene (46DMDBT) is often required for deep HDS of diesel fuels. It is generally agreed that the low reactivity of these compounds due to steric hindrance of the transition state particularly inhibits the C–S bond cleavage through DDS pathway. The reactivity of 46DMDBT can be improved by increasing the HYD function of a catalyst or using an acidic component (like zeolites) in the catalyst [45, 46]. The acidic component can enhance 46DMDBT hydroisomerization, which shifts one substituting group from the 4 or 6 position to the 3 or 7 position, where the steric hindrance can be relieved and HDS reactions can more readily occur [47]. Reaction network of 46DMDBT HDS is summarized in Figure 1.3.

HDS of benzonaphthothiophene (benzo[b]naphtho[2,3-d]thiophene, a four-ring aromatic compound) gave the reaction network that is similar to DBT (Figure 1.4). But the parallel HYD and HDS of benzonaphthothiophene progressed at nearly equal rates [48].

1.2.3. *Surface chemistry*

Sulfur anion vacancies have for a long time been considered as the active sites for HDS [49]. For example, the perpendicular adsorption of thiophene may occur on these coordinatively unsaturated sites, and bonding through the S atom. During HDS, the adsorbed thiophene molecule undergoes reactions to form hydrocarbons, leaving the S atom located at the original vacancy. This S atom is subsequently hydrogenated to H_2S and desorbed, regenerating the vacancy. When pre-HYD is needed, a flatwise adsorption is more likely on multivacancy centers [50], and HYD occurs via SH groups or hybrid species. Thus, a model of two different sites for HDS and HYD was proposed and further used to explain the interactions between HDS of thiophene and HDN of pyridine [51].

The surface adsorption study showed that equilibrium constants increased strongly in the order of Aromatic hydrocarbons < Sulfur compounds < Oxygen compounds < Nitrogen compounds. Consequently, HDS is strongly inhibited by the presence of nitrogen compounds. For example, the presence of quinoline significantly decreased the selectivity for cyclohexylbenzene and biphenyl derivatives formed via HYD and hydrogenolysis, respectively [52]. But HDN is only slightly influenced by the presence of sulfur compounds [53]. It's also known that HDS is severely inhibited by aromatic compounds, particularly diaromatics like naphthalene [54]. In addition, the hydroprocessing reactions are generally inhibited by reactants and products such as H_2S, NH_3, and H_2O.

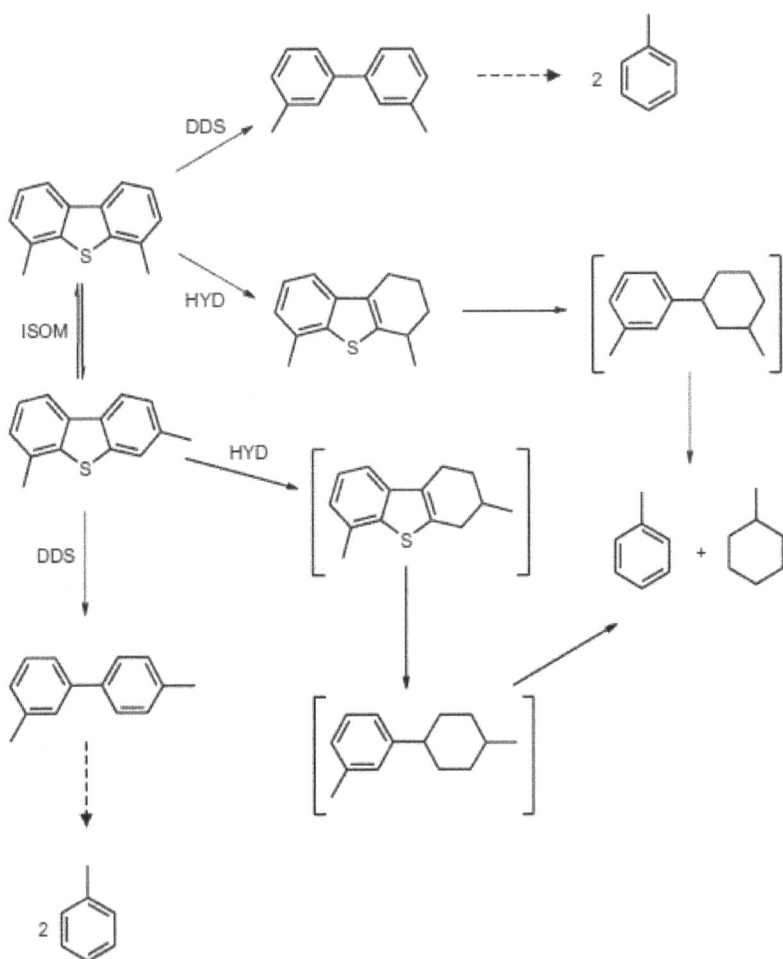

Figure 1.3.   Reaction network of HDS of 46DMDBT over acid-containing hydrotreating catalysts: isomerization (ISOM), DDS pathway (DDS), HYD. (Reprinted with permission from Ref. [45]. Copyright © 2003 Elsevier.)

1.3. Hydrodenitrogenation (HDN)

The nitrogen content in petroleum crudes is usually around 0.1 wt.%, but it may be as high as 1 wt.% in some cases. Processing heavy feeds has led to an increase in studies on the HDN of model compounds,

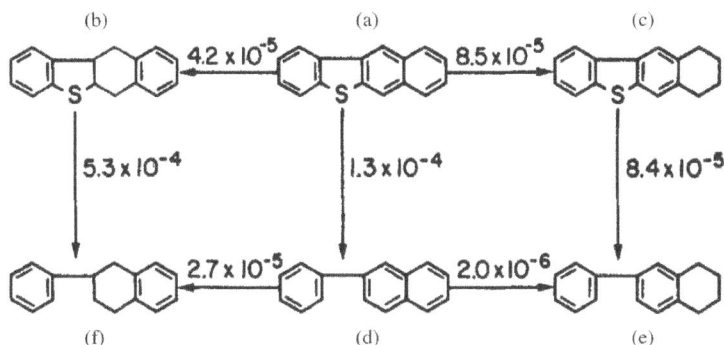

Figure 1.4. The reaction network for HDS of benzonaphthothiophene. The numbers next to the arrows are the pseudo-first-order rate constants in m^3/s kg catalyst. (a) Benzo[b] naphtho[2,3-d]thiophene, (b) 5a,6,11,11a-tetrahydrobenzo[b]naphtho[2,3-d]thiophene, (c) 7,8,9,10-tetrahydrobenzo[b]naphtho[2,3-d]thiophene, (d) 2-phenylnaphthalene, (e) 2-phenyltetralin, and (f) 6-phenyltetralin. (Reprinted with permission from Ref. [48]. Copyright © 1980 Wiley.)

which has been comprehensively reviewed by [53, 55, 56]. Nitrogen is generally concentrated in the cracked fractions or the fractions heavier than those containing sulfur. Approximately one third of these chemicals are present in the form of basic N compounds that contain the six-membered pyridinic ring, and the rest is mainly present in relatively nonbasic compounds containing the five-membered pyrrolic ring. These two groups of nitrogen heterocycles have different electronic configurations [55]. The unshared pair of electrons of N in six-membered ring compounds is not tied up in the π cloud and available for interacting with the acidic centers on the catalyst surface. The extra pair of electrons in N of a pyrrole ring are involved in the π cloud of the ring, and thus these nitrogen heterocycles may be primarily associated with the high electron density in the ring rather than with the nitrogen heteroatom. In terms of catalysts, nickel is generally a more active promoter for HDN than cobalt [57]. Some typical nitrogenated compounds are summarized in Table 1.3.

1.3.1. *Reactivities*

As aforementioned, HDS generally does not require a complete HYD of the ring prior to hydrogenolysis, while the HDN of heterocyclic

Table 1.3.   Some typical nitrogenated compounds.

Pyrrole   Indole   Carbazole

Pyridine   Quinoline

Acridine   Phenanthridine

Aniline   Diphenylamine

Porphyrin

nitrogenated compounds (even diphenylamine and aniline) via HYD is often inevitable [58, 59]. Although piperidine has an analogue structure to thiophene, the nitrogen removal from the piperidine ring is more difficult, which is mainly due to the thermodynamic constraints. The overall reaction rate may be partly governed by the equilibrium of the HYD of the N-containing ring [60]. Compared to HYD of aromatic ring, the N-containing ring is always hydrogenated faster [61].

In terms of relative reactivity for HYD of the N ring, Schulz *et al.* [53] gave following order:

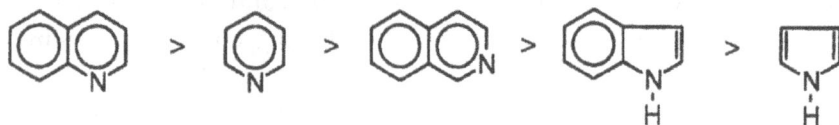

Quinoline > Pyridine > Isoquinoline > Indole > Pyrrole > Aromatic ring of Aniline > Aromatic ring. HYD of methyl-substituted quinolines can be greatly depressed by the methyl group adjacent to the N atom, whereas methyl groups at other positions enhanced the rate over that of quinoline. The reason can be explained as the synergistic effect of steric hindrance and the electron-withdrawing inductive effect.

1.3.2. *Reaction network*

Aliphatic amines are principal intermediates in ring-opening reactions of cyclic nitrogenated compounds, and often constituents of algae-derived fuels. Their reactivities are relatively high, and a complete conversion is possible below 300°C. Mechanism of HDN of pentylamine is shown in Figure 1.5. The heterolytic rupture leads to ammonia elimination and an olefin that could then be isomerized and hydrogenated. Both pentane and pentene-1 are primary products. The double-bond shift and amine dispro-portionation may take place too [62].

The aromatic ring of aniline is hydrogenated to cyclohexylamine. And then the conversion of cyclohexylamine is mainly through elimina-tion of NH_3 and simultaneous formation of cyclohexene (pathway 1), as shown in Figure 1.6. The reaction of hydrogenolysis of the $C(sp^2)$–N bond of aniline may occur only at temperatures above 300°C as a side reaction. N-phenylcyclo-hexylamine (3) is the predominant product at 250°C and only present in very small amounts at higher temperatures. The denitrogenation of anilines is a particularly slow reaction, so aniline is often used as a model compound for the testing of HDN catalyst activ-ity [53, 63].

Pyrrole denitrogenation involves a main slow HYD of the aromatic five-membered ring (1), cleavage of a C–N bond in pyrrolidine (2), and the further conversion of butylamine to butene and butane (Figure 1.7).

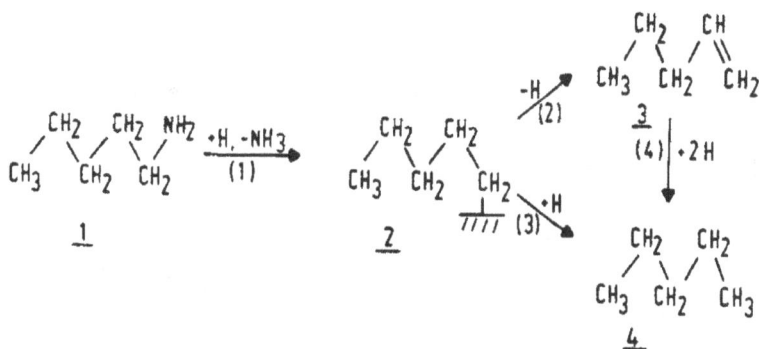

Figure 1.5. HDN of pentylamine. (Reprinted with permission of the authors from Ref. [53].)

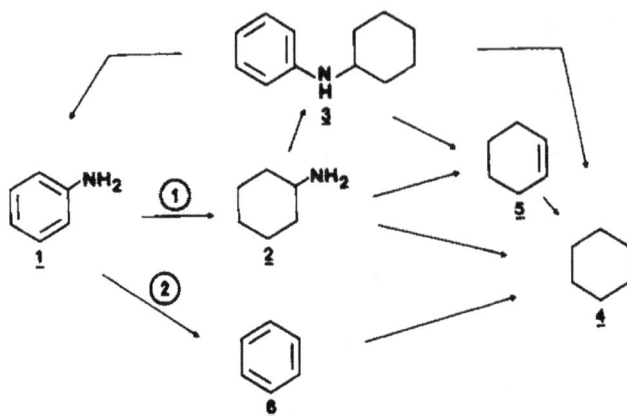

Figure 1.6. Reaction scheme for HDN of aniline. (Reprinted with permission from Ref. [59]. Copyright © 1987 Elsevier.)

The HDN process for pyrrole also involves reactions (1–4) of by-products formation [53].

Reaction network in HDN of indole is summarized in Figure 1.8. It follows the general pattern of reaction proposed for the HDN of N containing heterocyclics, i.e., HYD of the N ring followed by C–N bond breakage. The conversion of indole to 2,3-dihydroindole is reversible. By-products including aniline, toluidine, cyclohexane, methylcyclohexane, benzene, and toluene can form via the C–C bond rupture in 2,3-dihydroindole [64].

A preliminary investigation on carbazole denitrogenation indicated that the main product of carbazole HDN is bicyclohexyl [53]. Further studies observed the gradual HYD of carbazole involving at least four equilibrium systems, as shown in Figure 1.9 [65, 66]. The denitrogenated products include bicyclohexyl, 2-ethylbicyclo[4.4.0]decane, hexylcyclohexane, and a very small amount of cyclohexylbenzene, and the major hydrogenated compound was 1,2,3,4-tetrahydrocarbazole. The reaction scheme is summarized as follows: Carbazole is hydrogenated to tetrahydrocarbazole and further to perhydrocarbazole, which suffered C–N hydrogenolysis to form 1,1′-bicyclohexyl. At high temperatures, 1,1′-bicyclohexyl is either isomerized to 2-ethylbicyclo[4.4.0]decane or decomposed to hexylcyclohexane [67].

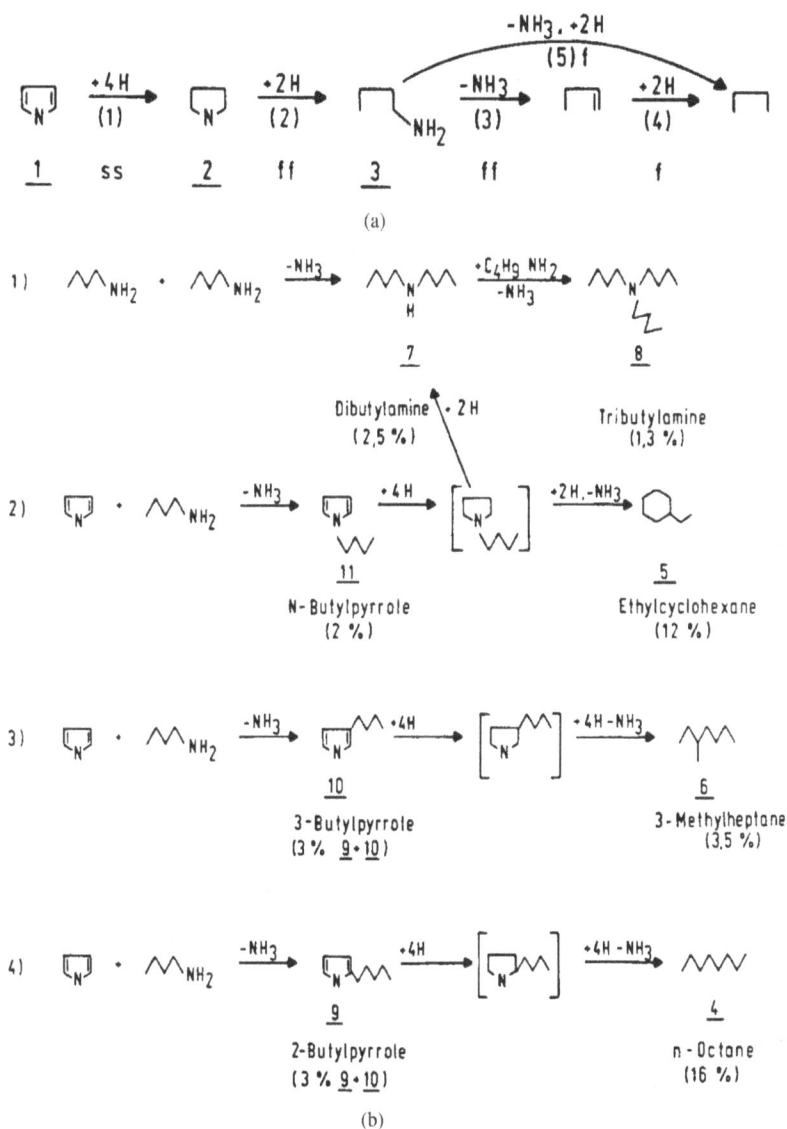

Figure 1.7. (a) HDN of pyrrole and (b) by-product formation. (Reprinted with permission of the authors from Ref. [53].)

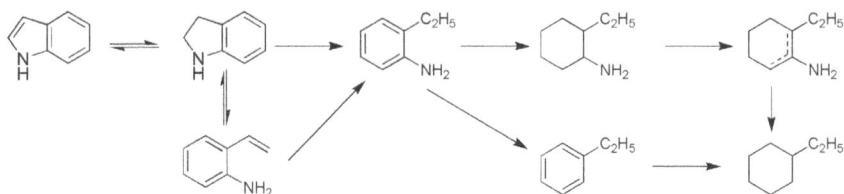

Figure 1.8.   Reaction network in HDN of indole. (Reproduced from Ref. [34] with permission from The Royal Society of Chemistry.)

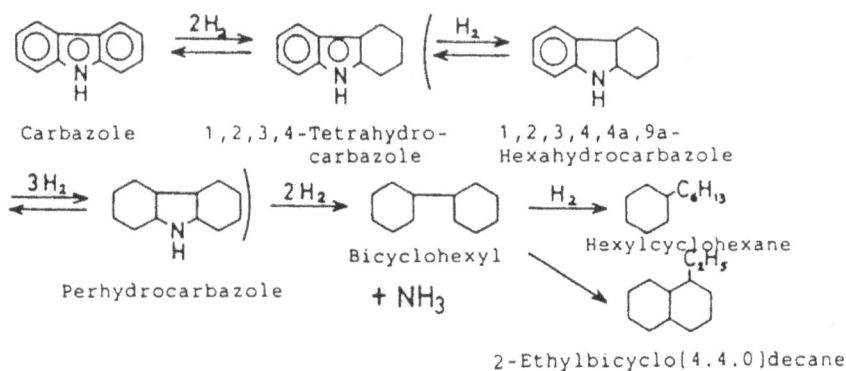

Carbazole   1,2,3,4-Tetrahydro-   1,2,3,4,4a,9a-
                        carbazole   Hexahydrocarbazole

Perhydrocarbazole   Bicyclohexyl   Hexylcyclohexane

+ NH₃

2-Ethylbicyclo(4.4.0)decane

Figure 1.9.   Reaction network of HDN of carbazole. (Adapted with permission from Ref. [67]. Copyright © 1993 Japan Petroleum Institute.)

The reaction mechanism for HYD of pyridine is shown in Figure 1.10. The reaction of pyridine to piperidine has often been regarded as a fast-reversible reaction. But equilibrium values are approached only at relatively high temperature (e.g., 350°C). The rates of the reactions 1 and 2 are similarly slow [68]. When the HDN is conducted under the hydrothermal conditions over a Pt/γ-Al$_2$O$_3$, intermediates including n-pentylamine, n-pentanol, and 1-pentene are present [69].

Reaction network for HDN of quinoline is shown in Figure 1.11. The HYD of quinoline to yield 1,2,3,4-tetrahydroquinoline (THQ1) is a very fast reaction, and can reach equilibrium at temperatures below 200°C. The reaction to 5,6,7,8-tetrahydroquinoline (THQ5) is a very slow reaction, which starts at about 350°C. This indicates a very much lower reactivity

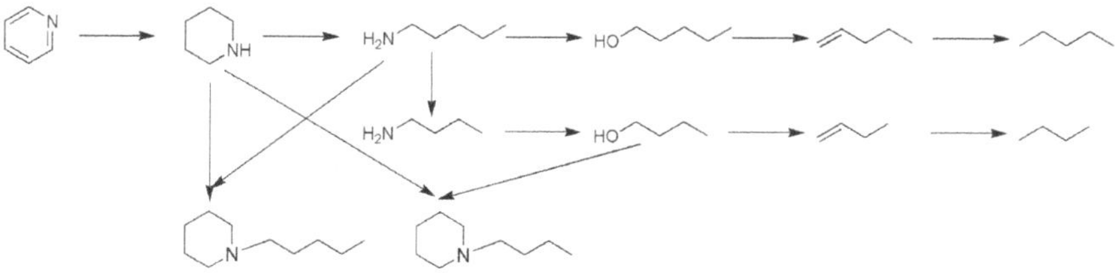

Figure 1.10. HDN of pyridine under hydrothermal conditions. (Reproduced with permission from Ref. [34]. Copyright © 2016 Royal Society of Chemistry.)

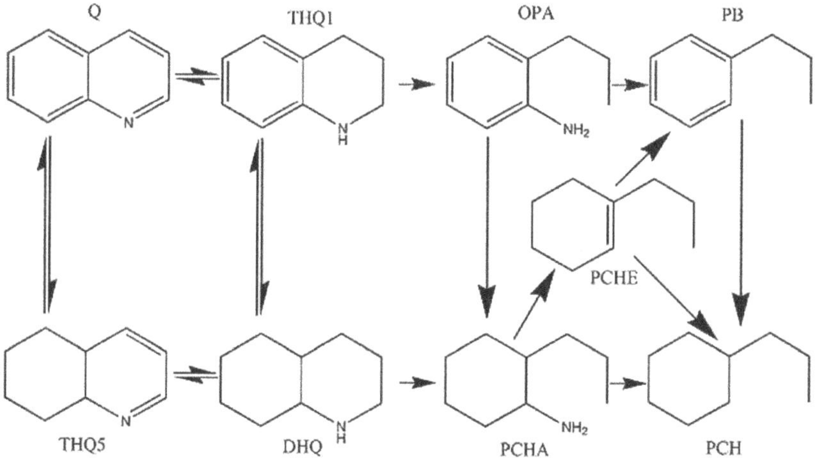

Figure 1.11. HDN reaction network of quinoline. Decahydroquinoline (DHQ), 1,2,3,4-tetrahydroquinoline (THQ1), 5,6,7,8-tetrahydroquinoline (THQ5), propylbenzene (PB), propylcyclohexane (PCH), 2-propyl-cyclohexylamine (PCHA), propylcyclohexene (PCHE) and ortho-propylaniline (OPA). (Reproduced with permission from Ref. [71]. Copyright © 2016 Royal Society of Chemistry.)

for the aromatic ring of quinoline. Both reactions to decahydroquinoline (DHQ) are very slow, however, reaction of THQ5 to DHQ starting at about 400°C is less slow. The ring opening reaction of THQ1 to OPA is a very slow irreversible reaction, happening only above 350°C. HYD of the aromatic ring of OPA needs about 400°C to form 2-propyl-cyclohexylamine (PCHA) that is unstable. Nitrogen is eliminated fast from PCHA as ammonia to yield propylcyclohexene (PCHE) [61, 70].

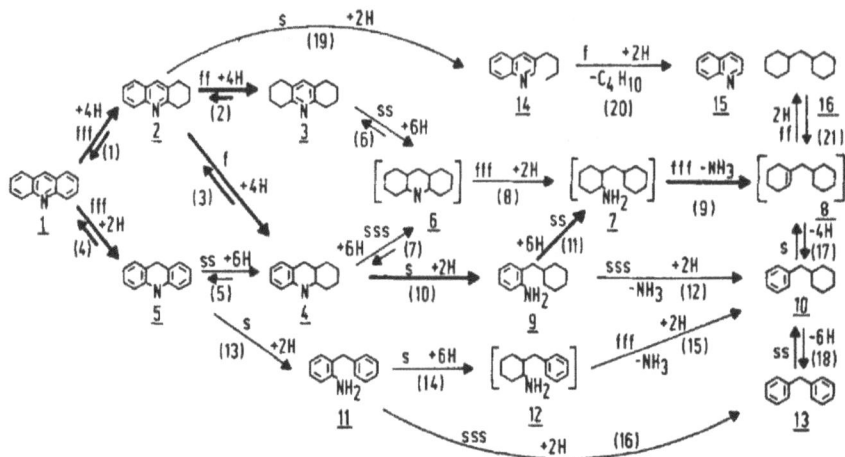

Figure 1.12. HDN of acridine (s: slow; f: fast). (Reprinted with permission of the authors from Ref. [53].)

Reaction network of HDN of acridine is shown in Figure 1.12 [53]. HYD of first two rings occur fast, and equilibrium is established at around 400°C. The C–N bond cleavage in 9, 10-dihydroacridine 5 happens at relatively low temperature, and is facilitated by weakening aniline type bonds due to the second aromatic ring. Overall, *o*-methylcyclohexylaniline 9 is a relatively stable intermediate, and its maximum concentration is attained only at the high temperature of 450°C. It is then hydrogenated to the very unstable saturated primary amine 7, which leads to other final products of cyclohexenylcyclohexylmethane 8, dicyclohexylmethane 16, cyclohexylphenylmethane 10, and diphenylmethane 13.

1.3.3. *Reaction sites*

It is generally assumed that N-cyclic compounds (like pyridine) adsorb perpendicular to the surface sites through the N lone-pair of electrons [72, 73]. Moreau and Geneste [74] proposed that the C–N bond rupture occurs via perpendicular adsorption through the N atom, and ring HYD happens via flatwise adsorption through the π-aromatic system. For N compounds with multiple rings, flatwise π-adsorption through the condensed aromatic ring system seems feasible. The energy of the perpendicular adsorption was lower than that of the flatwise adsorption.

It was proposed that HYD of N compounds and aromatics occurs at the same site, and hydrogenolysis happens at a different site. Yang and Satterfield suggested that the hydrogenolysis site is a Brønsted acid site, while sulfur vacancies are responsible for both HYD and hydrogenolysis reactions [75]. Rajagopal *et al.* reported that denitrogenation selectivity depends only on the concentration of Brønsted acid sites [76]. If a Brønsted site is involved in hydrogenolysis, it would appear to reside exclusively on the active phase itself.

Both the rim-edge model [77] and the (1010) edge plane model [78] of MoS_2 suggest that the C–N hydrogenolysis sites are located at the edges. After a N-ring HYD at the rim or corner, the molecule will change the adsorption mode on the surface to edge, which might result in a low HDN activity. If an SH group possessing Brønsted acid character can donate a proton to the adjacent coordinatively unsaturated sites (i.e., sulfur anion vacancies), a high HDN activity may be achieved.

Qu and Prins identified four catalytic sites for HDN over sulfided $NiMo/Al_2O_3$ [79]. The active site for $C(sp^3)$–N bond cleavage involves adjacent acid and base sites, where the site for $C(sp^2)$–N bond cleavage is highly unsaturated and accommodates aromatic amines in a flat adsorption mode. HYD of aromatics is preferred by high stacking of the MoS_2 slabs with more edge and corner sites, and HYD of alkenes is mainly on better dispersion and lower stacking of MoS_2 slabs.

1.3.4. *Mechanisms*

Nelson and Levy proposed mechanisms for C–N hydrogenolysis involving Brønsted acid sites via Hofmann elimination (reaction 1) or displacement reaction (reaction 2) (Figure 1.13) [80]. Both k_4 and k_6 were found to increase in the presence of H_2S, and it may be due to the increase in Brønsted acidity via dissociative adsorption of H_2S. Removal of the $-NH_2$ group from 2-propylaniline (IV) requires saturation of an aromatic ring.

Laine [81] proposed a mechanism for ring opening involving nucleophilic attack by H_2S on the metal-complexed ring, which resulted in enhanced HDN catalysis (Figure 1.14).

Portefaix *et al.* showed that the hydrogen atoms of the carbon in the γ position with respect to the nitrogen atom participate in the C–N bond cleavage, thus demonstrating a β-elimination Hofmann-type mechanism

$$(1)$$

$$(2)$$

$$H_2S \Leftrightarrow H^+ + {}^-HS \qquad (3)$$

$$(4)$$

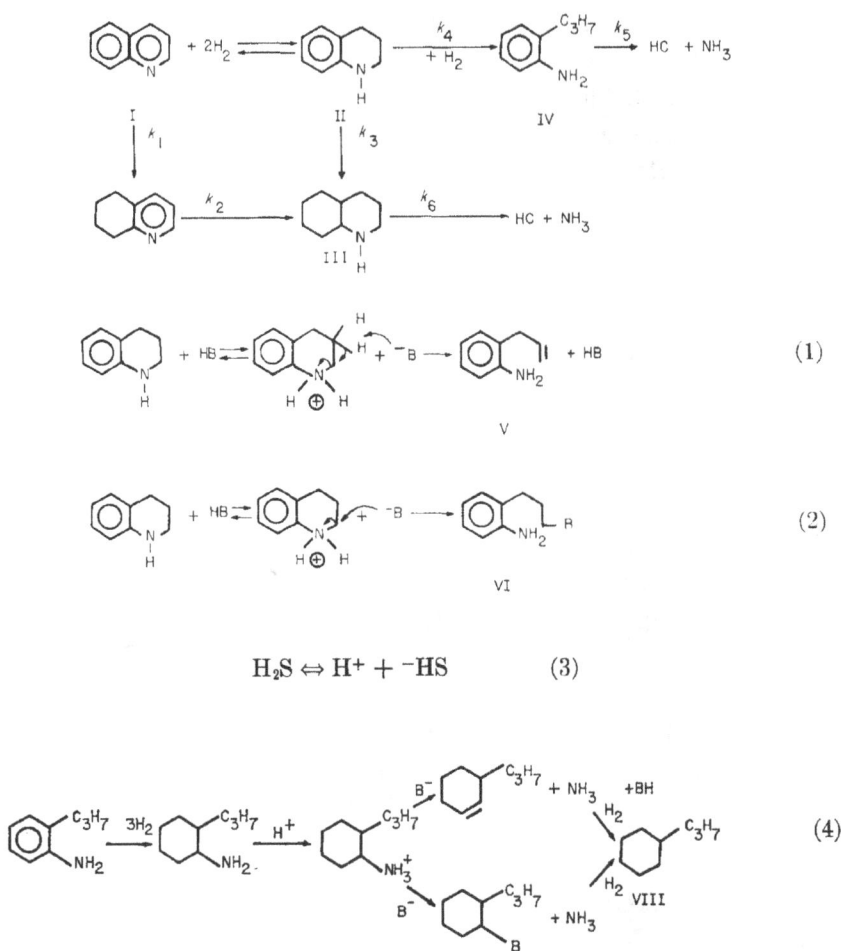

Figure 1.13. Mechanisms for C–N hydrogenolysis involving Brønsted acid sites. (Reprinted with permission from Ref. [80]. Copyright © 1979 Elsevier.)

[83]. As shown in Figure 1.15 [84], the first step of the carbon–nitrogen bond cleavage is the coordination by the nitrogen free electron pair of an amylamine molecule and a molybdenum atom having a coordination non-saturation. Next step involves activation of a hydrogen atom attached to the β carbon (C–H), and the activation of the hydrogen atom could be attributed to an S^{2-} group, an SH^- entity, or an adjacent Mo site. This hydrogen atom might attach to the bridged sulfur atom, and the electron

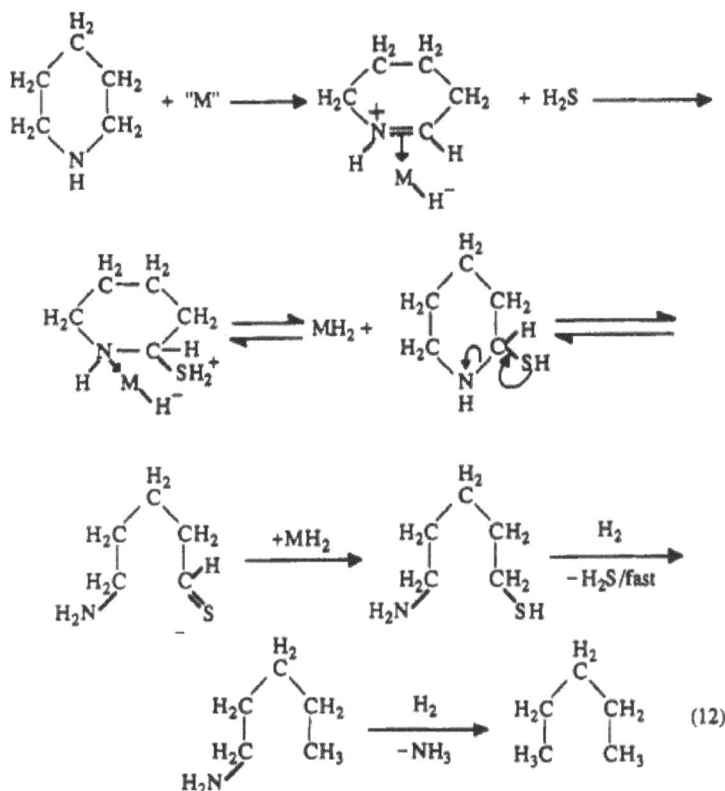

Figure 1.14. Mechanisms for C–N hydrogenolysis involving nucleophilic attack by H_2S. (Reprinted with permission from Ref. [82]. Copyright © 1992 American Chemical Society.)

pair shared between this atom and the molybdenum atom is responsible for the formation of a hydrosulfide. The anionic charge developed on the β carbon leads to the formation of a C=C bond, the C–N bond is cleaved, and the electron pair of this bond plays a part in the formation of a σ C–Mo bond. This process forms pentene, a NH_2 species bounded to the molybdenum atom, and a hydrosulfide (HS⁻) entity bridging two molybdenum atoms. Final products of pentane and NH_3 are formed via HYD and protonation, respectively.

Perot [56] suggested that HDN of 1,2,3,4-tetrahydroquinoline occurs either via the Hofmann elimination mechanism involving H⁺ as a cocatalyst (Figure 1.16(a)) or an S_N nucleophilic substitution by

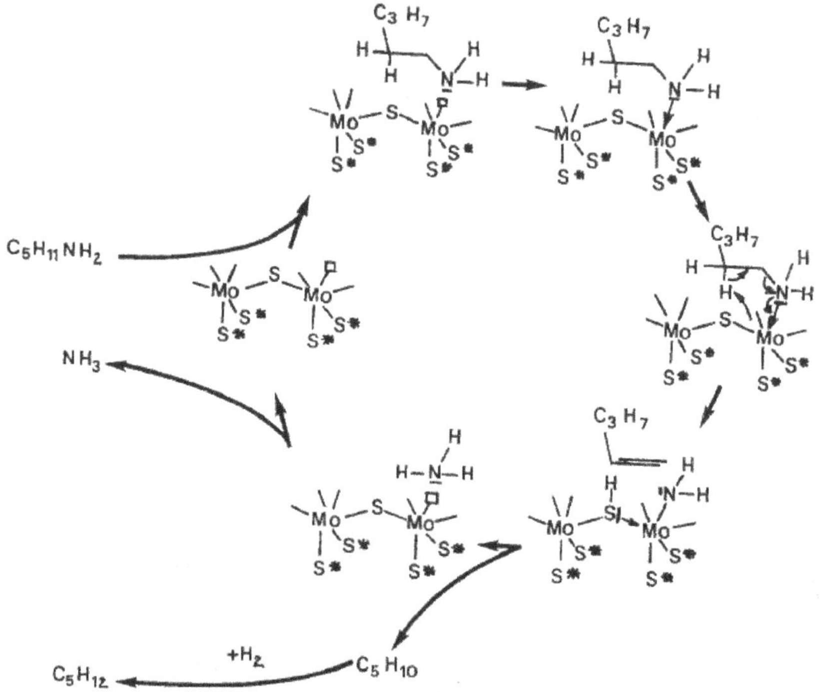

Figure 1.15. Mechanism proposed for C–N bond cleavage reaction during HDN of amylamine over a sulfided NiMo/Al$_2$O$_3$ catalyst. S* represents a sulfur atom of the bulk of the catalyst. (Reprinted with permission from Ref. [84]. Copyright © 1991 Springer.)

SH$^-$ followed by the hydrogenolysis of the C–S bond (Figure 1.16(b)). As shown in Figure 1.17, the HDN of 1,2,3,4-tetrahydroisoquinoline through an E$_2$ mechanism would lead to methyl-ethylcyclohexane, where its decomposition through an S$_N$ mechanism can lead to ortho-ethyltoluene [85].

The formation of HDN primary products may be also achieved via E$_1$ and S$_N$1 mechanisms (Figure 1.18). These reactions are initiated by a proton donated by the Brønsted acid site. The protonated amine releases NH$_3$ to form a stable carbocation (tertiary amines), which then reacts with a base (e.g., SH$^-$) to give either thiol (S$_N$1) or an alkene (E$_1$) [86]. Schwartz *et al.* explored different mechanisms via HDN of a series of isomeric amines, and concluded that the reaction occurred mainly through a β-elimination mechanism [87].

(a)

(b)

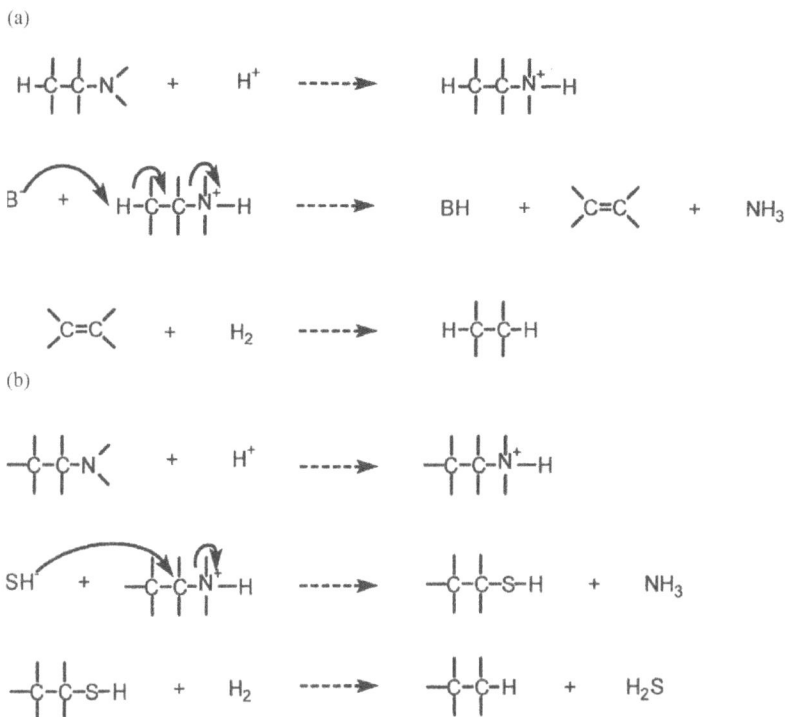

Figure 1.16. E_2 Hofmann-type elimination (a) and S_N nucleophilic substitution (b). (Reprinted with permission from Ref. [56]. Copyright © 1991 Elsevier.)

Interestingly, the CoMo catalysis of HDN is enhanced during the C–N bond breaking/ring-opening step of the saturated ring by the presence of H_2S and H_2O, which are normally catalyst poisons. The ring opening may be promoted through nucleophilic attack by H_2S, H_2O and NH_3 on the metal-complexed ring [81].

1.4. Hydrodeoxygenation (HDO)

Organic oxygen-containing compounds in petroleum feeds are present a relatively low content, mainly as phenolic compounds, ethers, and hetero-cyclics [88]. But the liquid bio-oil products generated from biomass are generally contain more than 10 wt.% oxygenated compounds, which makes these products highly unstable. Table 1.4 summarizes some typical compounds that require HDO.

(a)

(b)

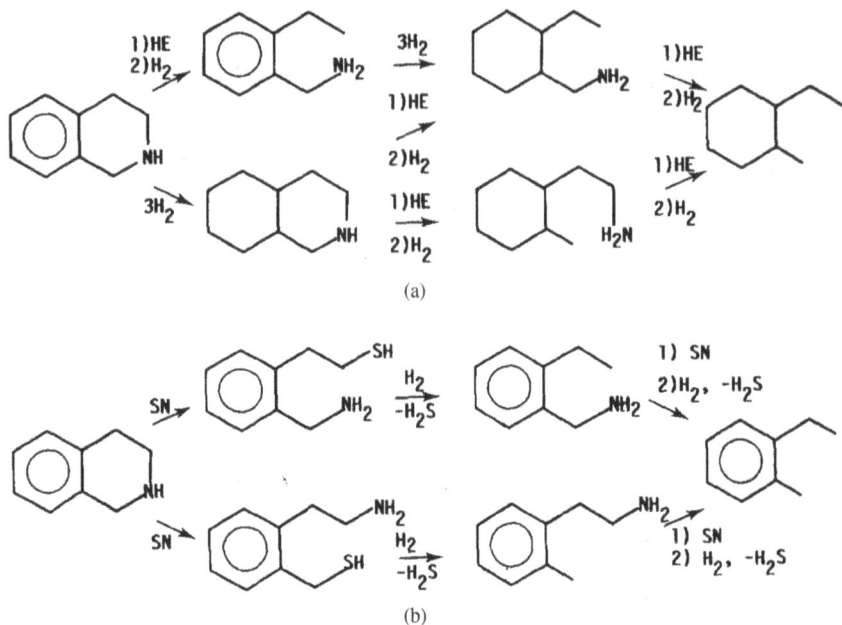

Figure 1.17. Mechanisms for HDN of 1,2,3,4-tetrahydroquinoline. (a) Hofmann-type elimination (HE) and (b) ethyltoluene formation through nucleophilic substitution (SN). (Reprinted with permission from Ref. [85]. Copyright © 1991 Elsevier.)

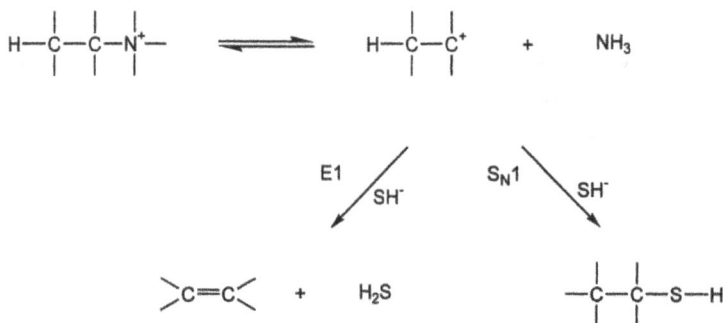

Figure 1.18. Formation of primary products during HDN of amines via E_1 and S_N1 mechanisms. (Reprinted with permission from Ref. [87]. Copyright © 1991 Elsevier.)

Table 1.4. Some typical compounds that require HDO.

Carboxylic acids	Aldehydes	Ketones	Ethers
Furan	Benzofuran	Dibenzofuran	
Phenol	*o*-Cresol	*m*-Cresol	*p*-Cresol
Anisole	Guaiacol	Naphthol	

1.4.1. Reactivities

Catalytic HDO over sulfided CoMo catalysts has been reviewed by Furimsky [89]. Relative stabilities of the oxygenated chemicals can be predicted according to the thermodynamic study on the equilibrium constants for transformations of model compounds. Although thermodynamic data could give relative trends of reactivities, the reactivities under hydrotreating conditions might be different because of the kinetic limitations.

Cronauer *et al.* [90] reported that the following order of reactivities of oxygen-containing compounds in the temperature range of 400–450°C:

Furans < Phenols < Ketones < Aldehydes < Alkyl ethers.

The reactivity of phenols depends on their structure. For example, anisole and guaiacol are more reactive than phenol and *o*-cresol [91], and the order of reactivities of ortho-substituted phenols was observed as

Tert-butyl phenol > Phenol > Methyl phenol > Ethyl phenol > Dimethyl phenol [92].

1.4.2. *Reaction pathways*

The HDO of phenols is known to proceed by the two reaction pathways shown in Figure 1.19 [93]. R can be hydrogen or an alkyl group (e.g., R = CH_3 for cresols).

The HDO of cresol over sulfided CoMo/γ-Al$_2$O$_3$ appears to go through two pathways (Figure 1.20): The fresh catalyst operating at lower temperatures (225–275°C) gives primarily a consecutive conversion

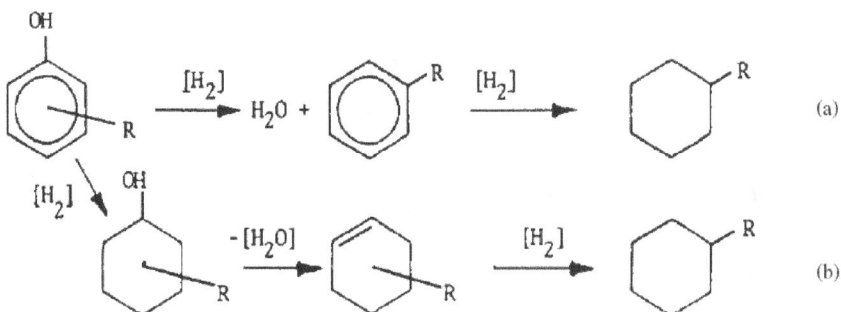

Figure 1.19. HDO of phenols. (a) Pathway 1 and (b) pathway 2. (Reprinted with permission from Ref. [93]. Copyright © 1983 Elsevier.)

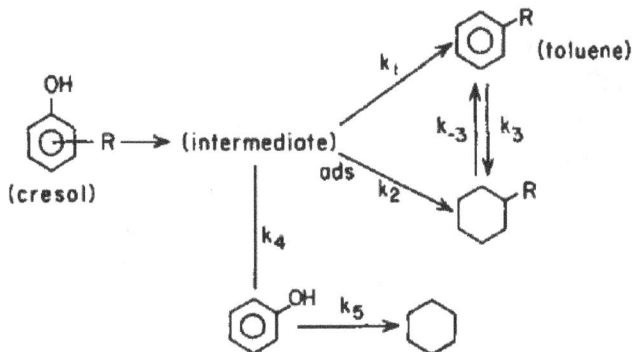

Figure 1.20. Reaction pathways for cresol HDO conversions. (Reprinted with permission from Ref. [93]. Copyright © 1983 Elsevier.)

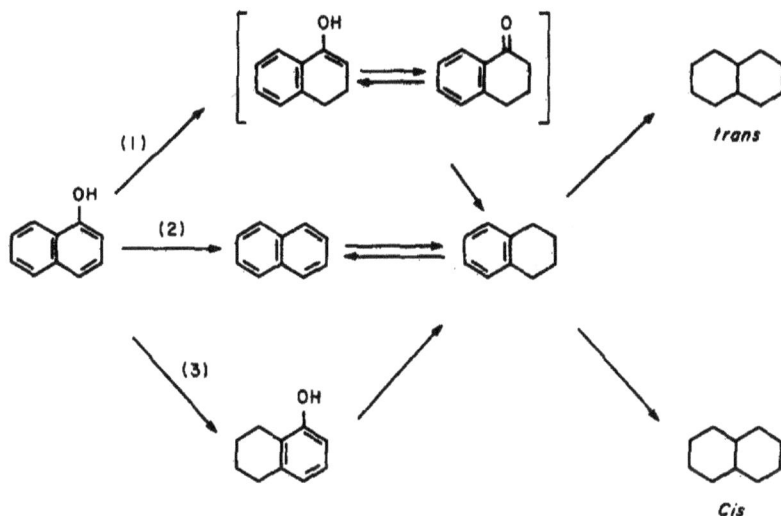

Figure 1.21. Reaction network for HDO of 1-naphthol catalyzed by sulfided NiMo/γ-Al$_2$O$_3$. (Reprinted with permission from Ref. [95]. Copyright © 1983 Elsevier.)

pattern (cresol → toluene → methylcyclohexane) with large k_1 and k_3 values. If a less active and aged catalyst is used at higher temperatures (350–400°C), it exhibits a parallel path behavior to toluene k_1 and methylcyclohexane k_2 from a common surface intermediate. Ring saturation prior to HDO is not necessary [93].

The reaction network for the HDO of 1-naphthol was determined as Figure 1.21. Li *et al.* reported that 1-tetralone, naphthalene, and 5,6,7,8-tetrahydro-1-naphthol are primary products with tetralin, *trans*-decalin, and *cis*-Decalin as minor products [94].

A study on HDO of furan at atmospheric pressure showed two routes (Figure 1.22): First, a transition state between the surface and furan molecule (state A) is formed; subsequently, the butadiene molecule is released as a primary product. In the second route, a partial HYD of the ring followed by ring opening forms a new transition state B, which will then decompose to either propylene or butenes [96]. Pratt and Christoverson proposed that 2,3-dihydrofuran is the intermediate, which will simultaneously yield C$_4$ compounds with H$_2$O and C$_3$ compounds with CO (Figure 1.23) [97].

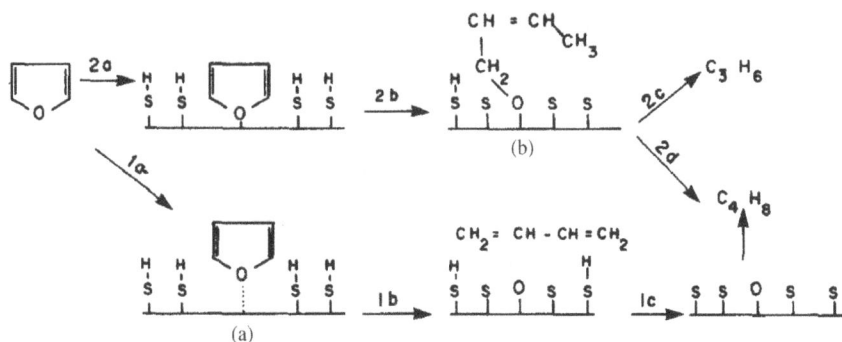

Figure 1.22. Tentative routes HDO of furan. Route 1 shows the formation and the removal of butadiene, and Route 2 is based on partial HYD of one double bond in the ring. (a) Transition state A and (b) transition state B. (Reprinted with permission from Ref. [96]. Copyright © 1983 Elsevier.)

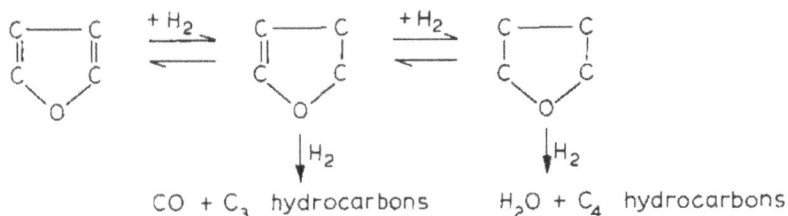

Figure 1.23. Proposed routes for HDO of furan. (Reprinted with permission from Ref. [97]. Copyright © 1983 Elsevier.)

HDO of benzofuran is summarized in Figure 1.24. The benzofuran reaction network includes initial HYD and hydrogenolysis to the oxygenated intermediates of 2,3-dihydrobenzofuran, *o*-ethyl phenol, and phenol. The major products formed from subsequent HDO of the intermediates are ethylbenzene, toluene, benzene, and ethylcyclohexane [98].

The reaction network for the HDO of dibenzofuran was proposed as Figure 1.25 [99]. The reaction intermediates detected include 1,2,3,4 tetrahydrodibenzofuran, biphenyl, cyclohexylbenzene, phenyl hexanol, bicyclohexyl, cyclohexyl-methyl cyclopentane, and cyclohexane. *O*-phenyl phenol, *o*-cyclohexyl phenol, cyclohexyl cyclohexanol, phenol, and benzene are not detected either due to their low concentrations or high

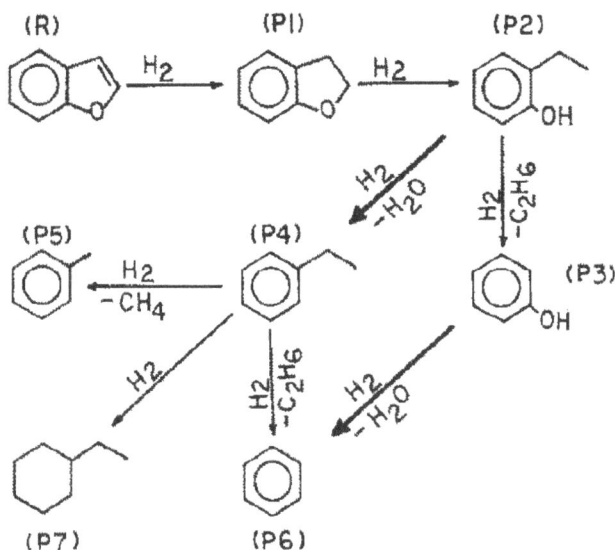

Figure 1.24. Proposed reaction network for the HDO of benzofuran. (Reprinted with permission from Ref. [98]. Copyright © 1988 Elsevier.)

reactivities. The presence of biphenyl indicates that a direct oxygen extrusion occurred prior to HYD of the adjacent benzene ring. The formation of phenyl hexanol indicates that 1,2,3,4 tetrahydrodibenzofuran can undergo the cleavage of C–O bond adjacent to the benzene ring.

The HDO of diphenyl ether results in a mixture of benzene, cyclohexane, and phenol (Figure 1.26). The reaction involves a simple C–O extrusion as the primary reaction step, with little tendency to prehydrogenate the aromatic ring.

The HDO of ketones can be relatively easy. Laurent and Delmon [100] showed that the carbonyl group of the model chemical of 4-methylacetophenone is catalyzed by sulfided CoMo and NiMo catalysts at temperatures higher than 200°C, and resulting in a primary product of ethyl methyl benzene with no intermediate products observed. Weisser and Landa [101] mentioned that the selectivity of the transformation of ketone to alcohol is lower with sulfides than with noble-metal catalysts, because dehydration reactions become dominant at temperatures higher than 250°C.

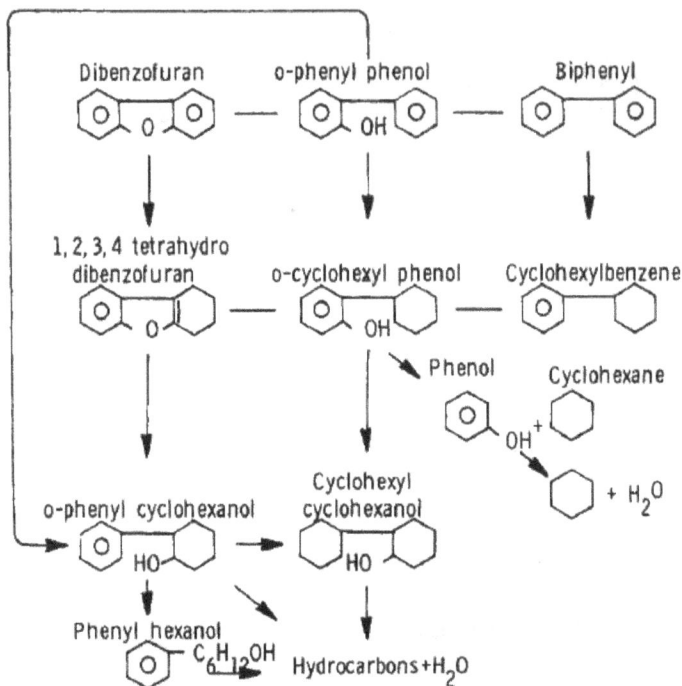

Figure 1.25.   Reaction network for the HDO dibenzofuran. (Reprinted with permission from Ref. [99]. Copyright © 1981 Wiley.)

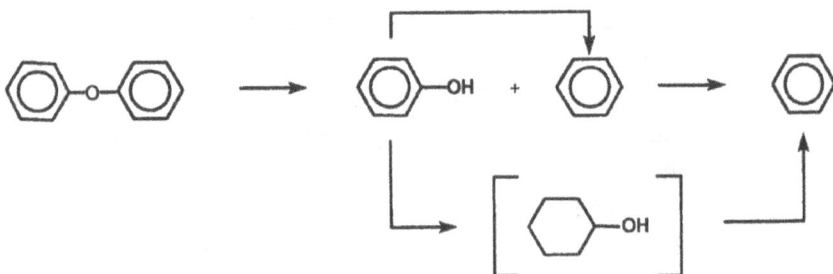

Figure 1.26.   Reaction pathways for HDO of diphenyl ether. (Reprinted with permission from Ref. [1]. Copyright © 1996 Springer.)

HDO of carboxylic groups are more refractory to deoxygenation than carbonyl groups. For example, di-ethyl decanedioate with two carboxylic groups only reacts at temperatures near 300°C in presence of CoMo and NiMo catalysts, and is almost exclusively converted into hydrocarbons with slight degradation [100]. Two main reaction pathways exist: one is the carboxylic group HYD and the other is the decarboxylation reaction. The reactivity of pure (e.g., decanoic acid), is less than that of the corresponding carboxylic ester (e.g., ethyl decanoate), and carboxylic acid gives a slightly higher yield of decarboxylated products. During HDO of the carboxylic ester, the carboxylic acid behaves as an intermediate product and does not accumulate in the reaction medium.

Reaction pathways for HDO of guaiacol (GUA) is summarized in Figure 1.27. Guaiacol with two types of oxygenated functional groups (i.e., phenolic and methoxyl groups) is often used as the model compound of lignin [102–105]. There are two categories of heterogeneous catalysts commonly used for HDO of GUA: (1) CoMo and NiMo series and (2) noble metals. Conventional CoMo and NiMo series catalysts are mainly used at temperatures around 280°C under pressures of about 3–7 MPa, and the major products are catechol, phenol, benzene, toluene, and cyclohexane. With CoMo and NiMo catalysts, GUA HDO began with demethylation, demethoxylation and deoxygenation, followed by benzene ring saturation [106]. Meanwhile, noble-metal catalysts, such as Ru, Pd, Rh [106] and Pt [107], have been identified as the most active selective-HYD catalysts within the low temperature range [108–110]. The major products are 2-methoxycyclohexanol, cyclohexanol and phenol, due to HYD of GUA's benzene ring followed by demethoxylation and dehydroxylation.

In summary, the HDO of chemicals often requires pre-HYD. As in the case of HDS and HDN, it may involve partially saturated intermediates, which may be carried out according to the reaction conditions to the aromatic compound or the fully saturated product.

1.4.3. *Reaction sites*

The behaviors of O and S heteroatoms have some similarities. Fox example, the electron-withdrawing effects of O and S atoms from corresponding heterorings are similar according to the resonance energies of the rings. Furimsky proposed that the active site for HDO is the same as that for HDS [89], where the O compound can adsorb on an S anion vacancy in a position perpendicular to the surface (Figure 1.28).

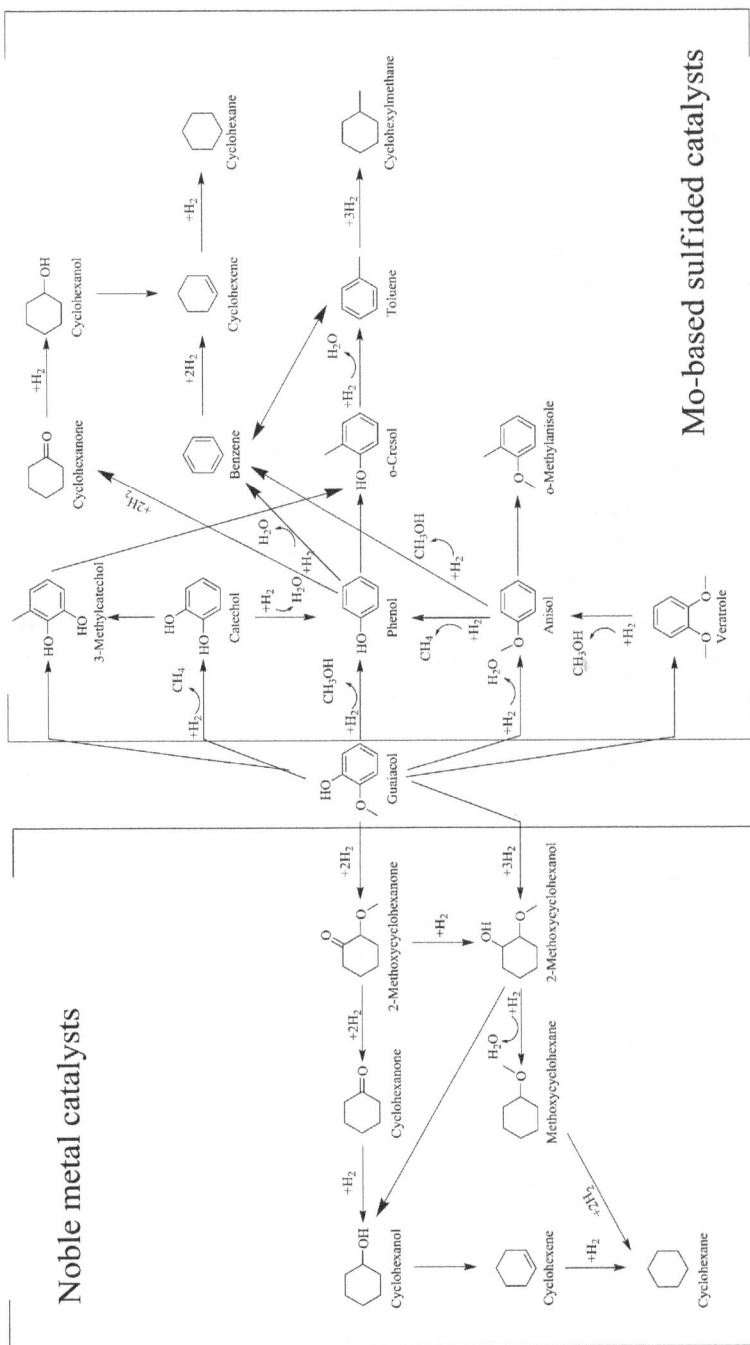

Figure 1.27.   Reaction pathways for guaiacol HYD over noble-metal catalysts and Mo-based sulfide catalysts. (Reprinted with permission from Ref. [111]. Copyright © 2016 Elsevier.)

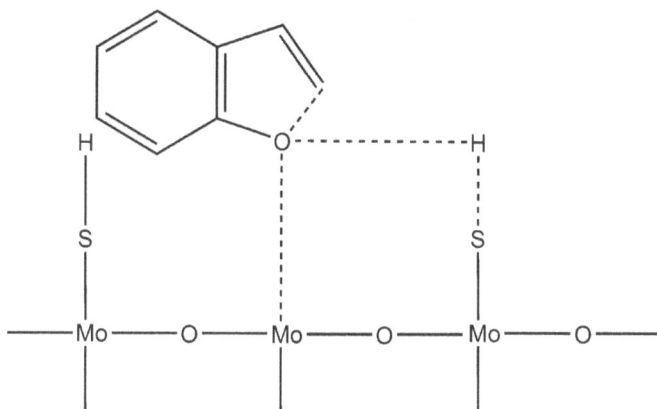

Figure 1.28. Hypothetical model of active site for HDO reactions.

Vogelzang *et al.* suggested that two kinds of catalytic sites are incorporated on the surface with one catalyzing direct C–O bond cleavage and the other for aromatic ring HYD [95]. These sites are susceptible to inhibition by bonding of H_2O. Studies on HDO of the substituted phenols supported this model, and found that addition of ammonia and H_2S could suppress both activities [112]. A σ-bonding through the O atom is suggested to lead to direct oxygen extrusion, and π-bonding of the ring leads to pre-HYD.

1.5. Hydrogenation (HYD)

HYD processes are necessary to remove unwanted olefins and diolefins from crude oils that tend to polymerize in the products, and convert polyaromatics to hydroaromatic compounds in the gasoline or diesel range to reduce coking. Meanwhile, HYD or partial HYD is sometimes required for other hydroprocessing reactions like HDO and HDN. Table 1.5 summarizes some typical compounds that require HYD processes.

HYD of olefins may happen at atmospheric pressure and 300°C. The reactivity generally decreases with the increase of olefin chain length and the number of substituents on the double bond [113].

HYD of aromatic hydrocarbons are reversible, with equilibrium conversions of hydrocarbons often being less than 100% under practical

Table 1.5.   Some typical compounds that require HYD processes.

Olefins — Ethylene, 1-Octene

Aromatics — Benzene, Tetralin, Biphenyl, Naphthalene, Anthracene, Phenanthrenes, Pyrene

processing conditions. The hydrotreating reactions are exothermic, and thus the removal of heteroatoms is usually not limited by thermodynamics except aromatics HYD. The thermodynamic limitations for HYD of aromatics are important at high temperatures and low hydrogen partial pressures [114]. Typically, HYD of aromatics requires a hydrogen pressure higher than that of olefins [115], due to resonance stabilization of the conjugated system and equilibrium constraints of the reactions. Reaction networks for HYD of benzene, biphenyl, naphthalene, and 2-phenylnaphthalene are summarized in Figure 1.29. The kinetics results showed that the reactivities of naphthalene and 2-phenylnaphthalene are one order of magnitude greater than those of benzene and biphenyl, which have approximately the same reactivities. Generally, the rate of HYD increases with the number of aromatic rings present.

Rosal *et al.* evaluated the reactivities of naphthalene, acenaphthene, phenanthrene, fluoranthene, anthracene, and pyrene [116], which showed an order of Anthracene > Fluoranthene > (Phenanthrene ≈ Naphthalene ≈

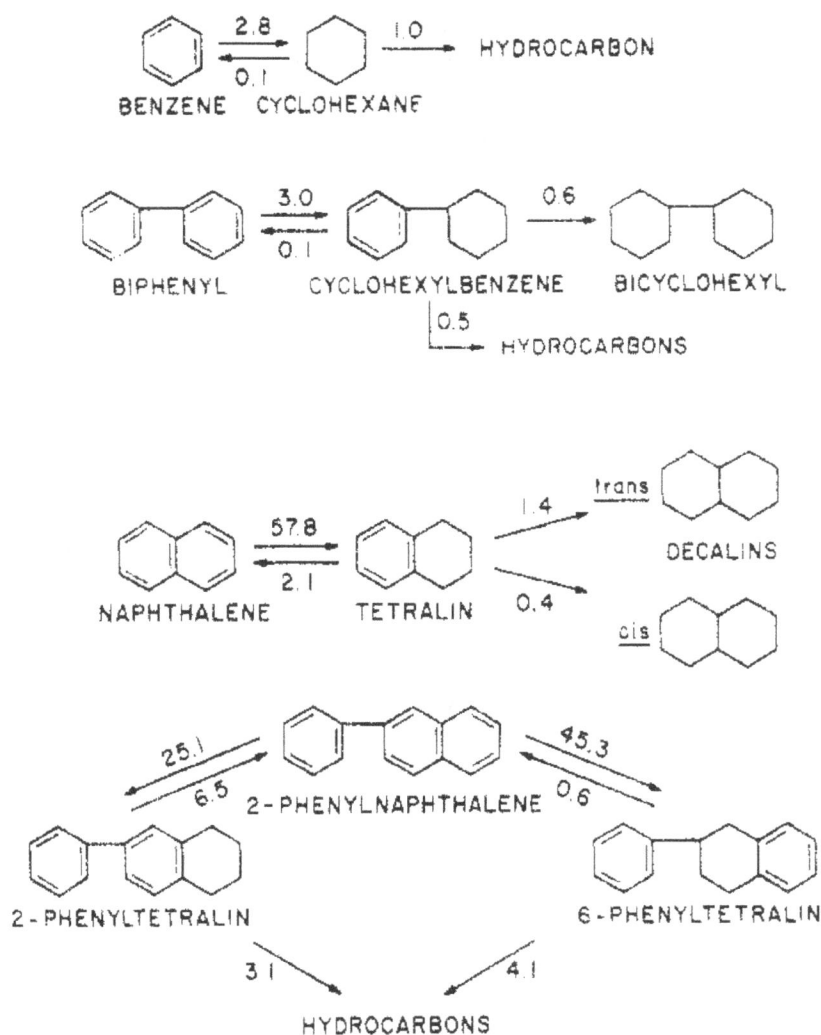

Figure 1.29. Reaction networks for HYD of benzene, biphenyl, naphthalene, and 2-phenylnaphthalene over sulfided CoMo/γ-Al$_2$O$_3$ at 325°C and 7.6 MPa. The numbers are the pseudo-first-order rate constants in cubic meters per kilogram of catalyst per second. (Reprinted with permission from [115]. Copyright © 1981 American Chemical Society.)

Pyrene $\approx$ Acenaphthene). Studies on aromatic model compounds gave a similar conclusion that the reactivities for HYD of one ring follow

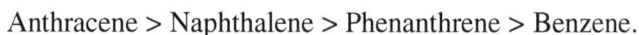

Anthracene > Naphthalene > Phenanthrene > Benzene.

The reactivity difference between two three-ring aromatics of phenanthrene and anthracene is due to the lower electron density in phenanthrene, which leads to a lower adsorption. For an aromatic compound with multiple rings, HYD of the outer ring occurs first, and the remaining rings become more resistant to HYD [74]. The decreasing reactivity can also be attributed to the resonance stabilization of rings in the molecule. Moreover, substituents on the benzene ring increase the HYD activity as

Ethylbenzene > Toluene ≈ *o*-xylene ≈ *p*-xylene > Benzene [117].

It is generally believed that aromatic hydrocarbons adsorb on catalytic surfaces through π-bonding of the ring [118]. The HYD activity of aromatic compounds is correlated to π-complexation, whose stability is related to the highest occupied molecular orbital and inversely proportional to the ionization potential. Stronger acidic functions on the catalyst surface are found to be conducive to stronger π-complex bonds and consequently to higher selectivity for the HYD reactions. In addition, the steric effect due to nonplanarity of the aromatic structure affects the reactivity too.

HYD of olefins and aromatics might occur at different sites. The olefin forms a σ-bond with the catalyst metal, while aromatics are more likely to be weakly π-bonded. An aromatic ring is assumed to adsorbed parallelly to the edge plane of MoS_2, which requires at least a six-vacancy center [119].

1.6. Hydrocracking

Hydrocracking catalysts possess dual functions including HYD and the cracking functions. Cracking reactions often require protonic (Brønsted) acid sites on the support and involve carbenium-ion intermediates. During hydrocracking, significant skeletal rearrangements occur, besides C–C bond hydrogenolysis. Conventional hydrotreating catalysts like sulfided $CoMo/Al_2O_3$ catalyst do not have sufficient protonic acidity and gave no skeletal rearrangements even under a high temperature of 400°C [120].

Conventional hydrocracking catalysts consist of the active metal sulfides and a support of silica–alumina, zeolites or alumina modified by F, P, or B. Because of the strong acid function, hydrocracking reactions involve skeletal isomerization, cracking, cyclization, alkylation, and dehydrogenation, besides HYD, HDS, HDN, and HDO [121].

1.7. Hydrodemetallization (HDM)

The metal components reported in the crude oil are predominantly vanadium (V) and nickel (Ni), and some oils also contain meaningful (>10 $\mu g \cdot g^{-1}$) amounts of other metals, such as iron (Fe) [122]. These compounds occur as porphyrinic and nonporphyrinic structural types (Figures 1.30 and 1.31) with a wide boiling range of 350 to >650°C. Hydrodemetallation mainly concerns the removal of nickel and vanadium. These organometallic compounds have high molecular weights ranging from several hundreds to 100,000 with the diameters from 2×10^{-9} to 10^{-7} m [123]. These dimensions approach or exceed catalyst pore size and contribute to fouling of catalysts and process equipment.

Typically, during the process of resid hydrotreating, metal compounds undergo hydrodemetallation (HDM) reactions simultaneously with HDS, HDN, and HDO reactions [124]. Model chemicals of Ni-etioporphyrin and Ni-tetra(3-methylphenyl)porphyrin have been used to study the HDM kinetics and reaction pathways [125]. As shown in Figure 1.32, the mechanism involves a partial HYD of the metalloporphyrin to a chlorin, which is followed by a sequential hydrogenolysis step [126]. More detailed information can be found in Ref. [124].

1.8. Hydroprocessing catalyst models

Typical industrial HDS catalysts consist of either Co or Ni promoted Mo or W that is supported on alumina [7]. The content of these metals in the catalyst is usually 1–4% for Co and Ni, 8–16% for Mo, and 12–25% for W. There are several models including the monolayer model, the intercalation model, and the contact synergy model that were developed to illustrate the structure and function of these catalysts. So far, the most favorable model is the Co–Mo–S model, which was developed by Topsoe and based on direct *in situ* physicochemical measurements [127–131].

The Co–Mo–S phase is shown to be MoS_2-like (slabs) structures with the promoter atoms located at the edges in five-fold coordinated sites

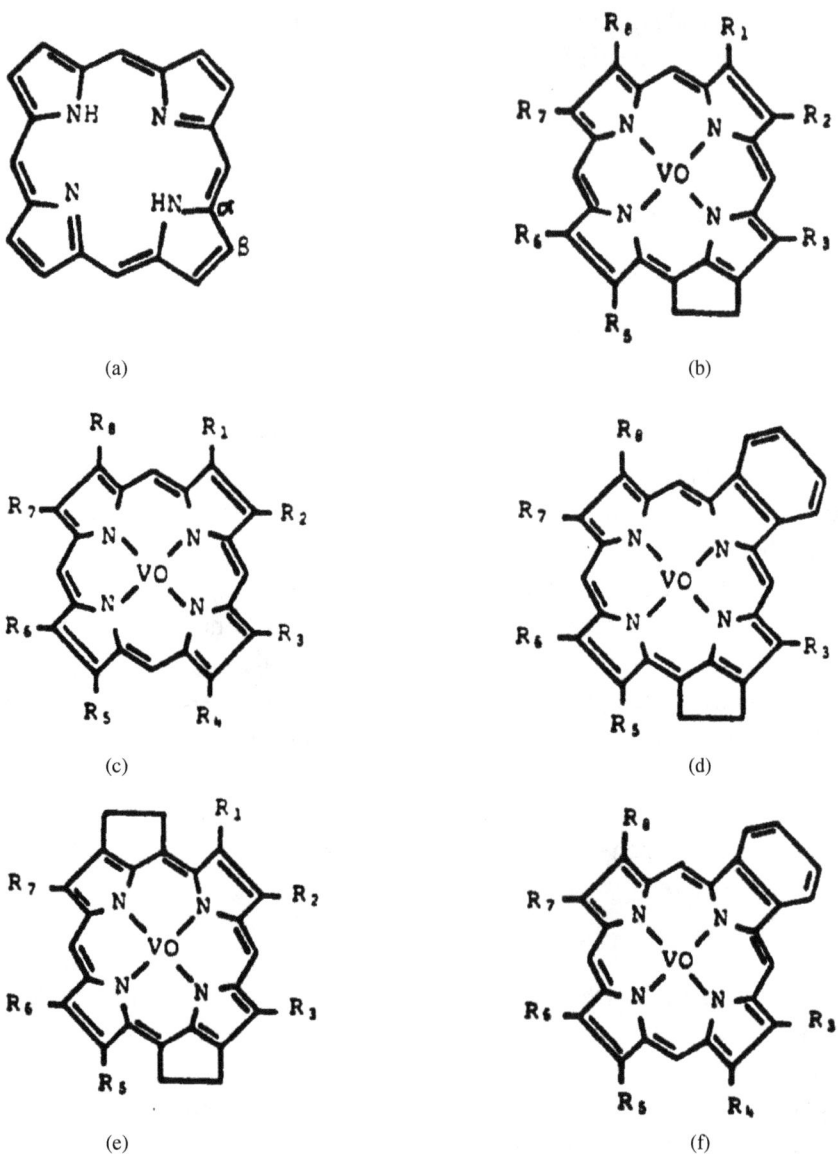

(a)

(b)

(c)

(d)

(e)

(f)

Figure 1.30. Porphyrin skeletal structure and metalloporphyrins found in petroleum: (a) porphine, (b) deoxophylloerythroetioporphyrin (DPEP), (c) etioporphyrin, (d) Rhodo-DPEP, (e) Di-DPEP, and (f) Rhodo-etioporphyrin. (Reprinted with permission from Ref. [124]. Copyright © 1988 Elsevier.)

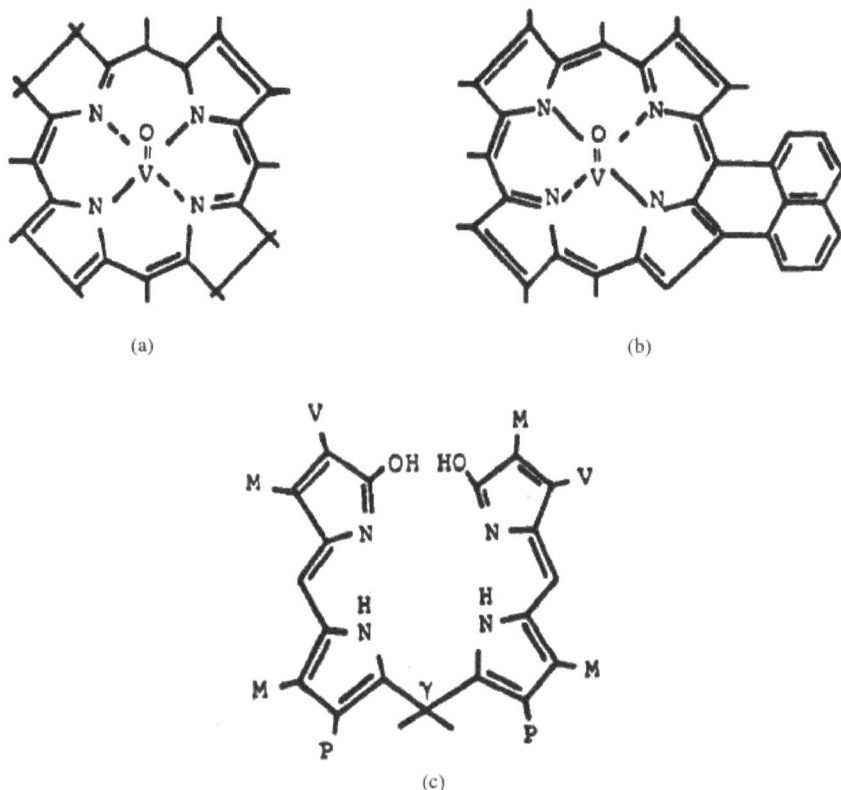

Figure 1.31. Nonporphyrinic metals: (a) vanadyl hydroporphyrin, (b) vanadyl arylpor-
phyrin (highly aromatic porphyrin), and (c) porphyrin-degraded product (bilirubin).
(Reprinted with permission from Ref. [124]. Copyright © 1988 Elsevier.)

(tetragonal pyramidal-like geometry) at the edge planes of MoS_2. The allover Co–Mo–S structure can be considered as a family of structures with a wide range of Co concentrations, ranging from pure MoS_2 up to essentially full coverage of the MoS_2 edges by Co. The Co atoms in Co–Mo–S may not have identical properties due to different edge-site geometries, Co–Co interactions and changes in sulfur coordination. Calculations of the standard molar free enthalpy of formation of slabs of MoS_2 and WS_2 indicate that the edge location (decoration) of promoters increases the stability of the slabs [132]. For alumina-supported catalysts, the single slab structures (i.e., type I Co–Mo–S) interact strongly with the support,

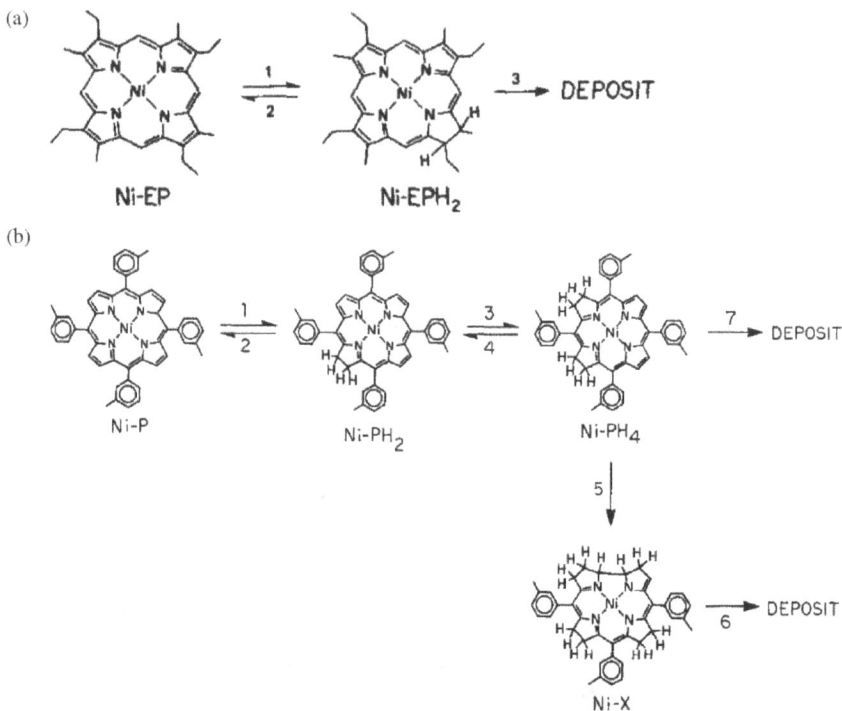

Figure 1.32.   Demetallation reaction pathway for (a) Ni-etioporphyrin (Ni-EP) and (b) Ni-tetra(3-methylphenyl) porphyrin (Ni-P). (Reprinted with permission from Ref. [125]. Copyright © 1985 Elsevier.)

probably via Mo–O–Al linkages located at the edges. For the multiple slab form (i.e., type-II Co–Mo–S), these interactions are small [133, 134]. In carbon-supported catalysts, where the support interactions are weaker, the single-slab structures may also exhibit type-II Co–Mo–S behavior.

The *in situ* studies via Mossbauer and IR spectroscopy have shown that most of the catalytic activity is linked to the presence of the promoter atoms in promoted Mo and W phases including Co–Mo–S, Fe–Mo–S, Ni–Mo–S, Co–W–S, Ni–W–S, and Fe–W–S. In the typical alumina-supported CoMo catalysts, other phases such as Co_9S_8 and $Co:Al_2O_3$ representing Co in the alumina lattice may exist too (Figure 1.33) [135].

For the alumina-supported NiMo catalysts, nickel is present in three forms after sulfidation: Ni_3S_2 crystallites on the support, nickel atoms

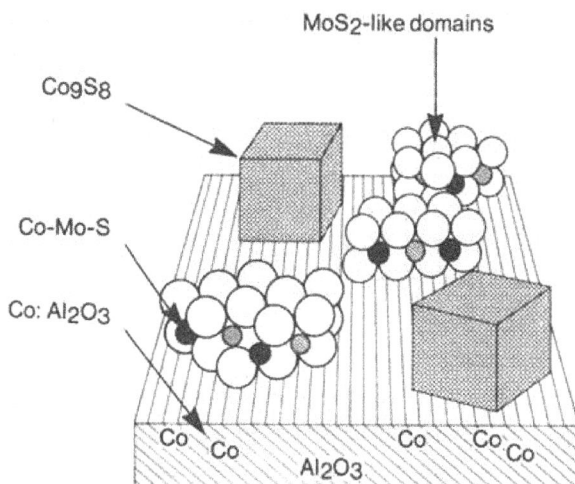

Figure 1.33. Schematic drawing showing Co–Mo–S and other Co sulfide structures on alumina and Co in the alumina. (Reprinted with permission from Ref. [136]. Copyright © 1984 Taylor & Francis.)

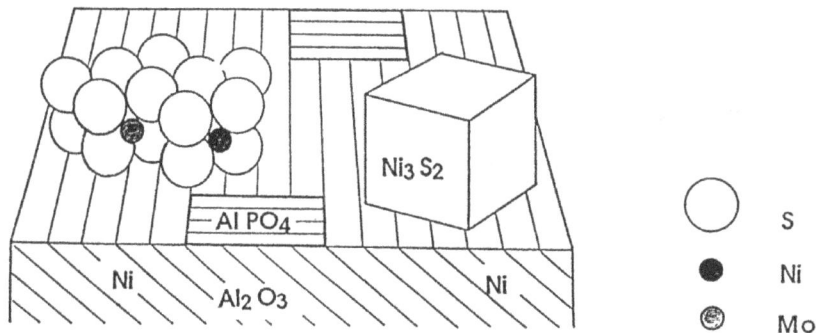

Figure 1.34. Three forms of nickel present in a sulfided Ni–Mo/Al$_2$O$_3$ catalyst. (Reprinted with permission from Ref. [57]. Copyright © 2001 Elsevier.)

adsorbed on the edges of MoS$_2$ crystallites (i.e., Ni–Mo–S phase), and nickel cations (Ni^{2+}) at octahedral or tetrahedral sites in the γ-Al$_2$O$_3$ lattice (Figure 1.34).

In terms of active sites, it's commonly assumed that the active sites in a hydrotreating catalyst are the molybdenum atoms at the surfaces of active phases such as Co–Mo–S and MoS_2 with at least sulfur anion vacancies to allow the reacting molecule to form a chemical bond with the molybdenum atom. Because sulfur atoms in the basal planes of MoS_2 are much more difficult to remove than those at edges and corners, reactive molybdenum atoms are predominantly present at edges and corners [57]. However, $Mo(W)S_2$ crystals supported on microroughened TiO_2 surface can form curved structures with radii of curvature ranging from 2 to 5 nm, which leads to the formation of unique catalytically active sites in the basal plane of MoS_2 [137].

In summary, CoMo catalysts are more selective for sulfur removal, and giving relatively low hydrogen consumption. NiMo catalysts are more active for HDN and HYD, but give rise to higher hydrogen consumption. Consequently, NiMo catalysts are often preferred for treating unsaturated feeds. The NiW catalysts have the highest activity for aromatic HYD at low H_2S partial pressures and are also active for hydrocracking. But tungsten is more expensive, and its industrial use is therefore limited. It's also shown that addition of Pd nanoparticles as a second active site for H_2 activation can partially saturate the heterocyclic ring, leading to a significant reduction in H_2 consumption and milder reaction conditions [138].

1.9. Closing remarks

The developments in bioderived fuel technologies bring up new questions and challenges, which require major developments within HDO and HDN catalyst technologies. In the meantime, the challenges that the refining industry needs to face to meet the new legislation and market demands still exist. The progress in fundamental hydroprocessing studies on support effect, catalyst morphology, reaction pathways, and reaction kinetics is expected to continue to provide new opportunities for the development of improved commercial catalysts for both traditional refining industry and new bioderived fuel upgrading.

Recommended reading

Furimsky, E. (1983). Chemistry of catalytic hydrodeoxygenation. *Catalysis Reviews*, 25(3), 421–458. DOI: 10.1080/01614948308078052.

Furimsky, E. and Massoth, F. E. (2005). Hydrodenitrogenation of petroleum. *Catalysis Reviews*, 47(3), 297–489. DOI: 10.1081/cr-200057492.

Girgis, M. J. and Gates, B. C. (1991). Reactivities, reaction networks, and kinetics in high-pressure catalytic hydroprocessing. *Industrial & Engineering Chemistry Research*, 30(9), 2021–2058. DOI: 10.1021/ie00057a001.

Langlois, G. E. and Sullivan, R. F. (1970). Chemistry of Hydrocracking. In: *Refining Petroleum for Chemicals*, American Chemical Society, pp. 38–67. DOI: 10.1021/ba-1970-0097.ch003.

Moreau, C. and Geneste, P. (1990). Factors affecting the reactivity of organic model compounds in hydrotreating reactions. In: *Theoretical Aspects of Heterogeneous Catalysis*, J. B. Moffat, ed., Springer, Dordrecht, pp. 256–310. DOI: 10.1007/978-94-010-9882-3_7.

Prado, G. H. C., Rao, Y., and de Klerk, A. (2017). Nitrogen removal from oil: a re-view. *Energy & Fuels*, 31(1), 14–36. DOI: 10.1021/acs.energyfuels.6b02779.

Prins, R. (2001). Catalytic hydrodenitrogenation. *Advances in Catalysis*, 46, 399–464. DOI: 10.1016/S0360-0564(02)46025-7.

Quann, R. J., Ware, R. A., Hung, C.-W., and Wei, J. (1988). Catalytic hydrodemetallation of petroleum. *Advances in Chemical Engineering*, 14, 95–259. DOI: 10.1016/S0065-2377(08)60101-5.

Schulz, H., Schon, M., and Rahman, N. M. (1986). Hydrogenative denitrogenation of model compounds as related to the refining of liquid fuels. *Studies in Surface Science and Catalysis*, 27, 201–255. DOI: 10.1016/S0167-2991(08)65352-5.

Stanislaus, A., Marafi, A., and Rana, M. S. (2010). Recent advances in the science and technology of ultra low sulfur diesel (ULSD) production. *Catalysis Today*, 153(1–2), 1–68. DOI: 10.1016/j.cattod.2010.05.011.

Topsøe, H., Clausen, B. S., and Massoth, F. E. (1996). Hydrotreating catalysis. In: *Catalysis: Science and Technology*, J. R. Anderson and M. Boudart, eds., Springer, Berlin, pp. 1–269. DOI: 10.1007/978-3-642-61040-0_1.

Acknowledgment

The author would like to thank Dr. Rui Li for his proofreading and suggestions.

References

1. Topsøe, H., Clausen, B. S., and Massoth, F. E. (1996). Hydrotreating catalysis. In: *Catalysis: Science and Technology*, J. R. Anderson and M. Boudart, eds., Springer, Berlin, pp. 1–269.

2. Jechura, J. (2016). Hydroprocessing: Hydrotreating & Hydrocracking. http://inside.mines.edu/~jjechura/Refining/08_Hydroprocessing.pdf.

3. Donath, E. E. (1956). Coal-hydrogenation vapor-phase catalysts. *Advances in Catalysis*, 8, 239–292.

4. Bergius, F. (1913). Production of hydrogen from water and coal from cellulose at high temperatures and pressures. *Journal of the Society of Chemical Industry*, 32(9), 462–467.

5. Luck, F. (1991). A review of support effects on the activity and selectivity of hydrotreating catalysts. *Bulletin des Sociétés Chimiques Belges*, 100(11–12), 781–800.

6. Wikipedia (2017). Friedrich Bergius. https://en.wikipedia.org/wiki/Friedrich_Bergius.

7. Byrns, A. C., Bradley, W. E., and Lee, M. W. (1943). Catalytic desulfurization of gasolines by cobalt molybdate process. *Industrial & Engineering Chemistry*, 35(11), 1160–1167.

8. TransportPolicy.net (2016). US: Fuels: Diesel and Gasoline. http://transportpolicy.net/index.php?title=US:_Fuels:_Diesel_and_Gasoline.

9. Stanislaus, A., Marafi, A., and Rana, M. S. (2010). Recent advances in the science and technology of ultra low sulfur diesel (ULSD) production. *Catalysis Today*, 153(1–2), 1–68.

10. Processing Magazine (2014). Refining catalyst market worth over $6 billion. https://www.processingmagazine.com/refinery-catalysts-market-worth-over-6-billion/.

11. Song, C. (2003). An overview of new approaches to deep desulfurization for ultra-clean gasoline, diesel fuel and jet fuel. *Catalysis Today*, 86(1–4), 211–263.

12. Dufresne, P. (2007). Hydroprocessing catalysts regeneration and recycling. *Applied Catalysis A: General*, 322, 67–75.

13. Berl, E. (1944). Production of oil from plant material. *Science*, 99(2573), 309–312.

14. Hurd, C. D. (1929). *Pyrolysis of Carbon Compounds*, Chemical Catalog Co., New York.

15. Elliott, D. C. (2007). Historical developments in hydroprocessing bio-oils. *Energy & Fuels*, 21(3), 1792–1815.

16. Xiu, S., Shahbazi, A., and Zhang, B. (2011). Biorefinery processes for biomass conversion to liquid fuel. In: *Biofuel's Engineering Process Technology*, M. A. dos Santos Bernardes, ed., InTech, Rijeka (open access).

17. Zhang, B., Huang, H. J., and Ramaswamy, S. (2012). A kinetics study on hydrothermal liquefaction of high-diversity grassland perennials. *Energy Sources, Part A: Recovery, Utilization, and Environmental Effects*, 34(18), 1676–1687.

18. Zhang, B., Yang, C., Moen, J., Le, Z., Hennessy, K., Wan, Y., Liu, Y., Lei, H., Chen, P., and Ruan, R. (2010). Catalytic conversion of microwave-assisted pyrolysis vapors. *Energy Sources, Part A: Recovery, Utilization, and Environmental Effects*, 32(18), 1756–1762.

19. Bridgwater, A. V. and Peacocke, G. V. C. (2000). Fast pyrolysis processes for biomass. *Renewable and Sustainable Energy Reviews*, 4(1), 1–73.

20. Amen-Chen, C., Pakdel, H., and Roy, C. (2001). Production of monomeric phenols by thermochemical conversion of biomass: a review. *Bioresource Technology*, 79(3), 277–299.

21. Mortensen, P. M., Grunwaldt, J. D., Jensen, P. A., Knudsen, K. G., and Jensen, A. D. (2011). A review of catalytic upgrading of bio-oil to engine fuels. *Applied Catalysis A: General*, 407(1–2), 1–19.

22. Huber, G. W., Iborra, S., and Corma, A. (2006). Synthesis of transportation fuels from biomass: chemistry, catalysts, and engineering. *Chemical Reviews*, 106(9), 4044–4098.

23. Elliott, D. C., Hart, T. R., Neuenschwander, G. G., Rotness, L. J., and Zacher, A. H. (2009). Catalytic hydroprocessing of biomass fast pyrolysis bio-oil to produce hydrocarbon products. *Environmental Progress & Sustainable Energy*, 28(3), 441–449.

24. Arun, N., Sharma, R. V., and Dalai, A. K. (2015). Green diesel synthesis by hydrodeoxygenation of bio-based feedstocks: strategies for catalyst design and development. *Renewable and Sustainable Energy Reviews*, 48, 240–255.

25. Srivastava, A. and Prasad, R. (2000). Triglycerides-based diesel fuels. *Renewable and Sustainable Energy Reviews*, 4(2), 111–133.

26. Zhang, B., Wang, L., Hasan, R., and Shahbazi, A. (2014). Characterization of a native algae species Chlamydomonas debaryana: strain selection, bioremediation ability, and lipid characterization. *BioResouces*, 9(4), 6130–6140.

27. Yang, C., Zhang, B., Cui, C., Wu, J., Ding, Y., and Wu, Y. (2016). Standards and protocols for characterization of algae-based biofuels. *Trends in Renewable Energy*, 2(2), 56–60.

28. Picardo, M., de Medeiros, J., Monteiro, J., Chaloub, R., Giordano, M., and de Queiroz Fernandes Araújo, O. (2013). A methodology for screening of

microalgae as a decision making tool for energy and green chemical process applications. *Clean Technologies and Environmental Policy*, 15(2), 275–291.

29. Davis, R., Kinchin, C., Markham, J., Tan, E., Laurens, L., Sexton, D., Knorr, D., Schoen, P., and Lukas, J. (2014). Process Design and Economics for the Conversion of Algal Biomass to Biofuels: Algal Biomass Fractionation to Lipid-and Carbohydrate-Derived Fuel Products (No. NREL/TP-5100-62368). National Renewable Energy Laboratory (NREL), Golden, CO.

30. Laurens, L. M. L., Nagle, N., Davis, R., Sweeney, N., van Wychen, S., Lowell, A., and Pienkos, P. T. (2015). Acid-catalyzed algal biomass pre-treatment for integrated lipid and carbohydrate-based biofuels production. *Green Chemistry*, 17(2), 1145–1158.

31. Jones, S. B., Zhu, Y., Snowden-Swan, L. J., Anderson, D., Hallen, R. T., Schmidt, A. J., Albrecht, K., and Elliott, D. C. (2014). *Whole algae hydrothermal liquefaction: 2014 state of technology*, Pacific Northwest National Laboratory (PNNL), Richland, WA (US).

32. Jones, S. B., Zhu, Y., Anderson, D. M., Hallen, R. T., Elliott, D. C., Schmidt, A., Albrecht, K., Hart, T., Butcher, M., and Drennan, C. (2014). *Process design and economics for the conversion of algal biomass to hydrocarbons: whole algae hydrothermal liquefaction and upgrading*, Pacific Northwest National Laboratory.

33. Vardon, D. R., Sharma, B. K., Scott, J., Yu, G., Wang, Z., Schideman, L., Zhang, Y., and Strathmann, T. J. (2011). Chemical properties of biocrude oil from the hydrothermal liquefaction of Spirulina algae, swine manure, and digested anaerobic sludge. *Bioresource Technology*, 102(17), 8295–8303.

34. Yang, C., Li, R., Cui, C., Liu, S., Qiu, Q., Ding, Y., and Wu, Y. (2016). Catalytic hydroprocessing of microalgae-derived biofuels: a review. *Green Chemistry*, 18(13), 3684–3699.

35. Nag, N. K., Sapre, A. V., Broderick, D. H., and Gates, B. C. (1979). Hydrodesulfurization of polycyclic aromatics catalyzed by sulfided CoO-$MoO_3\gamma$-Al_2O_3: the relative reactivities. *Journal of Catalysis*, 57(3), 509–512.

36. Kilanowski, D. R., Teeuwen, H., de Beer, V. H. J., Gates, B. C., Schuit, G. C. A., and Kwart, H. (1978). Hydrodesulfurization of thiophene, benzothiophene, dibenzothiophene, and related compounds catalyzed by sulfided $CoO\cdot MoO_3\gamma$-Al_2O_3: low-pressure reactivity studies. *Journal of Catalysis*, 55(2), 129–137.

37. Kabe, T., Ishihara, A., and Tajima, H. (1992). Hydrodesulfurization of sulfur-containing polyaromatic compounds in light oil. *Industrial & Engineering Chemistry Research*, 31(6), 1577–1580.

38. Hargreaves, A. E. and Ross, J. R. H. (1979). An investigation of the mechanism of the hydrodesulfurization of thiophene over sulfided Co-Moγ-Al$_2$O$_3$ catalysts. *Journal of Catalysis*, 56(3), 363–376.

39. McCarty, K. F. and Schrader, G. L. (1987). Deuterodesulfurization of thiophene: an investigation of the reaction mechanism. *Journal of Catalysis*, 103(2), 261–269.

40. Ruette, F. and Ludeña, E. V. (1981). Molecular orbital calculations of the hydrodesulfurization of thiophene over a Mo·Co catalyst. *Journal of Catalysis*, 67(2), 266–281.

41. Startsev, A. N., Burmistrov, V. A., and Yermakov, Y. I. (1988). Sulphide catalysts on silica as a support. *Applied Catalysis*, 45(2), 191–207.

42. Houalla, M., Nag, N. K., Sapre, A. V., Broderick, D. H., and Gates, B. C. (1978). Hydrodesulfurization of dibenzothiophene catalyzed by sulfided CoO-MoO$_3$γ-Al$_2$O$_3$: the reaction network. *AIChE Journal*, 24(6), 1015–1021.

43. Singhal, G. H., Espino, R. L., and Sobel, J. E. (1981). Hydrodesulfurization of sulfur heterocyclic compounds. *Journal of Catalysis*, 67(2), 446–456.

44. Mijoin, J., Pérot, G., Bataille, F., Lemberton, J. L., Breysse, M., and Kasztelan, S. (2001). Mechanistic considerations on the involvement of dihydrointermediates in the hydrodesulfurization of dibenzothiophene-type compounds over molybdenum sulfide catalysts. *Catalysis Letters*, 71(3), 139–145.

45. Pérot, G. (2003). Hydrotreating catalysts containing zeolites and related materials — mechanistic aspects related to deep desulfurization. *Catalysis Today*, 86(1), 111–128.

46. Breysse, M., Djega-Mariadassou, G., Pessayre, S., Geantet, C., Vrinat, M., Pérot, G., and Lemaire, M. (2003). Deep desulfurization: reactions, catalysts and technological challenges. *Catalysis Today*, 84(3), 129–138.

47. Ding, L., Zheng, Y., Zhang, Z., Ring, Z., and Chen, J. (2006). Hydrotreating of light cycled oil using WNi/Al$_2$O$_3$ catalysts containing zeolite beta and/or chemically treated zeolite Y. *Journal of Catalysis*, 241(2), 435–445.

48. Sapre, A. V., Broderick, D. H., Fraenkel, D., Gates, B. C., and Nag, N. K. (1980). Hydrodesulfurization of benzo[b]naphtho[2,3-d]thiophene catalyzed by sulfided CoO-MoO$_3$/γ-Al$_2$O$_3$: the reaction network. *AIChE Journal*, 26(4), 690–694.

49. Amberg, C. H. (1974). Molybdenum in hydrodesulphurisation catalysts. *Journal of the Less Common Metals*, 36(1), 339–352.

50. Diez, R. P. and Jubert, A. H. (1993). A molecular orbital picture of thiophene hydrodesulfurization. Part 2. Thiophene adsorption. *Journal of Molecular Catalysis*, 83(1), 219–235.

51. Satterfield, C. N., Modell, M., and Mayer, J. F. (1975). Interactions between catalytic hydrodesulfurization of thiophene and hydrodenitrogenation of pyridine. *AIChE Journal*, 21(6), 1100–1107.

52. Turaga, U. T., Ma, X., and Song, C. (2003). Influence of nitrogen compounds on deep hydrodesulfurization of 4,6-dimethyldibenzothiophene over Al_2O_3- and MCM-41-supported Co-Mo sulfide catalysts. *Catalysis Today*, 86(1–4), 265–275.

53. Schulz, H., Schon, M., and Rahman, N. M. (1986). Hydrogenative denitrogenation of model compounds as related to the refining of liquid fuels. *Studies in Surface Science and Catalysis*, 27, 201–255.

54. Duayne Whitehurst, D., Isoda, T., and Mochida, I. (1998). Present state of the art and future challenges in the hydrodesulfurization of polyaromatic sulfur compounds. *Advances in Catalysis*, 42, 345–471.

55. Ho, T. C. (1988). Hydrodenitrogenation catalysis. *Catalysis Reviews*, 30(1), 117–160.

56. Perot, G. (1991). The reactions involved in hydrodenitrogenation. *Catalysis Today*, 10(4), 447–472.

57. Prins, R. (2001). Catalytic hydrodenitrogenation. *Advances in Catalysis*, 46, 399–464.

58. Finiels, A., Geneste, P., Moulinas, C., and Olivé, J. L. (1986). Hydroprocessing of secondary amines over $NiW-Al_2O_3$ Catalyst. *Applied Catalysis*, 22(2), 257–262.

59. Geneste, P., Moulinas, C., and Olivé, J. L. (1987). Hydrodenitrogenation of aniline over $Ni-W/Al_2O_3$ catalyst. *Journal of Catalysis*, 105(1), 254–257.

60. Cocchetto, J. F., and Satterfield, C. N. (1976). Thermodynamic equilibria of selected heterocyclic nitrogen compounds with their hydrogenated derivatives. *Industrial & Engineering Chemistry Process Design and Development*, 15(2), 272–277.

61. Cocchetto, J. F. and Satterfield, C. N. (1981). Chemical equilibriums among quinoline and its reaction products in hydrodenitrogenation. *Industrial & Engineering Chemistry Process Design and Development*, 20(1), 49–53.

62. Sonnemans, J. and Mars, P. (1974). The mechanism of pyridine hydrogenolysis on molybdenum-containing catalysts. *Journal of Catalysis*, 34(2), 215–229.

63. Katzer, J. R. and Sivasubramanian, R. (1979). Process and catalyst needs for hydrodenitrogenation. *Catalysis Reviews*, 20(2), 155–208.
64. Stern, E. W. (1979). Reaction networks in catalytic hydrodenitrogenation. *Journal of Catalysis*, 57(3), 390–396.
65. Nagai, M., Masunaga, T., and Hanaoka, N. (1988). Hydrodenitrogenation of carbazole on a molybdenum/alumina catalyst. Effects of sulfiding and sulfur compounds. *Energy & Fuels*, 2(5), 645–651.
66. Nagai, M. and Miyao, T. (1992). Activity of alumina-supported molybdenum nitride for carbazole hydrodenitrogenation. *Catalysis Letters*, 15(1), 105–109.
67. Nagai, M. (1993). Inhibiting effect of sulfur and oxygen containing compounds on carbazole hydrodenitrogenation on a NiMo/Al$_2$O$_3$ catalyst. *Journal of the Japan Petroleum Institute*, 36(6), 502–506.
68. Satterfield, C. N., Modell, M., and Wilkens, J. A. (1980). Simultaneous catalytic hydrodenitrogenation of pyridine and hydrodesulfurization of thiophene. *Industrial & Engineering Chemistry Process Design and Development*, 19(1), 154–160.
69. Duan, P. and Savage, P. E. (2011). Catalytic hydrothermal hydrodenitrogenation of pyridine. *Applied Catalysis B: Environmental*, 108–109(0), 54–60.
70. Satterfield, C. N. and Cocchetto, J. F. (1981). Reaction network and kinetics of the vapor-phase catalytic hydrodenitrogenation of quinoline. *Industrial & Engineering Chemistry Process Design and Development*, 20(1), 53–62.
71. Shao, M., Cui, H., Guo, S., Zhao, L., and Tan, Y. (2016). Preparation and characterization of NiW supported on Al-modified MCM-48 catalyst and its high hydrodenitrogenation activity and stability. *RSC Advances*, 6(66), 61747–61757.
72. Kishi, K. and Ikeda, S. (1969). Ultraviolet study for the adsorption of pyridine and 2,2′-bipyridyl on evaporated metal films. *The Journal of Physical Chemistry*, 73(8), 2559–2564.
73. Fransen, T., van der Meer, O., and Mars, P. (1976). Surface structure and catalytic activity of a reduced molybdenum oxide-alumina catalyst. 1. The adsorption of pyridine in relation with the molybdenum valence. *The Journal of Physical Chemistry*, 80(19), 2103–2107.
74. Moreau, C. and Geneste, P. (1990). Factors affecting the reactivity of organic model compounds in hydrotreating reactions. In: *Theoretical Aspects of Heterogeneous Catalysis*, W. N. Lipscomb, ed., Springer, Dordrecht, pp. 256–310.
75. Yang, S. H. and Satterfield, C. N. (1983). Some effects of sulfiding of a NiMoAl$_2$O$_3$ catalyst on its activity for hydrodenitrogenation of quinoline. *Journal of Catalysis*, 81(1), 168–178.

76. Rajagopal, S., Grimm, T. L., Collins, D. J., and Miranda, R. (1992). Denitrogenation of piperidine on alumina, silica, and silica-aluminas: the effect of surface acidity. *Journal of Catalysis*, 137(2), 453–461.

77. Daage, M. and Chianelli, R. R. (1994). Structure-function relations in molybdenum sulfide catalysts: the "rim-edge" model. *Journal of Catalysis*, 149(2), 414–427.

78. Kasztelan, S., Wambeke, A., Jalowiecki, L., Grimblot, J., and Bonnelle, J. P. (1990). Site structure sensitivity of diene hydrogenation and isomerization reactions on $MOS_2/\gamma\text{-}Al_2O_3$ catalysts. *Journal of Catalysis*, 124(1), 12–21.

79. Qu, L. and Prins, R. (2003). Different active sites in hydrodenitrogenation as determined by the influence of the support and fluorination. *Applied Catalysis A: General*, 250(1), 105–115.

80. Nelson, N. and Levy, R. B. (1979). The organic chemistry of hydrodenitrogenation. *Journal of Catalysis*, 58(3), 485–488.

81. Laine, R. M. (1983). Comments on the mechanisms of heterogeneous catalysis of the hydrodenitrogenation reaction. *Catalysis Reviews*, 25(3), 459–474.

82. Hadjiloizou, G. C., Butt, J. B., and Dranoff, J. S. (1992). Catalysis and mechanism of hydrodenitrogenation: the piperidine hydrogenolysis reaction. *Industrial & Engineering Chemistry Research*, 31(11), 2503–2516.

83. Portefaix, J. L., Cattenot, M., Guerriche, M., Thivolle-Cazat, J., and Breysse, M. (1991). Conversion of saturated cyclic and noncyclic amines over a sulphided NiMo/Al$_2$O$_3$ catalyst: mechanisms of carbon — nitrogen bond cleavage. *Catalysis Today*, 10(4), 473–487.

84. Portefaix, J. L., Cattenot, M., Guerriche, M., and Breysse, M. (1991). Mechanism of carbon-nitrogen bond cleavage during amylamine hydrodenitrogenation over a sulphided NiMo/Al$_2$O$_3$ catalyst. *Catalysis Letters*, 9(1), 127–132.

85. Vivier, L., Dominguez, V., Perot, G., and Kasztelan, S. (1991). Mechanism of C-N bond scission. Hydrodenitrogenation of 1,2,3,4-tetrahydroquinoline and of 1,2,3,4-tetrahydroisoquinoline. *Journal of Molecular Catalysis*, 67(2), 267–275.

86. Furimsky, E. and Massoth, F. E. (2005). Hydrodenitrogenation of petroleum. *Catalysis Reviews*, 47(3), 297–489.

87. Schwartz, V., da Silva, V. T., and Oyama, S. T. (2000). Push–pull mechanism of hydrodenitrogenation over carbide and sulfide catalysts. *Journal of Molecular Catalysis A: Chemical*, 163(1), 251–268.

88. Petrakis, L., Ruberto, R. G., Young, D. C., and Gates, B. C. (1983). Catalytic hydroprocessing of SRC-II heavy distillate fractions. 1. Preparation of the fractions by liquid chromatography. *Industrial & Engineering Chemistry Process Design and Development*, 22(2), 292–298.

89. Furimsky, E. (1983). Chemistry of catalytic hydrodeoxygenation. *Catalysis Reviews*, 25(3), 421–458.

90. Cronauer, D. C., Jewell, D. M., Shah, Y. T., and Modi, R. J. (1979). Mechanism and kinetics of selected hydrogen transfer reactions typical of coal liquefaction. *Industrial & Engineering Chemistry Fundamentals*, 18(2), 153–162.

91. Bredenberg, J. B. S., Huuska, M., Räty, J., and Korpio, M. (1982). Hydrogenolysis and hydrocracking of the carbon-oxygen bond. *Journal of Catalysis*, 77(1), 242–247.

92. Furimsky, E., Mikhlin, J. A., Jones, D. Q., Adley, T., and Baikowitz, H. (1986). On the mechanism of hydrodeoxygenation of ortho substituted phenols. *The Canadian Journal of Chemical Engineering*, 64(6), 982–985.

93. Odebunmi, E. O. and Ollis, D. F. (1983). Catalytic hydrodeoxygenation. *Journal of Catalysis*, 80(1), 56–64.

94. Li, C. L., Xu, Z. R., Cao, Z. A., Gates, B. C., and Petrakis, L. (1985). Hydrodeoxygenation of 1-naphthol catalyzed by sulfided Ni-Mo/γ-Al$_2$O$_3$: reaction network. *AIChE Journal*, 31(1), 170–174.

95. Vogelzang, M. W., Li, C. L., Schuit, G. C. A., Gates, B. C., and Petrakis, L. (1983). Hydrodeoxygenation of 1-naphthol: activities and stabilities of molybdena and related catalysts. *Journal of Catalysis*, 84(1), 170–177.

96. Furimsky, E. (1983). The mechanism of catalytic hydrodeoxygenation of furan. *Applied Catalysis*, 6(2), 159–164.

97. Pratt, K. C. and Christoverson, V. (1983). Hydrogenolysis of furan over nickel — molybdenum catalysts. *Fuel Processing Technology*, 8(1), 43–51.

98. Edelman, M. C., Maholland, M. K., Baldwin, R. M., and Cowley, S. W. (1988). Vapor-phase catalytic hydrodeoxygenation of benzofuran. *Journal of Catalysis*, 111(2), 243–253.

99. Krishnamurthy, S., Panvelker, S., and Shah, Y. T. (1981). Hydrodeoxygenation of dibenzofuran and related compounds. *AIChE Journal*, 27(6), 994–1001.

100. Laurent, E. and Delmon, B. (1994). Study of the hydrodeoxygenation of carbonyl, carboxylic and guaiacyl groups over sulfided CoMo/γ-Al$_2$O$_3$ and NiMo/γ-Al$_2$O$_3$ catalysts. *Applied Catalysis A: General*, 109(1), 77–96.

101. Weisser, O. and Landa, S. (1973). *Sulphide Catalysts, Their Properties and Applications*, Pergamon Press, Oxford.

102. Bui, V. N., Laurenti, D., Delichère, P., and Geantet, C. (2011). Hydrodeoxygenation of guaiacol: Part II: Support effect for CoMoS catalysts on HDO activity and selectivity. *Applied Catalysis B: Environmental*, 101(3–4), 246–255.

103. Bykova, M. V., Ermakov, D. Y., Kaichev, V. V., Bulavchenko, O. A., Saraev, A. A., Lebedev, M. Y., and Yakovlev, V. A. (2012). Ni-based sol–gel catalysts as promising systems for crude bio-oil upgrading: guaiacol hydrodeoxygenation study. *Applied Catalysis B: Environmental*, 113–114(0), 296–307.

104. Ghampson, I. T., Sepúlveda, C., Garcia, R., Radovic, L. R., Fierro, J. L. G., de Sisto, W. J., and Escalona, N. (2012). Hydrodeoxygenation of guaiacol over carbon-supported molybdenum nitride catalysts: effects of nitriding methods and support properties. *Applied Catalysis A: General*, 439–440(0), 111–124.

105. Bui, V. N., Laurenti, D., Afanasiev, P., and Geantet, C. (2011). Hydrodeoxygenation of guaiacol with CoMo catalysts. Part I: Promoting effect of cobalt on HDO selectivity and activity. *Applied Catalysis B: Environmental*, 101(3–4), 239–245.

106. Lin, Y.-C., Li, C.-L., Wan, H.-P., Lee, H.-T., and Liu, C.-F. (2011). Catalytic hydrodeoxygenation of guaiacol on Rh-based and sulfided CoMo and NiMo catalysts. *Energy & Fuels*, 25(3), 890–896.

107. Runnebaum, R., Nimmanwudipong, T., Limbo, R., Block, D., and Gates, B. (2012). Conversion of 4-methylanisole catalyzed by Pt/γ-Al$_2$O$_3$ and by Pt/SiO$_2$-Al$_2$O$_3$: reaction networks and evidence of oxygen removal. *Catalysis Letters*, 142(1), 7–15.

108. Lin, S. D. and Song, C. (1996). Noble metal catalysts for low-temperature naphthalene hydrogenation in the presence of benzothiophene. *Catalysis Today*, 31(1–2), 93–104.

109. Sakanishi, K., Ohira, M., Mochida, I., Okazaki, H., and Soeda, M. (1989). The reactivities of polyaromatic hydrocarbons in catalytic hydrogenation over supported noble metals. *Bulletin of the Chemical Society of Japan*, 62(12), 3994–4001.

110. Smith, G. V. and Notheisz, F. (1999). *Heterogeneous Catalysis in Organic Chemistry*, Academic Press, San Diego.

111. Li, R., Shahbazi, A., Wang, L., Zhang, B., Hung, A. M., and Dayton, D. C. (2016). Graphite encapsulated molybdenum carbide core/shell nanocomposite for highly selective conversion of guaiacol to phenolic compounds in methanol. *Applied Catalysis A: General*, 528, 123–130.

112. Gevert, B. S., Otterstedt, J. E., and Massoth, F. E. (1987). Kinetics of the HDO of methyl-substituted phenols. *Applied Catalysis*, 31(1), 119–131.

113. Uchytil, J., Jakubíčková, E., and Kraus, M. (1980). Hydrogenation of alkenes over a cobalt-molybdenum-alumina catalyst. *Journal of Catalysis*, 64(1), 143–149.

114. Girgis, M. J. and Gates, B. C. (1991). Reactivities, reaction networks, and kinetics in high-pressure catalytic hydroprocessing. *Industrial & Engineering Chemistry Research*, 30(9), 2021–2058.

115. Sapre, A. V. and Gates, B. C. (1981). Hydrogenation of aromatic hydrocarbons catalyzed by sulfided cobalt oxide-molybdenum oxide/α-aluminum oxide. Reactivities and reaction networks. *Industrial & Engineering Chemistry Process Design and Development*, 20(1), 68–73.

116. Rosal, R., Diez, F. V., and Sastre, H. (1992). Catalytic hydrogenation of multiring aromatic hydrocarbons in a coal tar fraction. *Industrial & Engineering Chemistry Research*, 31(4), 1007–1012.

117. Dufresne, P., Bigeard, P. H., and Billon, A. (1987). New developments in hydrocracking: low pressure high-conversion hydrocracking. *Catalysis Today*, 1(4), 367–384.

118. Nag, N. K. (1984). On the mechanism of the hydrogenation reactions occurring under hydroprocessing conditions. *Applied Catalysis*, 10(1), 53–62.

119. Johnson, K. H. (1977). Spin-orbital electronegativity, the Xα method, and reactivity at transition-metal interfaces. *International Journal of Quantum Chemistry*, 12(S11), 39–60.

120. Muralidhar, G., Massoth, F. E., and Shabtai, J. (1984). Catalytic functionalities of supported sulfides. *Journal of Catalysis*, 85(1), 44–52.

121. Langlois, G. E. and Sullivan, R. F. (1970). Chemistry of hydrocracking. In: *Refining Petroleum for Chemicals*, American Chemical Society, Washington, DC, pp. 38–67.

122. Prado, G. H. C., Rao, Y., and de Klerk, A. (2017). Nitrogen removal from oil: a review. *Energy & Fuels*, 31(1), 14–36.

123. Oyekunle, L. O. and Hughes, R. (1987). Catalyst deactivation during hydrodemetalization. *Industrial & Engineering Chemistry Research*, 26(10), 1945–1950.

124. Quann, R. J., Ware, R. A., Hung, C.-W., and Wei, J. (1988). Catalytic hydrodemetallation of petroleum. *Advances in Chemical Engineering*, 14, 95–259.

125. Ware, R. A. and Wei, J. (1985). Catalytic hydrodemetallation of nickel porphyrins. *Journal of Catalysis*, 93(1), 100–121.

126. Mitchell, P. C. H. (1990). Hydrodemetallisation of crude petroleum: fundamental studies. *Catalysis Today*, 7(4), 439–445.

127. Topsoee, H., Clausen, B. S., Topsoee, N. Y., and Pedersen, E. (1986). Recent basic research in hydrodesulfurization catalysis. *Industrial & Engineering Chemistry Fundamentals*, 25(1), 25–36.

128. Clausen, B. S., Topsoe, H., Candia, R., Villadsen, J., Lengeler, B., Als-Nielsen, J., and Christensen, F. (1981). Extended x-ray absorption fine structure study of the cobalt-molybdenum hydrodesulfurization catalysts. *The Journal of Physical Chemistry*, 85(25), 3868–3872.

129. Topsøe, H., Clausen, B. S., Candia, R., Wivel, C., and Mørup, S. (1981). In situ Mössbauer emission spectroscopy studies of unsupported and supported sulfided Co-Mo hydrodesulfurization catalysts: evidence for and nature of a Co-Mo-S phase. *Journal of Catalysis*, 68(2), 433–452.

130. Wivel, C., Candia, R., Clausen, B. S., Mørup, S., and Topsøe, H. (1981). On the catalytic significance of a Co-Mo-S phase in Co-MoAl$_2$O$_3$ hydrodesulfurization catalysts: combined in situ Mössbauer emission spectroscopy and activity studies. *Journal of Catalysis*, 68(2), 453–463.

131. Topsøe, N.-Y. and Topsøe, H. (1983). Characterization of the structures and active sites in sulfided Co-MoAl$_2$O$_3$ and Ni-MoAl$_2$O$_3$ catalysts by NO chemisorption. *Journal of Catalysis*, 84(2), 386–401.

132. Toulhoat, H. and Kasztelan, S. (1988). In: *Proceedings of the 9th International Congress on Catalysis*, M. J. Phillips and M. Ternan, eds., Chemical Institute of Canada, Ottawa, p. 152.

133. Topsøe, H., Clausen, B. S., Topsøe, N.-Y., and Zeuthen, P. (1989). Progress in the design of hydrotreating catalysts based on fundamental molecular insight. *Studies in Surface Science and Catalysis*, 53, 77–102.

134. Topsøe, H. and Clausen, B. S. (1986). Active sites and support effects in hydrodesulfurization catalysts. *Applied Catalysis*, 25(1), 273–293.

135. Topsøe, H. (2007). The role of Co–Mo–S type structures in hydrotreating catalysts. *Applied Catalysis A: General*, 322, 3–8.

136. Topsøe, H. and Clausen, B. S. (1984). Importance of Co-Mo-S type structures in hydrodesulfurization. *Catalysis Reviews*, 26(3–4), 395–420.

137. Datye, A. K., Srinivasan, S., Allard, L. F., Peden, C. H. F., Brenner, J. R., and Thompson, L. T. (1996). Oxide supported MoS$_2$ catalysts of unusual morphology. *Journal of Catalysis*, 158(1), 205–216.

138. Bachrach, M., Marks, T. J., and Notestein, J. M. (2016). Understanding the hydrodenitrogenation of heteroaromatics on a molecular level. *ACS Catalysis*, 6(3), 1455–1476.

Chapter 2

Stabilization of Bio-oil to Enable Its Hydrotreating to Produce Biofuels

Huamin Wang

Pacific Northwest National Labortatory, Richland, WA, USA
Huamin.wang@pnnl.gov

Abstract

Fast pyrolysis is considered to be the simplest and most cost-effective approach to produce liquid oil (bio-oil) from biomass. Bio-oil is not suitable to substitute for petroleum as high-quality fuels and significant upgrading such as hydrotreating is required to remove oxygen, add hydrogen, and rearrange the carbon backbone of bio-oil. However, the grand challenge in bio-oil hydrotreating technology is bio-oil instability, which limits the lifetime of catalyst and operation. To enable a sustainable and economically viable process for bio-oil hydrotreating, it is vital to develop effective technologies for stabilizing bio-oils. This chapter will be devoted to bio-oil stabilization. The current understating of the major cause of bio-oil instability, condensation of reactive species such as sugar, aldehydes, ketones, and phenolics, is elucidated. The reported physical and chemical methods for bio-oil stabilization are summarized in detail, with a specific focus on bio-oil catalytic hydrogenation for stabilization. The impact of stabilization on bio-oil hydrotreating is discussed as well.

Keywords: Fast pyrolysis, Bio-oil instability, Bio-oil stabilization, Catalytic hydrogenation

2.1. Introduction

Lignocellulosic biomass, because of its characteristic of being renewable, abundantly available, and low-cost, is considered to be the only sustainable resource for production of fuels and chemicals [1–5]. Biochemical and thermochemical processes have been developed to convert lignocellulosic biomass into fuels and chemicals via different intermediate products [1–5]. In general, it is believed that thermochemical process, including fast pyrolysis, gasification, and liquefaction, are more flexible in terms of feed and products [6–8]. Among the thermochemical processes listed above, fast pyrolysis is considered to be the simplest and most cost-effective approach to produce liquid oil from biomass [1–9]. Fast pyrolysis treats biomass in the absence of oxygen at the temperature of 400–600°C with rapid ramping to thermally decompose the lignocellulose and, after the subsequent rapid quenching, produces pyrolysis oil (bio-oil) as a complex mixture of oxygenates. Fast pyrolysis is highly compatible with a large variety of feedstock and retains most of carbon and energy of biomass in the produced bio-oil [10–12].

Bio-oil can directly be combusted to produce heat and power [13]. However, it is not suitable to substitute for petroleum as a high-quality fuel because some properties of bio-oil, such as high oxygen and water content, poor stability, and corrosiveness, present serious problems [13, 14]. Therefore, significant upgrading is required to remove oxygen, add hydrogen, and rearrange carbon backbone for bio-oil to produce liquid transportation fuels which have similar properties as petroleum fuels. Hydrotreating, one of the key process in modern oil refining, is also the most efficient and common approach to upgrade bio-oil by hydrodeoxygenation (HDO), hydrogenation (HYD), and hydrocracking and therefore has attracted the most attention in recent decades [4, 5, 7, 15]. Supported CoMo- and NiMo-based sulfide catalysts, which served as industrial hydrotreating catalysts in refining for decades, have been primarily used in HDO of bio-oil and showed good performance without much specific modification.

However, the grand challenge in bio-oil hydrotreating technology is the instability of bio-oil, which will cause catalyst fouling and even hydrotreating reactor plugging within 1 or 2 days during the processing of bio-oil under conventional hydrotreating conditions [4, 7]. The thermal and chemical instability of bio-oil is a result of fast pyrolysis during which thermodynamic equilibrium cannot be attained because of short pyrolysis residence time and rapid quenching. The properties of bio-oil will change even at storage temperature and such change accelerates significantly as temperature increases [16, 17]. This makes any treatment of bio-oil at high temperatures extremely challenging. Therefore, to arrive at a sustainable and economically viable process for bio-oil upgrading, it is vital to develop effective technologies for stabilizing bio-oils. Several bio-oil stabilization approaches have been reported and some methods have shown to be very promising for producing thermally stable bio-oil for high-temperature catalytic treatment.

This chapter is devoted to bio-oil stabilization with a specific focus on catalytic bio-oil HYD. The major cause of bio-oil instability, the reported physical and chemical methods for bio-oil stabilization and the consequence of such stabilization on their upgrading will be discussed in detail in the following sections.

2.2. Instability of bio-oil

Bio-oil is mixture of chemicals with a wide range of size and functional groups derived from lignocellulosic biomass feedstock. Its composition is highly dependent on the variety of feedstocks and pyrolysis conditions. Lignocellulosic biomass is generally composed of cellulose (28–55%), hemicellulose (17–35%), and lignin (17–35%), which are all oxygen-containing organic polymers as structured portion, and some other minor components such as minerals [5]. During fast pyrolysis, biomass decomposes to produce vapors, aerosols, and chars, which then condenses during cooling to produce bio-oil, uncondensed gas, and char. Each component of biomass, cellulose, hemicellulose, and lignin decomposes at different temperatures and forms different species in bio-oil. Bio-oil, as shown in Table 2.1, has an overall elemental composition in dry basis similar to the biomass feedstock [18]. In general, bio-oil has significant amount of water and oxygen compared to conventional crude oil. The oxygen in bio-oil is in various forms as oxygen-containing function

Table 2.1. Comparison of some properties of biomass feedstock, pyrolysis bio-oil, and conventional fuel oil.

	Pine feedstock [18]	Pine bio-oil [18]	Conventional fuel oil [13, 20]
Elementary analysis, dry basis (wt.%)			
Carbon	49.6	45.0	85
Hydrogen	5.9	7.8	11–13
Oxygen (calculated)	44.3	47.1	0.1–1.0
Sulfur	<0.1	<0.1	1.0–1.8
Nitrogen	0.2	0.08	0.1
Water (wt.%)	—	21.1	0.02–0.1
Viscosity @ 25°C (cP)	—	83.4	180
Density (kg/m^3)	—	1.15–1.25	0.9–1.0

groups in more than 400 oxygenate species, including acids, esters, alcohols, ketones, aldehydes, sugars, furans, phenolics, hydrocarbons, and high molecular weight species [19]. The concentration of these species depends on the feedstock and pyrolysis condition and significantly impacts the stability of bio-oil.

Typical woody fast pyrolysis bio-oils are not thermally and chemically stable as conventional fossil fuels. Even during storage or transportation of bio-oil at near 20°C, there is change of physical and chemical properties such as phase separation, viscosity increase, molecular weight increase, acidity change, and water content change [16, 17]. Unlike fuel stability and storage stability that can be tested by ASTM standards, there is no standardized methods for measuring bio-oil stability. Accelerated aging, for instance at 80–100°C for 24 h, is broadly used to test thermal stability of bio-oil by comparing properties of bio-oil before and after the treatment by various analyses [21, 22]. The analytical methods used include viscosity measurement, density measurement, water content by Karl Fischer titration, acid number by acid titration, carbonyl content by chemical titration, molecular weight test by gel permeation chromatography, and other typical analytical spectroscopy methods such as Fourier transform infrared spectroscopy (FTIR), gas chromatography/mass spectrometry (GC/MS) and ^{1}H and ^{13}C nuclear magnetic resonance (NMR). As reported by Czernik *et al.* [21], the aging of bio-oil at 37, 60, and 90°C resulted in relatively similar chemical change. Similar change of viscosity (from 152 to ~250 cP) were obtained after aging the bio-oil at 37°C for 56 days, 60°C for 4 days, and 90°C for 6 h. It indicates the feasibility of using accelerated aging to evaluate stability of bio-oil at temperature range of 20–120°C. Besides the viscosity, other properties of bio-oil also changed during aging. As reported by Meng *et al.* [16], characterization of fresh and aged bio-oil (80°C for 24 h) from pine showed very similar elemental composition and, however, increased water content (20–24 wt.%), viscosity (16–56 cP), molecular weight (*Mw* from 340 g/mol to 660 g/mol), and acidity (TAN from 59 mg KOH/g to 69 mg KOH/g). This change was believed to be because of the condensation reaction occurring between the reactive species in bio-oil, especially aldehydes and sugar. As reported by Oasmma *et al.* [19], the chemical change of bio-oil in accelerated aging; the content of aldehydes, ketones, sugars, and phenolics decreased and that of high molecular weight species increased. Alsbou and Helleur [23] reported that the largest change of a birch wood bio-oil after aging at 80°C was the reduction in the amount of olefinic containing compounds together with the reduction in aldehyde and hydroxyl carbon. The possible reactions that occurred during bio-oil aging include condensation of aldehydes and ketones, phenol (pyrolytic lignin)-aldehydes reaction, conversion of sugars to humins, alcohol etherification, and carboxylic acid and alcohol esterification. Meng *et al.* [16] explored the reaction mechanism causing bio-oil instability by aging bio-oil and its fractions and some model compound tests and proposed the following condensation reaction of sugar and lignin fractions: sugar decomposition and condensation forming humins, aldol condensation of furfural and ketone, acid-catalyzed lignin condensation, radical-initiated lignin condensation, and phenol glycolaldehyde coupling. Some representative reactions are shown in Figure 2.1. To stabilize bio-oil for storage, these reactions must be slowed down, and the reported approach for this will be discussed in the following sections.

The process used to convert bio-oil to a more feasible fuel normally requires a temperature much higher than that used for accelerated aging

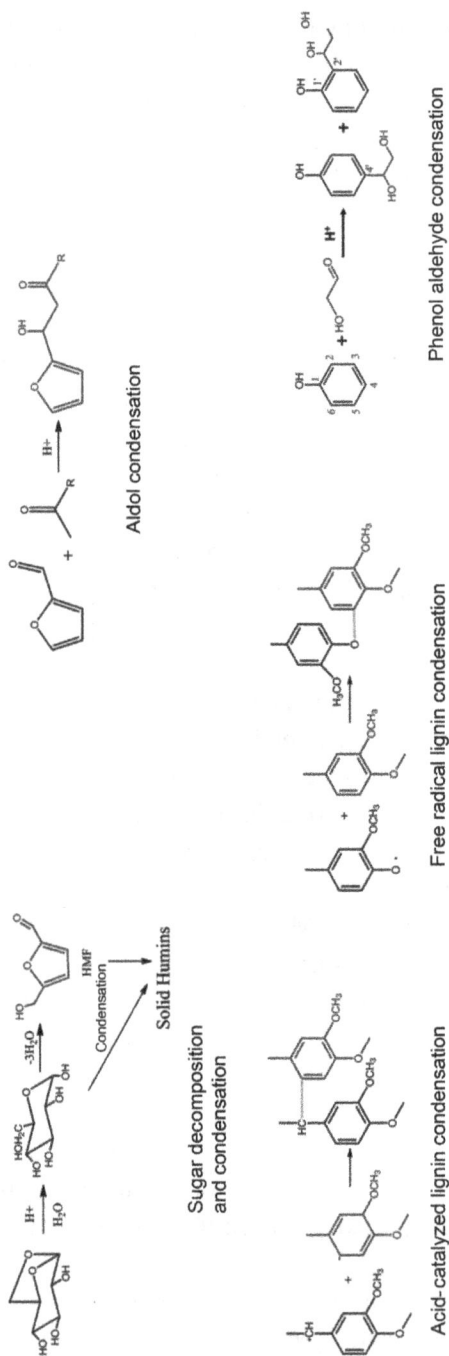

Figure 2.1. Proposed condensation reactions of typical reactive species in bio-oil. (Redrawn with permission from Ref. [16]. Copyright © 2014 American Chemical Society.)

shown above [4, 5, 7, 24]. Therefore the change of bio-oil properties during heating up bio-oils to a process temperature is much more significant and causes most challenges to the processing of bio-oil. For instance, hydrotreating of bio-oil by HDO is the most efficient and common approach to produce fuel range hydrocarbons. HDO of bio-oil requires a reaction temperate of around 300–420°C [4, 5, 7]. Early tests of direct hydrotreating of raw bio-oil clearly demonstrated that bio-oil is not suitable for hydrotreating as it resulted in heavy product plugging the reactor system and encapsulating catalysts within 24 h of operation [4]. There is lack of knowledge on the change of properties of bio-oil and the properties of formed heavy species at a temperature similar to these process temperatures. At high temperatures, one should expect a mechanism similar to that at a temperature typically used for accelerated aging with a much different rate. However, it is still questionable if one can extrapolate the differences in bio-oils during aging test at 80°C to their different polymerization trends and products during hydrotreating test at 300–400°C. Hu *et al.* [25] studied the condensation/polymerization reactions in bio-oil during heating of the bio-oil to 190°C. A surrogate mixture, including sugar, acids, aldehydes, ketones, furans, and phenols, was used to understand their contribution to the condensation/polymerization of bio-oil. Glucose played a key role for the polymer formation due to its decomposition to reactive compounds with multiple hydroxyl groups, carbonyl groups, or conjugated π bonds. The sugar derivatives, including furfural, hydroxyl aldehyde, and hydroxyl acetone, were also found to be reactive toward polymerization. The carboxylic acids were shown to be the catalysts for polymerization, and formic acid was much more efficient to catalyze polymerization than acetic acid. The phenolic compounds also promoted the acid-catalyzed reactions. These reactions are very similar to those presented in Figure 2.1, showing the similarity in the change of bio-oil properties at a much wider temperature range.

In summary, condensation/polymerization of reactive species such aldehydes, ketones, sugars, and phenolics are believed to cause the instability of bio-oil. In order to address this, physical methods, such as solvents' addition, and chemical methods such as esterification or mild HYD were developed to stabilize bio-oil. In the following sections, these methods will be discussed in detail.

2.3. Bio-oil stabilization by hydrotreating

Hydrotreating of bio-oil at a moderate temperature could convert active species to more stable products, such as aldehydes to alcohols or hydro-carbons, and therefore stabilize bio-oil. It has been considered as the most effective method for stabilizing bio-oil until now and has attracted the most attention [4, 5, 7]. A two-step process was developed almost three decades ago and includes a low-temperature hydrotreating step to stabilize the bio-oil followed by a high-temperature hydrodeoxygenation/hydroc-racking step for oxygen removal [4, 5, 7, 15]. The low-temperature hydro-treating step has been changed extensively regarding catalyst and reaction temperature used in recent years. The general trend is that more reactive catalysts for HYD at lower temperatures are used for the stabilization step. For instance, conventional hydrotreating catalyst, such as supported CoMo sulfide, was used at ~200–300°C when the two-step process was developed, supported Ru sulfide catalyst was then used at ~160–250°C, and, in the most recent years, noble-metal catalyst in reduced form was used to stabilize bio-oil at ~120–160°C. By using more reactive catalysts and lower temperature for the stabilization step, the lifetime of catalyst and operation of both the stabilization step and the second step were sig-nificantly improved. Such a trend indicates that condensation/polymeriza-tion reactions of reactive species in bio-oil are very rapid at high temperatures. Therefore the reaction to stabilize bio-oil by using a catalyst must be fast enough to surpass the condensation/polymerization reactions at a relatively low temperature at which the latter reactions are slower. Next, several examples will be given to show how bio-oil is stabilized by using different catalysts and conditions and how this stabilization enables bio-oil stabilization and high-temperature processing.

The early work done by Elloitt and Baker has used a CoMo sulfide catalyst for low-temperature (273°C) hydrotreating of a woody bio-oil [4]. The results showed that the bio-oil was significantly upgraded and ther-mal stability was improved. Chemical composition analysis by GC-MS indicated that carbonyl components were reduced and the olefinic side chains were saturated. Some HYD of the aromatic rings also occurred. The produced bio-oil could be distilled batchwise without coke formation in the pot, indicating an enhancement of stability of bio-oil. Based on these results, the two-step process was developed and then demonstrated to convert the bio-oil to fuel-range hydrocarbons. However, such a process still suffers from char formation and even reactor plugging problems at longer time on stream. For instance, after the testing using CoMo sulfide catalyst at 273°C, a large amount of char material was found in the reactor and the catalyst pellets carried a high carbon loading (7–13%). It indicates that the HYD activity of CoMo sulfide was not sufficient to compete with the condensation/polymerization reaction in this temperature range.

To further promote the stabilization step, noble-metal catalysts, which showed good performance for HYD at low temperatures, were then considered. Sulfided ruthenium (RuS_x) has been used for the HYD of sugars and polyols without the high methane production of Ru metal [26, 27]. RuS_x was used as the first step catalyst, together with promoted sulfide Mo-based catalyst, for bio-oil hydrotreating. The use of RuS_x also allows the nonisothermal operation by placing RuS_x and Mo sulfide catalysts in a single reactor with different temperature zones, RuS_x at around 170°C and promoted Mo sulfide catalysts at around 400°C. The use of RuS_x significantly extended the time on stream of bio-oil hydrotreating at ~0.19 liquid hourly space velocity, which enabled the steady-state operation to have reliable yield data and sufficient fuel product for evaluation [26]. However, catalyst bed plugging was always observed at 90–99 h on stream with typical bio-oil [27]. It resulted from carbon fouling of the catalyst at the point where the temperature was increased to the second step temperature, indicating the RuS_x was not able to fully stabilize bio-oil.

Noble-metal catalysts in their reduced form with their excellent low-temperature HYD ability were then considered for bio-oil stabilization. Bio-oil model compounds studied by Elliott and Hart [29, 30] used Pd/C and Ru/C catalysts in a batch reactor at 150–300°C. The results showed that at lower temperatures (<200°C), HYD of furfural and guaiacol to their saturated products occurs on both catalysts and acetic acid could only be hydrogenated over Pd at a higher temperature (300°C). Ru appeared to be a more active catalyst for HYD of these oxygenates in aqueous phase than Pd. HYD of individual components in bio-oil, such as sugars, aldehydes, ketones, and acids, over a reduced noble-metal catalyst such as Ru have been widely reported, although without the purpose of

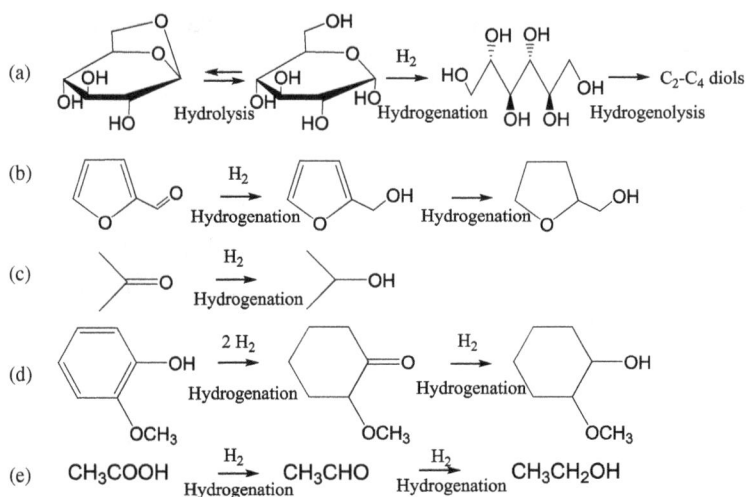

Figure 2.2. HYD of representative compounds (a) levoglucosan and glucose, (b) furfural, (c) acetone, (d) guaiacol, and (e) acetic acid over reduced noble-metal catalysts. (Redrawn with permission from Ref. [28]. Copyright © 2016 American Chemical Society.)

bio-oil stabilization. For instance, D-glucose could be hydrogenated over Ru/C at 100–180°C with the primary product of sorbitol and byproducts such as C_3–C_6 sugar alcohols as shown in Figure 2.2 (a) [31, 32]. Levoglucosan could be converted over Ru/C catalyst at 125–160°C via levoglucosan hydrolysis to glucose followed by glucose HYD (Figure 2.2 (a)) [33]. As discussed above, a fast HYD of furfural to tetrahydrofurfuryl alcohol could occur on Ru/C at 150°C (Figure 2.2 (b)) [22]. HYD of acetone to 2-propanol over Ni/SiO$_2$ catalyst at >100°C was reported (Figure 2.2 (c)) [34]. Phenol could be hydrogenated to cyclohenxanol at 200°C on a Pd/C catalyst [35] and guaiacol could be hydrogenated to methoxycyclohexanol at 150–200°C on an Ru/C catalyst (Figure 2.2(d)) [22]. Carboxylic acids, such as acetic acid, could be hydrogenated on an Ru catalyst at 160°C with a high selectivity to ethanol (Figure 2.2 (e)) [36, 37]. Apparently, these species with specific function groups could be HYD by Ru catalyst at a temperature from 100°C to 200°C with different reactivity and therefore temperature requirement to get high conversion.

Similar HYD reactions occurred during stabilization of actual bio-oil by HYD over Ru catalyst under conditions relevant to practical bio-oil hydrotreating processes. Wang *et al.* [28] conducted a detailed investigation of HYD of woody bio-oil over an Ru/TiO_2 catalyst at 120–160°C, 12.4 MPa, and 0.4 h^{-1} LHSV. HYD was the dominant reaction with very limited deoxygenation. Various components of the bio-oil, including sugars, aldehydes, ketones, alkenes, aromatics, and carboxylic acids, were hydrogenated and consequently chemical and physical properties of the bio-oil were significantly changed. Analysis of a series of sample with different HYD extent showed the following HYD reaction rate sequence: sugar conversion to sugar alcohols, followed by ketone and aldehyde conversion to alcohols, followed by alkene and aromatic HYD, and then followed by carboxylic acid HYD to alcohols. This sequence was in agreement with the reaction temperature required for hydrogenating model compounds as discussed above. They also reported the selectivity of hydrogen addition to the individual components, as shown in Figure 2.3. It indicated that HYD of carbonyls in sugar, aldehyde, and ketones consumed ~50% of the overall hydrogen added. Phenolic compounds were the second group of components that consumed hydrogen, and carboxylic acids consumed the least hydrogen. By HYD, the concentration of reactive

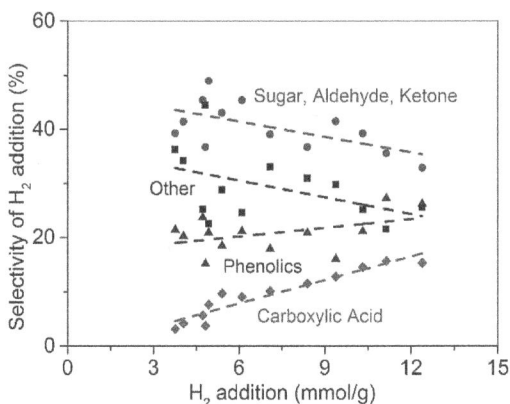

Figure 2.3. Selectivity of hydrogen additions to different components of bio-oil at different hydrogen additions. (Adapted with permission from Ref. [28]. Copyright © 2016 American Chemical Society.)

species such as aldehydes, ketones, and sugars, which could be titrated as carbonyl content in bio-oil, decreased significantly and therefore the stability of bio-oil was significantly improved. As reported by Mareifel *et al.* [38] from the same group, the low temperature HYD of bio-oil at 140°C over an Ru catalyst enabled the successful operation of RuS$_x$ and promoted sulfide Mo two-stage hydrotreating over 1,400 h, which was significantly higher than 100 h historically obtained without the HYD step.

Venderbosch *et al.* [39] conducted the HYD of bio-oil on Ru/C at 175–225°C. The hydrogenated bio-oil was analyzed by solvent fractionation method. The major observation was that aldehydes, ketones, and acids were converted and the sugar fraction was reduced considerably from 35 wt.% to 24 wt.%. Huber *et al.* [40, 41] conducted the low-temperature HYD of light oxygenates in the aqueous fraction of bio-oil using Ru catalysts. The major reactions included HYD of hydroxy-ketones into diols, ketones and aldehydes into mono-alcohols, levoglucosan to sorbitol, and phenols to cyclohexanol. The low-temperature HYD stabilized the bio-oil so that higher-temperature HYD could be attained to produce useful products (gasoline-cuts and diols).

Deactivation of HYD catalyst for bio-oil stabilization is considered to be a critical issue to be addressed [28, 42, 35]. There are several potential deactivation modes: catalyst fouling by carbonaceous species formed by condensation reaction of active species in bio-oil, catalyst poisoning by sulfur species, nitrogen species, or inorganics in bio-oil, and loss or structural change of active sites or support. Analysis of deactivated RuS$_x$ catalyst for bio-oil HYD showed the major deactivation mode was fouling of catalysts by condensation products of the active species, which was in agreement with the insufficient HYD activity of RuS$_x$ [42]. For catalyst in reduced form, sulfur poisoning was believed to one of the primary deactivation modes [28, 35]. A sulfur-poisoned catalyst could then result in slow HYD competing with the condensation reaction and further lead to the formation of carbonaceous species that could further deactivate catalyst by fouling. Leaching of metal in acidic bio-oil was also observed in some reported work [35]. Most of the catalysts for bio-oil HYD used hydrothermally stable material, such as carbon or TiO_2, as support to avoid the potential change of support, which could occur on certain type of materials such as Al_2O_3.

Besides the thermochemical HYD of bio-oil using H_2, catalytic transfer HYD and electrocatalytic hydrogenation (ECH) have also been used for bio-oil HYD. The aim was mainly to avoid the requirement for molecular hydrogen (which is often sourced from fossil fuels) and lower the temperature requirement for bio-oil HYD. Kannapu *et al.* [43] conducted the reduction of bio-oil and model compounds using transfer HYD from isopropanol with supported Ni–Cu catalyst at 300°C in a batch reactor. Ring HYD of phenolics and reduction of aldehydes to alcohols were observed, and bio-oils from tail gas reactive pyrolysis were upgraded to a product with a higher level of hydrogen. Xu *et al.* [44] reported the transfer HYD of bio-oil with different hydrogen donors using a Ni catalyst at 230°C in a batch reactor. The results showed that the hydrogen donors methanol, ethanol, and formic acid could all provide hydrogen for the transfer HYD with different product distributions. HYD of ketones, aldehydes, acids, and phenols were observed, and some other reactions such as esterification also occurred.

ECH is a novel alternative approach to stabilize bio-oil considering that it could occur at near room temperature at aqueous phase without external H_2 supply. Li *et al.* [45, 46] reported the HYD of model compounds and aqueous fraction of bio-oil using ECH at very mild conditions (≤80°C and ambient pressure) compared to hydrotreatment. Ru-based electrocatalysts were used as the catalytic cathode. HYD of aldehydes, ketones, and phenols to the corresponding alcohols were achieved with more than 80% carbon recovery into the liquid product of bio-oil aqueous fraction. The stability evaluation of the ECH-treated bio-oil fraction using accelerated aging test showed the much-improved stability. ECH provides a very promising and unique method for HYD of bio-oil or other biomass-derived intermediate, and more research is expected in the near future to address some challenges such as feedstock complexity and energy efficiency.

2.4. Other bio-oil stabilization approaches

Beside the hydrotreating, other chemical or physical approaches have been developed to stabilize bio-oil, with a primary focus of improving the storage and transport stability of bio-oil [17]. Esterification has been well studied for bio-oil stabilization, in which carboxylic acids and aldehydes in bio-oil could react with alcohols to produce esters and acetals. Lohitharn and Shanks [47] conducted model compound studies of organic acid esterification and reported that both esterification and acetalization occurred and acetalization of the aldehydes with ethanol was faster than acetic acid esterification at 50–70° C. Xiong *et al.* [48] used an ionic liquid as catalyst for bio-oil esterification with ethanol at room temperature. Organic acid was converted and the produced oil layer of bio-oil had improved heating value, higher pH, and lower viscosity. Wang *et al.* [49] also reported the significant decrease of acidity of bio-oil by esterification with methanol using ion-exchange resins as catalyst. Li *et al.* [50] reported the simultaneous esterification and acetalization of a woody bio-oil with methanol using Amberlyst-70 catalyst at 70–170°C. Light organic acids and aldehydes were converted to esters and acetals. Some acetals could decompose at higher operating temperatures (>110°C). However, as reported by Moens *et al.* [51], treating crude bio-oil with alcohols in the presence of a solid acid catalyst through these reactions was not sufficient for neutralizing and stabilizing bio-oils because of the equilibrium limitation in aqueous environment and inefficient conversion of carbonyl compounds (inherent hydrolytic instability of formed acetals). Nevertheless, the reduction of pH of bio-oil could mitigate the corrosiveness and slow down condensation reaction during storage.

Physical methods, such as solvents addition [17, 52] or char removal by filtration [17, 53, 54], were also tested to enhance the storage and transport stability of bio-oils. Oasmaa *et al.* [52] reported that the addition of alcohols retarded the aging reactions of bio-oil and methanol was the most effective one of the alcohols tested (methanol, ethanol, isopropanol). It was believed to be because of dilution of reactive species, especially the hydrophobic high molecular mass lignin-derived fraction, and reaction of alcohols with aldehydes, ketones, and sugars. Baldwin and Feik [53] applied hot gas filtration during fast pyrolysis to reduce the alkali and alkaline earth metals and total solids content in produced bio-oil. They found that bio-oil obtained by hot gas filtration with a ceramic filter element showed much-improved stability (10-fold reduced from the raw oil for the rate of viscosity increase) based on accelerated aging test. Similarly, Pattiya and Suttibak [54] found significant improved bio-oil stability measured by viscosity change after aging when a hot vapor filter was applied.

Catalytic fast pyrolysis (CFP) is a process that applied catalyst in pyrolysis reactor to react with pyrolysis vapor and there to produce upgraded bio-oil products [6, 55, 56]. It normally utilizes acidic catalysts such as zeolite to conduct reaction such as decarbonylation, decarboxylation, dehydration, HYD, aromatization, and ketonization/condensation. It could effectively remove reactive species normally found in regular bio-oil and therefore produce more stable bio-oil which could be directly upgraded by hydrotreating without the stabilization step. However, the major challenge of CFP was the lower bio-oil yield and fast catalyst deactivation. Significant work has been done recently on CFP, and several reviews have been published [6, 55, 56].

2.5. Conclusions

Instability of bio-oil is the major barrier in pyrolysis and upgrading pathway of biomass conversion to produce biofuels. In order to enable a sustainable and economically viable bio-oil hydrotreating process, effective technologies for bio-oil stabilization must be developed. The bio-oil instability is mainly because of condensation/polymerization reaction of reactive species such aldehydes, ketones, sugars, and phenolic in bio-oil. Therefore, to slow down these reactions or to remove these species, physical methods, such as solvents addition, and chemical methods such as esterification or mild HYD were developed to stabilize bio-oil. Bio-oil hydrotreating, especially HYD, has been demonstrated to stabilize bio-oil by hydrogenating the reactive species into a stable form such as aldehydes to alcohols and therefore enable the upgrading of stabilized bio-oil to produce a fuel-range hydrocarbon with long-term steady-state operation.

Successful stabilization of bio-oil by HYD has significantly moved forward toward understanding biofuel quality and the correlation between the source of the biomass and final biofuel quality; this has also improved the economic feasibility of the fast pyrolysis-upgrading technology. Catalyst development should be the key for the further improvement of bio-oil HYD technology with a focus on design of catalyst for conducting fast HYD of complex mixtures at low temperature and in an acidic aqueous environment. Both metal site, supports, and the microenvironment could be finely tuned. Low-cost metal catalysts should be considered to further lower the cost for the whole process, and catalyst regeneration protocols should be developed as well. Well-defined criteria to evaluate bio-oil stability regarding their processability at high temperature hydrotreating process should also be established.

References

1. Huber, G. W., Iborra, S., and Corma, A. (2006). Synthesis of transportation fuels from biomass: chemistry, catalysts, and engineering. *Chemical Reviews*, 106(9), 4044–4098.
2. Gallezot, P. (2012). Conversion of biomass to selected chemical products. *Chemical Society Reviews*, 41, 1538–1558.
3. Zhou, C.-H., Xia, X., Lin, C.-X., Tong, D.-S., and Beltramini, J. (2011). Catalytic conversion of lignocellulosic biomass to fine chemicals and fuels. *Chemical Society Reviews*, 40(11), 5588–5617.
4. Elliott, D. C. (2011). Historical developments in hydroprocessing bio-oils. *Energy & Fuels*, 21(3), 1792–1815.
5. Wang, H., Male, J., and Wang, Y. (2013). Recent advances in hydrotreating of pyrolysis bio-oil and its oxygen-containing model compounds. *ACS Catalysis*, 3(5), 1047–1070.
6. Liu, C., Wang, H., Karim, A. M., Sun, J., and Wang, Y. (2014). Catalytic fast pyrolysis of lignocellulosic biomass. *Chemical Society Reviews*, 43(22), 7594–7623.
7. Zacher, A. H., Olarte, M. V., Santosa, D. M., Elliott, D. C., and Jones, S. B. (2014). A review and perspective of recent bio-oil hydrotreating research. *Green Chemistry*, 16(2), 491–515.
8. Bulushev, D. A. and Ross, J. R. H. (2011). Catalysis for conversion of biomass to fuels via pyrolysis and gasification: a review. *Catalysis Today*, 171(1), 1–13.
9. Bridgwater, A. V. (2012). Review of fast pyrolysis of biomass and product upgrading. *Biomass and Bioenergy*, 38, 68–94.
10. Elliott, D. C., Baker, E. G., Beckman, D., Solantausta, Y., Tolenhiemo, V., Gevert, S. B., Hornell, C., Ostman, A., and Kjellstrom, B. (1990). Technoeconomic assessment of direct biomass liquefaction to transportation fuels. *Biomass*, 22(1–4), 251–269.

11. Mortensen, P. M., Grunwaldt, J. D., Jensen, P. A., Knudsen, K. G., and Jensen, A. D. (2011). A review of catalytic upgrading of bio-oil to engine fuels. *Applied Catalysis A: General*, 407(1–2), 1–19.
12. Agrawal, R. and Singh, N. R. (2009). Synergistic routes to liquid fuel for a petroleum-deprived future. *AIChE Journal*, 55(7), 1898–1905.
13. Czernik, S. and Bridgwater, A. V. (2004). Overview of applications of biomass fast pyrolysis oil. *Energy & Fuels*, 18(2), 590–598.
14. Oasmaa, A., Elliott, D. C., and Korhonen, J. (2010). Acidity of biomass fast pyrolysis bio-oils. *Energy & Fuels*, 24, 6548–6554.
15. Furimsky, E. (2004). Hydroprocessing challenges in biofuels production. *Catalysis Today*, 217, 13–56.
16. Meng, J., Moore, A., Tilotta, D., Kelley, S., and Park, S. (2014). Toward understanding of bio-oil aging: accelerated aging of bio-oil fractions. *ACS Sustainable Chemistry & Engineering*, 2(8), 2011–2018.
17. Yang, Z., Kumar, A., and Huhnke, R. L. (2015). Review of recent developments to improve storage and transportation stability of bio-oil. *Renewable and Sustainable Energy Reviews*, 50, 859–870.
18. Howe, D., Westover, T., Carpenter, D., Santosa, D., Emerson, R., Deutch, S., Starace, A., Kutnyakov, I., and Lukins, C. (2015). Field-to-fuel performance testing of lignocellulosic feedstocks: an integrated study of the fast pyrolysis–hydrotreating pathway. *Energy & Fuels*, 29(5), 3188–3197.
19. Oasmaa, A., Kuoppala, E., and Elliott, D. C. (2012). Development of the basis for an analytical protocol for feeds and products of bio-oil hydrotreatment. *Energy & Fuels*, 26(4), 2454–2460.
20. Furimsky, E. (2000). Catalytic hydrodeoxygenation. *Applied Catalysis A: General*, 199(2), 147–190.
21. Czernik, S., Johnson, D. K., and Black, S. (1994). Stability of wood fast pyrolysis oil. *Biomass and Bioenergy*, 7(1), 187–192.
22. Oasmaa, A. and Kuoppala, E. (2003). Fast pyrolysis of forestry residue. 3. Storage stability of liquid fuel. *Energy & Fuels*, 17(4), 1075–1084.
23. Alsbou, E. and Helleur, B. (2014). Accelerated aging of bio-oil from fast pyrolysis of hardwood. *Energy & Fuels*, 28(5), 3224–3235.
24. Lindfors, C., Paasikallio, V., Kuoppala, E., Reinikainen, M., Oasmaa, A., and Solantausta, Y. (2015). Co-processing of dry bio-oil, catalytic pyrolysis oil, and hydrotreated bio-oil in a micro activity test unit. *Energy & Fuels*, 29(6), 3707–3714.
25. Hu, X., Wang, Y., Mourant, D., Gunawan, R., Lievens, C., Chaiwat, W., Gholizadeh, M., Wu, L., Li, X., and Li, C.-Z. (2013). Polymerization on

heating up of bio-oil: a model compound study. *AIChE Journal*, 59(3), 888–900.

26. Elliott, D. C., Hart, T. R., Neuenschwander, G. G., Rotness, L. J., Olarte, M. V., Zacher, A. H., and Solantausta, Y. (2012). Catalytic hydroprocessing of fast pyrolysis bio-oil from pine sawdust. *Energy & Fuels*, 26(6), 3891–3896.

27. Montassier, C., Ménézo, J. C., Hoang, L. C., Renaud, C., and Barbier, J. (1991). Aqueous polyol conversions on ruthenium and on sulfur-modified ruthenium. *Journal of Molecular Catalysis*, 70(1), 99–110.

28. Wang, H., Lee, S.-J., Olarte, M. V., and Zacher, A. H. (2016). Bio-oil stabilization by hydrogenation over reduced metal catalysts at low temperatures. *ACS Sustainable Chemistry & Engineering*, 4(10), 5533–5545.

29. Elliott, D. C. and Hart, T. R. (2009). Catalytic hydroprocessing of chemical models for Bio-oil. *Energy & Fuels*, 23(1), 631–637.

30. Elliott, D. C., Hart, T. R., Neuenschwander, G. G., Rotness, L. J., and Zacher, A. H. (2009). Catalytic hydroprocessing of biomass fast pyrolysis bio-oil to produce hydrocarbon products. *Environmental Progress and Sustainable Energy*, 28(3), 441–449.

31. Crezee, E., Hoffer, B. W., Berger, R. J., Makkee, M., Kapteijn, F., and Moulijn, J. A. (2003). Three-phase hydrogenation of d-glucose over a carbon supported ruthenium catalyst — mass transfer and kinetics. *Applied Catalysis A: General*, 251(1), 1–17.

32. Lazaridis, P. A., Karakoulia, S., Delimitis, A., Coman, S. M., Parvulescu, V. I., and Triantafyllidis, K. S. (2015). D-Glucose hydrogenation/hydrogenolysis reactions on noble metal (Ru, Pt)/activated carbon supported catalysts. *Catalysis Today*, 257, 281–290.

33. Bindwal, A. B. and Vaidya, P. D. (2013). Kinetics of aqueous-phase hydrogenation of levoglucosan over Ru/C catalyst. *Industrial & Engineering Chemistry Research*, 52(50), 17781–17789.

34. Witsuthammakul, A. and Sooknoi, T. (2015). Selective hydrodeoxygenation of bio-oil derived products: ketones to olefins. *Catalysis Science & Technology*, 5(7), 3639–3648.

35. Zhao, C., He, J., Lemonidou, A. A., Li, X., and Lercher, J. A. (2011). Aqueous-phase hydrodeoxygenation of bio-derived phenols to cycloalkanes. *Journal of Catalysis*, 280(1), 8–16.

36. Wan, H., Chaudhari, R. V., and Subramaniam, B. (2013). Aqueous phase hydrogenation of acetic acid and its promotional effect on p-cresol hydrodeoxygenation. *Energy & Fuels*, 27(1), 487–493.

37. Olcay, H., Xu, L. J., Xu, Y., and Huber, G. W. (2010). Aqueous-phase hydrogenation of acetic acid over transition metal catalysts. *ChemCatChem*, 2(11), 1420–1424.

38. Olarte, M. V., Zacher, A. H., Padmaperuma, A. B., Burton, S. D., Job, H. M., Lemmon, T. L., Swita, M. S., Rotness, L. J., Neuenschwander, G. N., Frye, J. G., and Elliott, D. C. (2015). Stabilization of softwood-derived pyrolysis oils for continuous bio-oil hydroprocessing. *Topics in Catalysis*, 59(1), 55–64.
39. Venderbosch, R. H., Ardiyanti, A. R., Wildschut, J., Oasmaa, A., and Heeresb, H. J. (2010). Stabilization of biomass-derived pyrolysis oils. *Journal of Chemical Technology and Biotechnology*, 85(5), 674–686.
40. Sanna, A., Vispute, T. P., and Huber, G. W. (2015). Hydrodeoxygenation of the aqueous fraction of bio-oil with Ru/C and Pt/C catalysts. *Applied Catalysis B: Environmental*, 165, 446–456.
41. Bergem, H., Xu, R., Brown, R. C., and Huber, G. W. (2017). Low temperature aqueous phase hydrogenation of the light oxygenate fraction of bio-oil over supported ruthenium catalysts. *Green Chemistry* (in press).
42. Wang, H. and Wang, Y. (2015). Characterization of deactivated bio-oil hydrotreating catalysts. *Topics in Catalysis*, 59(1), 65–72.
43. Reddy Kannapu, H. P., Mullen, C. A., Elkasabi, Y., and Boateng, A. A. (2015). Catalytic transfer hydrogenation for stabilization of bio-oil oxygenates: reduction of p-cresol and furfural over bimetallic Ni–Cu catalysts using isopropanol. *Fuel Processing Technology*, 137, 220–228.
44. Xu, Y., Li, Y., Wang, C., Wang, C., Ma, L., Wang, T., Zhang, X., and Zhang, Q. (2017). In-situ hydrogenation of model compounds and raw bio-oil over Ni/CMK-3 catalyst. *Fuel Processing Technology*, 161, 226–231.
45. Li, Z., Garedew, M., Lam, C. H., Jackson, J. E., Miller, D. J., and Saffron, C. M. (2012). Mild electrocatalytic hydrogenation and hydrodeoxygenation of bio-oil derived phenolic compounds using ruthenium supported on activated carbon cloth. *Green Chemistry*, 14(9), 2540–2549.
46. Li, Z., Kelkar, S., Raycraft, L., Garedew, M., Jackson, J. E., Miller, D. J., and Saffron, C. M. (2015). A mild approach for bio-oil stabilization and upgrading: electrocatalytic hydrogenation using ruthenium supported on activated carbon cloth. *Green Chemistry*, 16(2), 844–852.
47. Lohitharn, N. and Shanks, B. H. (2009). Upgrading of bio-oil: effect of light aldehydes on acetic acid removal via esterification. *Catalysis Communications*, 11(2), 96–99.
48. Xiong, W.-M., Zhu, M.-Z., Deng, L., Fu, Y., and Guo, Q.-X. (2009). Esterification of organic acid in bio-oil using acidic ionic liquid catalysts. *Energy & Fuels*, 23(4), 2278–2283.
49. Wang, J.-J., Chang, J., and Fan, J. (2010). Upgrading of bio-oil by catalytic esterification and determination of acid number for evaluating esterification degree. *Energy & Fuels*, 24(5), 3251–3255.

50. Li, X., Gunawan, R., Lievens, C., Wang, Y., Mourant, D., Wang, S., Wu, H., Garcia-Perez, M., and Li, C.-Z. (2011). Simultaneous catalytic esterification of carboxylic acids and acetalisation of aldehydes in a fast pyrolysis bio-oil from mallee biomass. *Fuel*, 90(7), 2530–2537.

51. Moens, L., Black, S. K., Myers, M. D., and Czernik, S. (2009). Study of the neutralization and stabilization of a mixed hardwood bio-oil. *Energy & Fuels*, 23(5), 2695–2699.

52. Oasmaa, A., Kuoppala, E., Selin, J.-F., Gust, S., and Solantausta, Y. (2004). Fast pyrolysis of forestry residue and pine. 4. Improvement of the product quality by solvent addition. *Energy & Fuels*, 18(5), 1578–1583.

53. Baldwin, R. M. and Feik, C. J. (2013). Bio-oil stabilization and upgrading by hot gas filtration. *Energy Fuels*, 27(6), 3224–3238.

54. Pattiya, A. and Suttibak, S. (2013). Influence of a glass wool hot vapour filter on yields and properties of bio-oil derived from rapid pyrolysis of paddy residues. *Bioresource Technology*, 116, 107–113.

55. Dickerson, T. and Soria, J. (2013). Catalytic fast pyrolysis: a review. *Energies*, 6(1), 514–538.

56. Zheng, A., Jiang, L., Zhao, Z., Huang, Z., Zhao, K., Wei, G., and Li, H. (2013). Catalytic fast pyrolysis of lignocellulosic biomass for aromatic production: chemistry, catalyst and process. *Wiley Interdisciplinary Reviews: Energy and Environment*, 6(3), e234.

Chapter 3

Hydroprocessing Catalysts: Inexpensive Ni-Based Nonsulfided Catalysts

Jing Liu and Jiandu Lei*

Beijing Key Laboratory of Lignocellulosic Chemistry, Beijing Forestry University, Beijing 100083, China
Corresponding author: ljd2012@bjfu.edu.cn

Abstract

Conventional hydroprocessing catalysts are usually sulfided to retain their active forms. However, sulfuration may cause sulfur dioxide emissions, corrosion of equipment and sulfur residues in the products, since plant oils are free of sulfur compounds. The high price of noble metals also limits their industrial applications. Therefore, the nonsulfided nonnoble-metal catalysts, which can get rid of the presulfurization step and avoid the harm of sulfide to environment and human health, have received a lot of research interests. Recent development of less expensive metallic catalysts, like nickel (Ni)-based catalysts, has shown promising results. The purpose of this chapter is to review current developments in the species and application of inexpensive Ni-based nonsulfided hydroprocessing catalysts.

Keywords: Ni-based catalysts, Nonsulfided, Hydroprocessing

3.1. Introduction

Hydroprocessing is a well-known technology in the petroleum refining industry, which can be carried out either by hydrocracking technology or by the less severe hydrotreating technology [1]. Biofuels show promising potential in the manufacture of liquid fuels. An alternative technology for biofuels production technology, which employs the existing infrastructure of petroleum refineries, is the catalytic hydroprocessing of renewable oils, such as plant oil [2]. There are several issues that need to be addressed prior to hydroprocessing to be widely acceptable as a convenient route for the production of transportation fuel: (1) the process should be insensitive to the free fatty acid content of the feed, since the free fatty acid could vary from 1% to 25% depending on the source of feedstock, (2) hydroprocessing catalysts should be selective for desired transportation fuels, and (3) catalysts should be stable as well as recyclable [3]. Hydroprocessing of renewable oils is compatible with a petroleum refinery, and it could become a regular part of an oil refinery in the future [4].

So far NiMo-, CoMo-, Pt- and Pd-based catalysts with porous SiO_2–Al_2O_3 and zeolite supports have been used in the hydroprocessing processes [1, 5]. On one hand, presulfiding catalysts with H_2S promotes deoxygenation of fuels over both NiMo and CoMo catalysts, regenerates the sulfide sites, and protects the catalyst from desulfurization and deactivation [6]. However, the resulting fuels are contaminated by sulfur, which loses the advantage of sulfur-free fuels [7]. On the other hand, the sintering of noble metals as well as their high cost become obstacles to their industrial utilization (Figure 3.1(a)). Based on this view, there is an apparent need to develop an inexpensive nonsulfided catalyst for the hydroprocessing [8]. Zhao *et al.* [6] has reviewed nonsulfided metal catalysts for hydrodeoxygenation (HDO) of microalgae oil to green hydrocarbons. Kordulis *et al.* [9] also reviewed low cost Ni-based nonsulfide catalysts for the transformation of natural triglycerides and related compounds into green diesel. There is an increasing number of articles devoted to Ni-based nonsulfided catalysts, mainly during the last 4 y. Figure 3.1(b) illustrates the increasing rate of articles on the subject. The Ni-based nonsulfided catalysts had been proved quite promising in many hydroprocessing reactions.

(a)

(b)

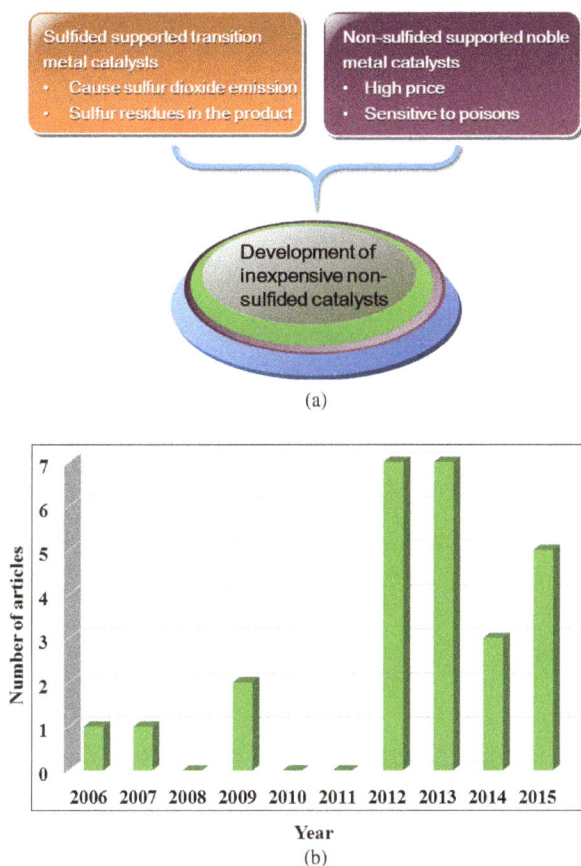

Figure 3.1. From the sulfided transition metal and nonsulfided noble-metal catalysts to inexpensive nonsulfided catalysts (a) and the articles reported for the Ni-based nonsulfided catalysts in the last decade (b).

3.2. Rare-earth metal supported Ni-based catalysts

3.2.1. *NiMoLa/Al$_2$O$_3$ and NiMoCe/Al$_2$O$_3$ catalyst*

The nonsulfided NiMoLa/Al$_2$O$_3$ and NiMoCe/Al$_2$O$_3$ catalyst were developed to produce renewable C$_{15}$–C$_{18}$ alkanes from the hydroprocessing of *Jatropha* oil [10, 11]. Table 3.1 shows that the maximum yield of C$_{15}$–C$_{18}$ alkanes of 78%, selectivity of 94% and conversion of 83% were obtained at 370°C, 3.5 MPa and 0.9 h^{-1}. The experimental results demonstrated that

Table 3.1. Hydroprocessing of *Jatropha* oil on different catalysts (370°C, 3.5 MPa, 0.9 h^{-1}) [10].

Catalyst	NiMo	Sulfided NiMo	NiMoLa(0.5)	NiMoLa(5.0)	NiMoLa(15.0)
<C$_{15}$	7.65	5.70	9.31	4.60	7.61
C$_{15}$–C$_{18}$	53.63	77.29	60.31	78.37	69.88
Other hydrocarbons and esters[a]	37.87	16.98	30.34	16.97	22.44
(C$_{15}$ + C$_{17}$)/ (C$_{16}$ + C$_{18}$)	0.97	0.73	1.35	1.46	2.30
Conversion	67.48	95.37	73.09	82.96	88.59
C$_{15}$–C$_{18}$ selectivity	79.48	81.04	82.51	94.47	78.88

Note: [a]Not specified hydrocarbons and esters of heavy fraction.

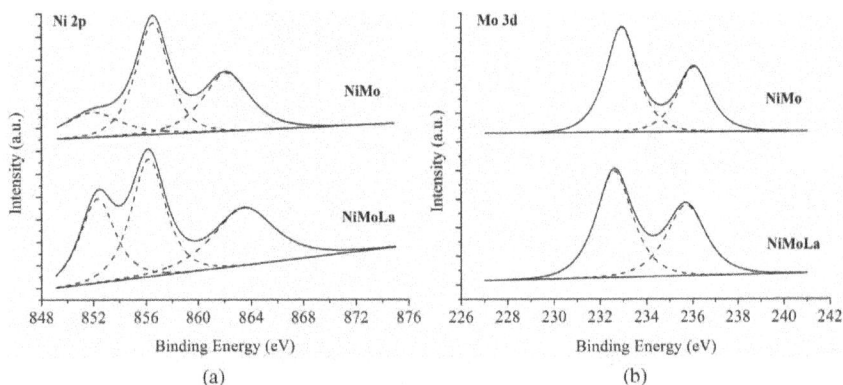

Figure 3.2. X-ray photoelectron spectroscopy (XPS) spectra of (a) Ni 2p and (b) Mo 3d levels of NiMo/Al$_2$O$_3$ catalyst and NiMoLa (5.0)/Al$_2$O$_3$ catalyst. (Reproduced with permission from Ref. [10]. Copyright © 2012 Royal Society of Chemistry.)

a suitable amount of rare-earth metal (5.0 wt.%) doping on the NiMo/Al$_2$O$_3$ catalyst presented stable catalytic performance (177 h) and enhanced *Jatropha* oil conversion as well as C$_{15}$–C$_{18}$ fraction selectivity.

From XPS results (see Figure 3.2 and Table 3.2), La loading can promote the reduction of Ni^{2+} and the oxidation of Mo^{4+}. This implied that

Table 3.2.   Relative peak areas and ratios of Ni and Mo in the XPS analysis [10].

Catalyst	Ni (%)			Mo (%)		
	Ni^0	Ni^{2+}	$[Ni^0]/[Ni^{2+}]$	Mo^{4+}	Mo^{6+}	$[Mo^{6+}]/[Mo^{4+}]$
NiMo	18.0	82.0	22.0	60.6	39.4	65.0
NiMoLa	28.2	71.8	39.3	57.6	42.4	73.6

Mo species might donate partial electrons to Ni oxide species, resulting in the electron transfer between Mo and Ni due to La loading. The temperature-programed desorption (TPD)–H_{ads} spectra showed that La markedly enhanced H_2 uptake as compared with $NiMo/Al_2O_3$ catalyst. This is because the electron density on Ni oxide species increased after La was added to $NiMo/Al_2O_3$ catalyst, as confirmed by XPS results, leading to the formation of more activating centers. Then the hydrogenation (HYD) of $NiMo/Al_2O_3$ catalyst was improved by adding La. As shown in thermogravimetric analysis (TGA) results; the amount of carbon deposited on $NiMoLa(5.0)/Al_2O_3$ catalyst (1.93%) was lower than that on $NiMo/Al_2O_3$ catalyst (3.16%). Compared with the reaction of 10 h, the carbon residue of $NiMoLa/Al_2O_3$ catalyst (2.20%) after reaction of 177 h was only increased by 0.27%. It can be deduced that adding La to NiMo catalysts could decrease carbon residues to some extent.

However, the conversion over $NiMoLa/Al_2O_3$ catalyst was still lower than sulfided $NiMo/Al_2O_3$ catalyst at the temperature from 280°C to 400°C. Thus, in order to improve the catalytic activity and increase the conversion of *Jatropha* oil, we are searching for other additives and supports, such as heteropolyacid (HPA) and porous materials.

3.2.2.   *Effect of reducing catalyst coke by La loading*

During the hydroprocessing reaction, the carbonaceous compound (coke) will deposit on the catalyst. As the amount of coke increases, it will cover the active centers and block the pores of catalysts, leading to deactivation in the long-term reaction [12]. In Section 3.2.1, adding La to Ni-based catalysts could decrease carbon residues. In order to study the mechanism of reducing coke formation, La was loaded into NiW/Al_2O_3,

NiW/nanohydroxyapatite (nHA) and NiW/HY catalysts for hydroprocessing of *Jatropha* oil [13].

The species and the amount of coke on the hydroprocessing catalyst were measured by FT-IR, solid state ^{13}C NMR and TGA. It was demonstrated that the coke was constituted of polyaromatic hydrocarbons and the La loading on catalyst had a very positive effect on coke reduction. As shown in Figure 3.3, the amount of coke on NiW/Al$_2$O$_3$, NiW/nHA and NiW/HY catalysts decreased, respectively, by 0.94%, 1.19%, 1.91% after 8 h and 1.66%, 1.89%, 3.78% after 180 h. The catalytic stability and the catalyst lifetime were also improved by La loading.

3.3. HPA supported Ni-based catalysts

3.3.1. *Effect of supports on the hydroprocessing*

HPAs have attracted considerable attention as efficient catalysts because of their unique composition, which includes heteropolyanions and countercations [14, 15]. The phosphotungstic acid (PTA or HPW) as one of the HPAs supplies the metal sites (W oxide species) and also provides acid sites for hydroprocessing. Ni and PTA supported on different supports (Al$_2$O$_3$, ZIF-8, and nHA) have been investigated [16–18], and the effect of supports on the hydroprocessing of *Jatropha* oil has also been discussed.

The Ni–PTA/Al$_2$O$_3$ catalyst was prepared by one-pot synthesis of Ni/Al$_2$O$_3$ with the coprecipitation method and then impregnating Ni/Al$_2$O$_3$ with PTA solution. Results of Brunauer–Emmett–Teller (BET) surface area measurement (see Table 3.3 and Figure 3.4) and NH$_3$–TPD (see Table 3.4) indicated that the pore structure, physicochemical adsorption properties, and acidity of supports and catalysts can affect hydroprocessing catalytic activity. Scanning electron microscopy with energy dispersive X-ray analysis (SEM-EDX), transmission electron microscopy (TEM), and X-ray diffraction (XRD) results showed that Ni and W species were distributed on the surface of the synthetic Ni–PTA/Al$_2$O$_3$ catalyst evenly and no aggregates were observed. This led to the formation of a more active center. XPS clearly showed the existence of assembled PTA on Ni/Al$_2$O$_3$ catalyst. Ni and W metal oxide/hydroxides not only protected

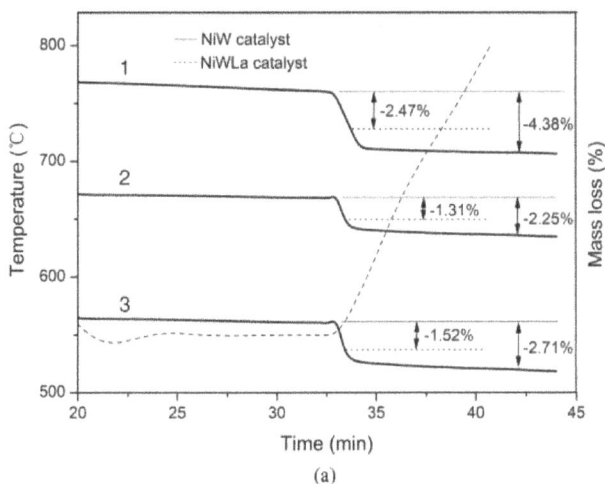

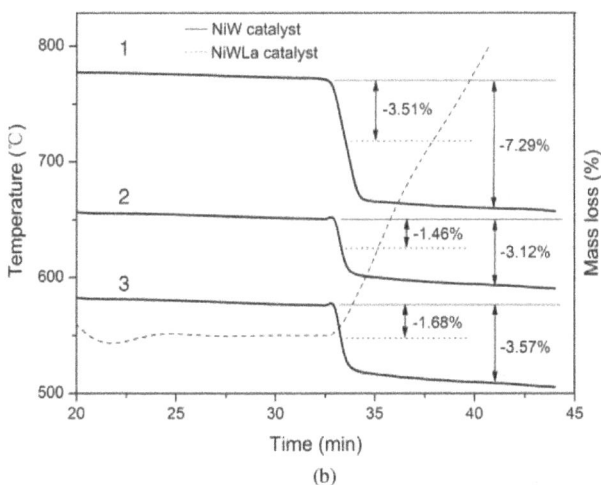

Figure 3.3. TGA profiles of (1) NiW/HY and NiWLa/HY (2) NiW/nHA and NiWLa/nHA (3) NiW/Al$_2$O$_3$ and NiWLa/Al$_2$O$_3$ after use for (a) 8 h and (b) 180 h at 360°C, 3MPa, H$_2$/oil (v/v) = 600, LHSV = 2 h^{-1}. (Reproduced with permission from Ref. [13]. Copyright © 2015 Royal Society of Chemistry.)

the amorphous structure of the catalyst, aiding in HYD activity, but also enhanced the dispersion of active sites. The maximum conversion of *Jatropha* oil (98.5 wt.%) and selectivity of the C$_{15}$–C$_{18}$ alkanes fraction (84.5 wt.%) occurred at 360°C, 3.0 MPa, 0.8 h^{-1}. It exhibited a higher

Table 3.3. Textural properties of different Al_2O_3 samples [16].

Samples	Specific surface areas (m²/g)	Total pore volume (cm³/g)	Average pore diameter (nm)
Commercial Al_2O_3 grain	207	0.38	5.18
Commercial Al_2O_3 powder	257	0.42	4.84
Synthetic Al_2O_3 support	296	0.43	6.32
Synthetic Ni–PTA/Al_2O_3 catalyst	232	0.44	7.49

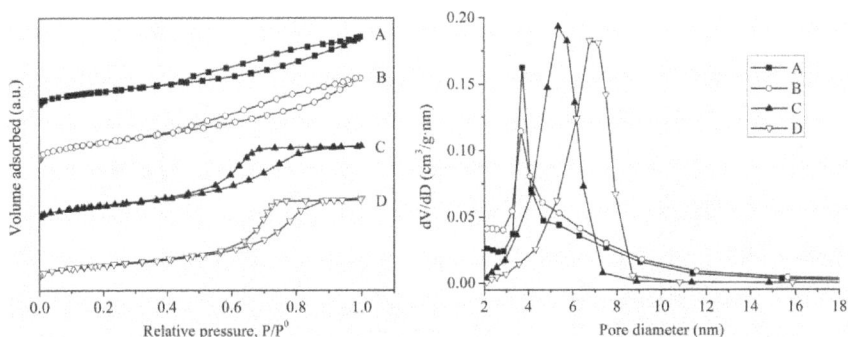

Figure 3.4. Nitrogen adsorption–desorption isotherms and pore size distribution of A — commercial Al_2O_3 grain, B — commercial Al_2O_3 powder, C — synthetic Al_2O_3 support, and D — synthetic Ni–PTA/Al_2O_3 catalyst. (Adapted with permission from Ref. [16]. Copyright © 2015 Creative Commons BY license.)

Table 3.4. Acidity of PTA, synthetic support, and catalysts from NH_3–TPD [16].

Surface acidity (mmol/g NH_3)	Weak acidity (100–300°C)	Strong acidity (300–800°C)
Pure PTA	0.02	1.71
Synthetic Al_2O_3	1.19	0.29
Ni–PTA(10)	0.65	0.67
Ni–PTA(30)	0.53	1.22
Ni–PTA(50)	0.51	1.06

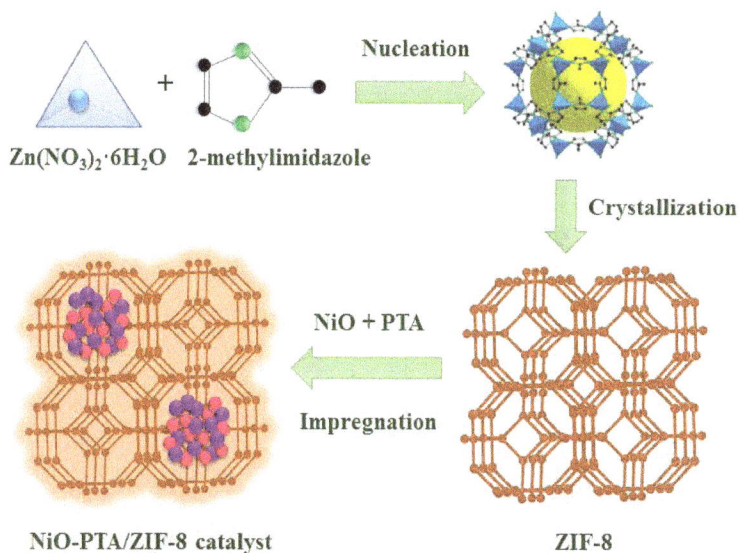

Figure 3.5. Schematic diagram of synthesis of NiO–PTA/ZIF-8 catalyst. (Adapted with permission from Ref. [17]. Copyright © 2016 Creative Commons BY license.)

catalytic activity than the Ni–PTA catalyst supported with commercial Al_2O_3.

ZIF-8, a subfamily of metal organic frameworks (MOFs), is a microporous material with uniform small pores [19, 20]. Figure 3.5 displays the synthetic procedure of NiO–PTA/ZIF-8 catalyst. Compared with the Ni–PTA/commercial Al_2O_3 catalyst, the selectivity of C_{15}–C_{18} hydrocarbon increased over 36%, and catalytic efficiency increased 10 times over the NiO–PTA/ZIF-8 catalyst. The BET surface area of NiO–PTA/commercial Al_2O_3 was only 223 m^2/g, while that of the NiO–PTA/ZIF-8 catalyst was more than 600 m^2/g. Generally, the catalysts at a nanosize scale with a high surface area should provide the catalysts with higher activity. Thus, the NiO–PTA/ZIF-8 catalyst possessed more active sites for hydroprocessing of *Jatropha* oil than the NiO–PTA/commercial Al_2O_3 catalyst. Results are presented in Table 3.5. The kinetic behavior of two catalysts was examined with an assumption of pseudo-first-order reaction

Table 3.5.   Kinetic and thermodynamic parameters for triglyceride conversion of hydrocracking of *Jatropha* oil [17].

Catalysts	k_H $(h^{-1})^a$	E_H $(kJ/mol)^b$
NiO–PTA/Al$_2$O$_3$	0.01657	22.24
NiO–PTA/ZIF-8	0.02392	15.26

Note: [a]Reaction temperature 360°C, H$_2$ pressure 3.0 MPa, 9–36 h^{-1}.

[b]Reaction temperature 280–400°C, H$_2$ pressure 3.0 MPa, 9 h^{-1}.

kinetics. The relative pseudo-first-order rate constant (k_H) of NiO–PTA/ZIF-8 catalyst was 1.4-fold higher than that of the NiO–PTA/commercial Al$_2$O$_3$ catalyst under the same reaction condition, while the activation energy (E_H) of conversion of triglycerides over NiO–PTA/ZIF-8 was lower. This indicates that conversion of triglyceride was easier over NiO–PTA/ZIF-8, due to its higher hydroprocessing activity. This may be related to the metal active center (Ni and W phase). The NiO–PTA/ZIF-8 catalyst showed a high surface area, high metal content and uniform metal dispersion (proven by BET, XPS, and SEM-EDX results), leading to the formation of a more active center and then a marked increase in HYD activity.

Nanosized materials offer different morphology and surface area that could enhance reaction rates [21]. Ni–HPW/nHA catalyst was designed for hydroprocessing of *Jatropha* oil, the effect of phosphotungstic acid (HPW) amount on the isoparaffins and pour point of product oil were investigated. The Ni–HPW(30%)/nHA sample reached the relative high *Iso/n* ratio (1.64); the conversion was almost 100%; the yield of liquid product oil was 83.4% and the pour point was dropped to −28°C; the final product oil had favorable low-temperature fluidity.

Jin *et al.* [22] directly synthesized nonsulfided 8%Ni–50%Cs$_{1.5}$H$_{1.5}$PW/SiO$_2$ catalyst, which has shown the highest activity for the hydrocracking of *n*-decane and excellent tolerance to the sulfur and nitrogen compounds in the feedstock, superior to the impregnated catalyst and the industrial catalyst. The catalyst has higher surface area, better dispersion and reducibility of the active components, moderate acid strength, and suitable HYD–dehydrogenation function.

3.3.2. *Effect of Keggin-type HPAs on hydroprocessing*

To investigate the effect of Keggin-type HPAs on hydroprocessing of *Jatropha* oil, the Ni–HPAs/nHA catalysts were prepared with different Keggin-type HPAs loading: PTA, phosphomolybdic acid (PMA or HPMo), silicotungstic acid (STA or HSiW) and silicomolybdic acid (SMA or HSiMo) [23]. The acidity of catalysts at reaction temperature (360°C) was determined by solid-state ^{1}H NMR (see Figure 3.6 and Table 3.6); the acidity order was as follows: STA > PTA > SMA > PMA. The catalytic effect of isomerization was influenced by the acidity of catalyst [24–26]. Through the comparison of different Ni–HPAs/nHA catalysts, the Ni–STA/nHA catalyst produced more *Iso*-alkanes, achieving the highest *Iso/n* ratio (1.78) and lowest pour point (−32°C). The *Iso/n* ratio of liquid product was the same order with the acidity of the hydroprocessing catalyst at 360°C. Based on the characterization data of XPS and NH_3–TPD, the HPA increased the content of Ni^0 and offered protons for HYD; therefore the hydroprocessing activity was improved.

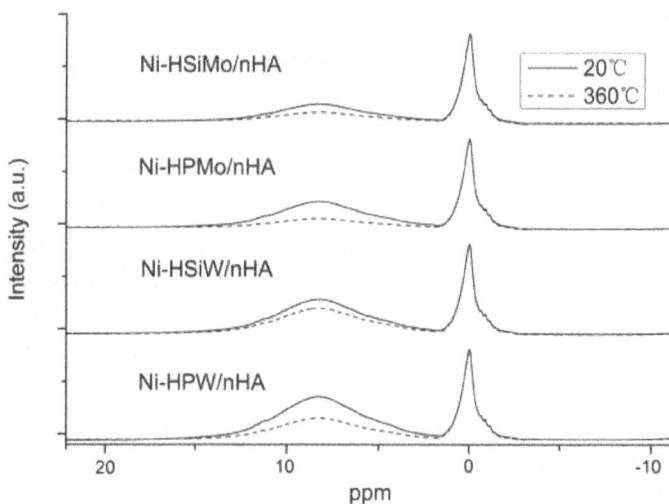

Figure 3.6. Solid-state ^{1}H NMR spectra of Ni–HPAs/nHA catalyst with different Keggin-type HPA loading at 20°C and 360°C. (Reproduced with permission from Ref. [23]. Copyright © 2015 Royal Society of Chemistry.)

Table 3.6.   Relative peak area of HPAs and hydroxyl group in Ni–HPAs/nHA catalyst with different Keggin-type HPA loading [23].

Catalyst samples	Relative peak area of HPAs (%)		Relative peak area of hydroxyl group (%)	
	20°C	360°C	20°C	360°C
Ni–HPW/nHA	100	51.8	100	100
Ni–HSiW/nHA	81.6	63.3	100	100
Ni–HPMo/nHA	64.1	24.8	100	100
Ni–HSiMo/nHA	48.7	32.5	100	100

3.4. Flower-like Ni–PTA catalyst

The flower-like Ni–PTA catalyst was obtained by flower-like NiO supported PTA by the impregnating method [27]. As a bimetallic catalyst containing solid acid with flower-like morphology and porous structure, flower-like Ni–PTA catalyst will potentially exhibit superior performance in hydroprocessing of plant oil without sulfuration. In the experiment, 10 g of the Ni–PTA/commercial Al_2O_3 catalyst was required to reach the similar catalytic activity (conversion and selectivity) as only 0.6 g of the flower-like Ni–PTA catalyst. Compared with the Ni–PTA/commercial Al_2O_3 catalyst, the catalytic efficiency was increased more than 15-fold by the flower-like Ni–PTA catalyst. The Ni–PTA catalyst with flower-like morphology possessed more metal atoms on edges and corners (proven by TEM and EDX). There is no support in the flower-like Ni–PTA catalyst, and the Ni content per unit of area in the flower-like Ni–PTA catalyst (19.02%) was much higher than that in the Ni–PTA/commercial Al_2O_3 catalyst (5.74%), revealed by XPS analysis. This likely led to the formation of more active sites (metal and acid centers) which would improve the catalytic activity. Additionally, mesoporous surfaces can promote the desorption of the products, followed by the inhibition of some side reactions, consistent with the SEM and BET results. Therefore, the morphology and flower-like architectures are important for efficient catalytic HYD. Figure 3.7 shows the SEM images of samples obtained at 180°C after 1 h, 3 h, 5 h, and 7 h, respectively and the morphology evolution was a function of the hydrothermal reaction

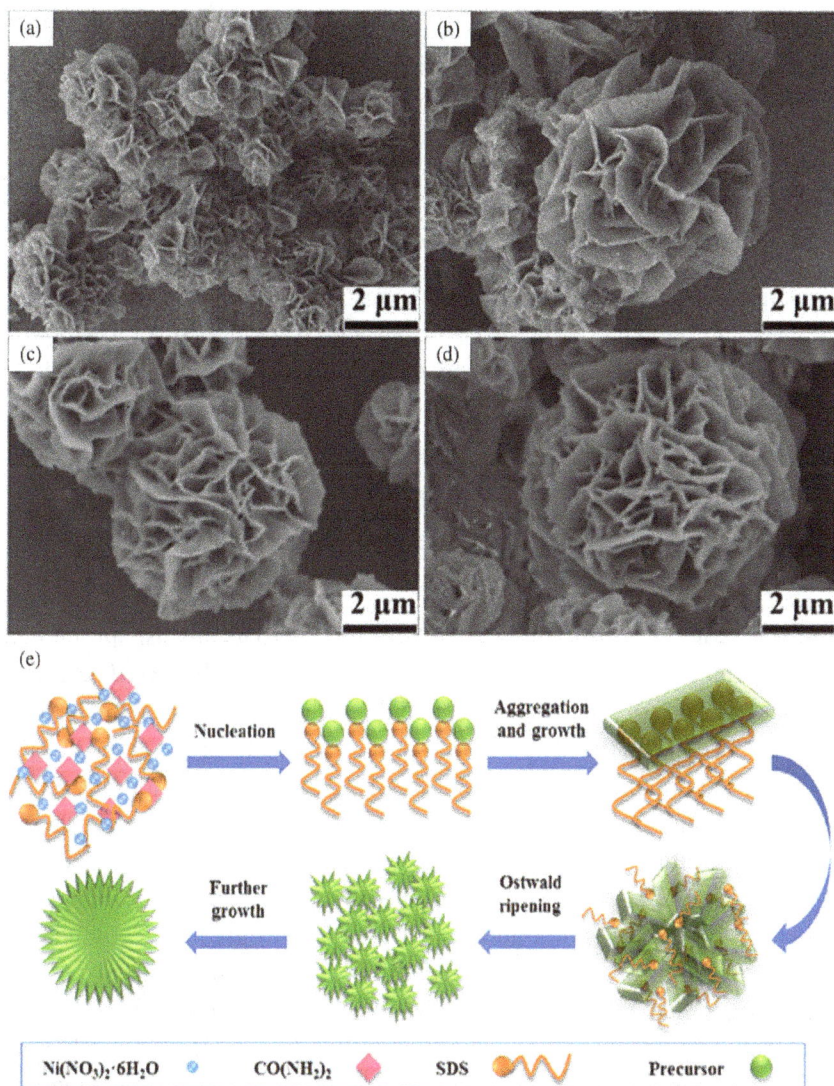

Figure 3.7. SEM images of the synthesized flower-like Ni(OH)$_2$ with reaction time of (a) 1 h, (b) 3 h, (c) 5 h, (d) 7 h at 180°C, and (e) schematic illustration for the formation mechanism of flower-like Ni(OH)$_2$ architectures. (Adapted with permission from Ref. [27]. Copyright © 2015 Creative Commons BY license.)

time. This process is consistent with the previously reported so-called two-stage growth process, which involves a fast nucleation of amorphous primary particles followed by a slow aggregation and crystallization of primary particles [28].

3.5. Ni-based catalysts supported by other supports

The supports can be sorted into four classes: (1) carbon; (2) solid Brønsted acids including H-beta zeolite (HBEA) and H-zeolite Socony Mobil-5 (HZSM-5) zeolites (H is a positively charged hydrogen ion); (3) refractory oxides such as SiO_2, Al_2O_3, and MgO; and (iv) the reducible oxides such as ZrO_2, TiO_2, and CeO_2 [6].

3.5.1. *Carbon nanotube supported Ni-based catalysts*

Carbon nanotube (CNT) was a suitable support compared with Al_2O_3 support and zeolite Socony Mobil-5 (ZSM-5), owing to its intrinsic properties such as chemical inertness, and thermal stability [29]. Zhou *et al.* [30] has synthesized a series of nonsulfided NiMo/CNTs catalysts with HNO_3-pretreated CNTs as the support, which was evaluated for the HYD of bio-oil model compounds (acetic acid, furfural, and hydroxyacetone) and raw bio-oil. Results showed that reduced (10 + 10) wt.% NiMo/CNT catalyst exhibited the highest activity. A synergistic effect was observed in the HYD process of a model compound mixture, which led to the aldol condensation between furfural and hydroxyacetone to give a C_8 compound, which could give valuable hydrocarbons after deep HYD. After the upgrading process, both the pH and the hydrogen content of bio-oil increased, and the oxygen content decreased, which were all desirable for the further application of upgraded bio-oil. They also prepared nonprecious transition metal and nonsulfided NiCo/CNT catalysts for the conversion of guaiacol to cyclohexanol, which was successful at reaction temperature 200°C over a reaction time 10 h. NiCo/CNT catalysts exhibited high catalytic activity for the HDO of guaiacol. The guaiacol conversion rate is up to 93% with a high selectivity towards cyclohexanol (about 85%) catalyzed by (10 + 10) wt.% NiCo/CNT.

3.5.2. *Zeolites supported Ni-based catalysts*

Chen *et al.* [31] developed a nonsulfide and nonnoble-metal $NiMo_{3-x}$/SAPO-11 catalyst for the one step deoxygenation and isomerization of methyl laurate to hydrocarbons that are identical to diesel fuel. In their study, $NiMo_{3-x}$/SAPO-11 was prepared and its performance was investigated after reduction pretreatment at various temperatures (400–550°C). The results showed the $NiMo_{3-x}$/SAPO-11 catalyst reduced at 450°C possessed a good balance between its deoxygenation activity and isomerization selectivity and exhibited a synergetic effect on HDO due to the presence of nickel metal and Mo^{4+} species. Zhang *et al.* [32] prepared Ni/HZSM-5 with different ratio of Si/Al and Ni/Al_2O_3 through wet impregnating with $Ni(NO_3)_2 \cdot 6H_2O$ and calcining at 450°C. Phenol conversion was the highest (91.8%) at 240°C in the presence of HZSM-5 (Si/Al = 38) loaded with 10% Ni. It showed higher activity than Ni/HZSM-5 (Si/Al = 50) and Ni/Al_2O_3 catalysts for phenol HDO. Their work proved that the properties of bio-oil, particularly the hydrogen content and the acidity can be effectively improved through hydrotreating.

3.5.3. *Complex oxide supported Ni-based catalysts*

Zhang *et al.* [33] synthesized SiO_2–ZrO_2 (SZ) complex oxide by precipitation method with different ratios of Si/Zr, and then impregnating with $Ni(NO_3)_2 \cdot 6H_2O$. The Ni/SZ catalysts were evaluated for HDO using guaiacol as the model compound. Guaiacol conversion reached the maximum (conversion of 100% and selectivity of above 98%) at 300°C and 5.0 MPa H_2 pressure in the presence of Ni/SZ-3. SZ oxide possesses amphoteric character, which promotes the excellent recyclability and anticoking performance of Ni/SZ-3 catalyst in the HDO reaction of model phenolic compounds. When HDO reaction was carried out with real lignin-derived phenolic compounds under the optimal conditions determined for guaiacol, the total yield of hydrocarbons was 62.81%. The HDO products have a high octane number and would be the most desirable components for fungible liquid transportation fuel.

3.6. Conclusions

Ni-based catalysts are usually sulfided to retain their active form. However, these catalysts are very susceptible to poisoning by atmospheric sulfur and the eventual S-contamination of the end product when using the sulfided catalysts. Although noble-metal catalysts can be used without sulfuration, their application might be limited because of their high cost. Thus, it is imperative to exploit possible alternatives for the sulfided catalysts or noble-metal catalysts for the upgrading of biofuel and bio-oil.

Inexpensive nonsulfided hydroprocessing catalysts are bifunctional, i.e., acid sites which provide the cracking function and metal sites with a HYD function. The activity of hydroprocessing catalysts is affected by the nature of the support and metal's properties. On one hand, acid sites are usually from HPA, zeolites or complex oxide (e.g., SiO_2–ZrO_2) supports. The hydroprocessing catalysts with Keggin-type HPA could increase the isoalkane content of product oil and reduce the pour point, demonstrating the HPAs have potential to be important catalysts for hydroprocessing. The acidity order was as follows: silicotungstic acid > PTA > silicomolybdic acid > PMA. The hydroprocessing results of these catalysts were consistent with their acidity order; Ni–STA/nHA catalyst achieved the highest ratio of i/n-alkane and the lowest pour point. Moreover, a catalyst with a small pore diameter or low specific surface area would decrease its catalytic activity by restricting the access of reactants to the catalytic sites or decreasing the numbers of activity sites per unit area. Thus, HPAs are loaded on the support with macropores or high specific surface area. The Al_2O_3, ZIF-8, and nHA have been proved to be appropriate supports, which aid in diffusion and adsorption of reactants on the catalyst surface and increase the catalytic activity and help remove products from the catalyst surface and enhanced catalytic stability. On the other hand, metal sites are Ni, Mo, Co etc. (transition metals), and Ni is the most popular one. There is no support in the flower-like Ni–PTA catalyst, and the Ni content per unit of area in the flower-like Ni–PTA catalyst was much higher than that in the Ni–PTA/Al_2O_3 catalyst. Furthermore, loading of rare-earth metals, such as La, Ce, on catalysts could improve the HYD capacity to increase metal function, enhance the alkalinity of catalyst to suppress the aromatization process, and thus reduce the coke formation on the hydroprocessing catalyst.

Ni-based hydroprocessing catalysts showed good usability in the upgrading process of biofuel and bio-oil, which could provide possible alternatives for those sulfided catalysts and noble-metal catalysts. Development of an inexpensive nonsulfided catalyst, especially Ni-based catalyst, will be helpful for the energy, environment, and economy.

References

1. Sotelo-Boyás, R., Liu, Y., and Minowa, T. (2011). Renewable diesel production from the hydrotreating of rapeseed oil with Pt/zeolite and NiMo/Al$_2$O$_3$ catalysts. *Industrial & Engineering Chemistry Research*, 50(5), 2791–2799.
2. Bezergianni, S., Voutetakis, S., and Kalogianni, A. (2009). Catalytic hydrocracking of fresh and used cooking oil. *Industrial & Engineering Chemistry Research*, 48(18), 8402–8406.
3. Kumar, R., Rana, B. S., Tiwari, R., Verma, D., Kumar, R., Joshi, R. K., Garg, M. O., and Sinha, A. K. (2010). Hydroprocessing of Jatropha oil and its mixtures with gas oil. *Green Chemistry*, 12(12), 2232–2239.
4. Šimáček, P., Kubička, D., Šebor, G., and Pospíšil, M. (2009). Hydroprocessed rapeseed oil as a source of hydrocarbon-based biodiesel. *Fuel*, 88(3), 456–460.
5. Ishihara, A., Fukui, N., Nasu, H., and Hashimoto, T. (2014). Hydrocracking of soybean oil using zeolite–alumina composite supported NiMo catalysts. *Fuel*, 134(0), 611–617.
6. Zhao, C., Bruck, T., and Lercher, J. A. (2013). Catalytic deoxygenation of microalgae oil to green hydrocarbons. *Green Chemistry*, 15(7), 1720–1739.
7. Chang, J., Danuthai, T., Dewiyanti, S., Wang, C., and Borgna, A. (2013). Hydrodeoxygenation of guaiacol over carbon-supported metal catalysts. *ChemCatChem*, 5(10), 3041–3049.
8. Chen, N., Gong, S., and Qian, E. W. (2015). Effect of reduction temperature of NiMoO$_{3-x}$/SAPO-11 on its catalytic activity in hydrodeoxygenation of methyl laurate. *Applied Catalysis B*: *Environmental*, 174, 253–263.
9. Kordulis, C., Bourikas, K., Gousi, M., Kordouli, E., and Lycourghiotis, A. (2016). Development of nickel based catalysts for the transformation of natural triglycerides and related compounds into green diesel: a critical review. *Applied Catalysis B*: *Environmental*, 181, 156–196.
10. Liu, J., Liu, C., Zhou, G., Shen, S., and Rong, L. (2012). Hydrotreatment of Jatropha oil over NiMoLa/Al$_2$O$_3$ catalyst. *Green Chemistry*, 14(9), 2499–2505.

11. Liu, J., Fan, K., Tian, W., Liu, C., and Rong, L. (2012). Hydroprocessing of Jatropha oil over NiMoCe/Al$_2$O$_3$ catalyst. *International Journal of Hydrogen Energy*, 37(23), 17731–17737.

12. Al-Marshed, A., Hart, A., Leeke, G., Greaves, M., and Wood, J. (2015). Optimization of heavy oil upgrading using dispersed nanoparticulate iron oxide as a catalyst. *Energy & Fuels*, 29(10), 6306–6316.

13. Fan, K., Yang, X., Liu, J., and Rong, L. (2015). Effect of reducing catalyst coke by La loading in hydrocracking of Jatropha oil. *RSC Advances*, 5(42), 33339–33346.

14. Martin, A., Armbruster, U., and Atia, H. (2012). Recent developments in dehydration of glycerol toward acrolein over heteropolyacids. *European Journal of Lipid Science and Technology*, 114(1), 10–23.

15. Marci, G., Garcia-Lopez, E. I., and Palmisano, L. (2014). Heteropolyacid-based materials as heterogeneous photocatalysts. *European Journal of Inorganic Chemistry*, 2014(1), 21–35.

16. Liu, J., Lei, J., He, J., Deng, L., Wang, L., Fan, K., and Rong, L. (2015). Hydroprocessing of Jatropha oil for production of green diesel over non-sulfided Ni-PTA/Al$_2$O$_3$ catalyst. *Scientific Reports*, 5, 11327.

17. Liu, J., He, J., Wang, L., Li, R., Chen, P., Rao, X., Deng, L., Rong, L., and Lei, J. (2016). NiO-PTA supported on ZIF-8 as a highly effective catalyst for hydrocracking of Jatropha oil. *Scientific Reports*, 6, 23667.

18. Fan, K., Liu, J., Yang, X., and Rong, L. (2014). Hydrocracking of Jatropha oil over Ni-H$_3$PW$_{12}$O$_{40}$/nano-hydroxyapatite catalyst. *International Journal of Hydrogen Energy*, 39(8), 3690–3697.

19. Tang, J., Salunkhe, R. R., Liu, J., Torad, N. L., Imura, M., Furukawa, S., and Yamauchi, Y. (2015). Thermal conversion of core–shell metal–organic frameworks: a new method for selectively functionalized nanoporous hybrid carbon. *Journal of the American Chemical Society*, 137(4), 1572–1580.

20. Zhang, M., Gao, Y., Li, C., and Liang, C. (2015). Chemical vapor deposition of Pd(C3H5)(C5H5) for the synthesis of reusable Pd@ZIF-8 catalysts for the Suzuki coupling reaction. *Chinese Journal of Catalysis*, 36(4), 588–594.

21. Zhou, G., Hou, Y. Z., Liu, L., Liu, H. R., Liu, C., Liu, J., Qiao, H. T., Liu, W. Y., Fan, Y. B., Shen, S. T., and Rong, L. (2012). Preparation and characterization of NiW-nHA composite catalyst for hydrocracking. *Nanoscale*, 4(24), 7698–7703.

22. Jin, H., Guo, D., Sun, X., Sun, S., Liu, J., Zhu, H., Yang, G., Yi, X., and Fang, W. (2013). Direct synthesis, characterization and catalytic performance of non-sulfided Ni–CsxH3–xPW12O40/SiO2 catalysts for hydrocracking of n-decane. *Fuel*, 112, 134–139.

23. Fan, K., Liu, J., Yang, X., and Rong, L. (2015). Effect of Keggin-type heteropolyacids on the hydrocracking of Jatropha oil. *RSC Advances*, 5(47), 37916–37924.

24. Sun, J. M., Baylon, R. A. L., Liu, C. J., Mei, D. H., Martin, K. J., Venkitasubramanian, P., and Wang, Y. (2016). Key roles of lewis acid-base pairs on ZnxZryOz in direct ethanol/acetone to isobutene conversion. *Journal of the American Chemical Society*, 138(2), 507–517.

25. Gonçalves, J. C. and Rodrigues, A. E. (2016). Xylene isomerization in the liquid phase using large-pore zeolites. *Chemical Engineering & Technology*, 39(2), 225–232.

26. Rabaev, M., Landau, M. V., Vidruk-Nehemya, R., Goldbourt, A., and Herskowitz, M. (2015). Improvement of hydrothermal stability of Pt/SAPO-11 catalyst in hydrodeoxygenation–isomerization–aromatization of vegetable oil. *Journal of Catalysis*, 332, 164–176.

27. Liu, J., Chen, P., Deng, L., He, J., Wang, L., Rong, L., and Lei, J. (2015). A non-sulfided flower-like Ni-PTA catalyst that enhances the hydrotreatment efficiency of plant oil to produce green diesel. *Scientific Reports*, 5, 15576.

28. Wei, X. W., Zhou, X. M., Wu, K. L., and Chen, Y. (2011). 3-D flower-like NiCo alloy nano/microstructures grown by a surfactant-assisted solvothermal process. *CrystEngComm*, 13(5), 1328–1332.

29. Zhou, L. and Lawal, A. (2015). Evaluation of Presulfided NiMo/γ-Al2O3 for hydrodeoxygenation of microalgae oil to produce green diesel. *Energy & Fuels*, 29(1), 262–272.

30. Zhou, M., Tian, L., Niu, L., Li, C., Xiao, G., and Xiao, R. (2014). Upgrading of liquid fuel from fast pyrolysis of biomass over modified Ni/CNT catalysts. *Fuel Processing Technology*, 126, 12–18.

31. Chen, N., Gong, S., and Qian, E. W. (2015). Effect of reduction temperature of $NiMoO_{3-x}$/SAPO-11 on its catalytic activity in hydrodeoxygenation of methyl laurate. *Applied Catalysis B: Environmental*, 174–175, 253–263.

32. Zhang, X., Wang, T., Ma, L., Zhang, Q., and Jiang, T. (2013). Hydrotreatment of bio-oil over Ni-based catalyst. *Bioresource Technology*, 127, 306–311.

33. Zhang, X., Zhang, Q., Wang, T., Ma, L., Yu, Y., and Chen, L. (2013). Hydrodeoxygenation of lignin-derived phenolic compounds to hydrocarbons over Ni/SiO_2–ZrO_2 catalysts. *Bioresource Technology*, 134, 73–80.

Chapter 4

Catalytic Upgrading of Pinewood Pyrolysis Bio-oil over Carbon-Encapsulated Bimetallic Co–Mo Carbide and Sulfide Catalysts

Rui Li*, Qiangu Yan[†,§], Zhongqing Ma[‡], and Guangyao Li[‡]

*Department of Natural Resources and Environmental Design,
North Carolina A&T State University,
Greensboro, NC 27411, USA
[†]Department of Sustainable Bioproducts,
Mississippi State University,
Starkville, MS 39759, USA
[‡]School of Engineering, Zhejiang A&F University, Lin'an,
Hangzhou, Zhejiang, China
[§]Corresponding author: yanqiangu@gmail.com

Abstract

The objective of this study was to synthesize carbon-encapsulated bimetallic Co–Mo carbides and sulfides nanoparticles by carbothermal reduction (CR) of molybdenum-promoted biochar and post sulfiding. The bimetallic Co–Mo carbides and sulfides nanoparticles were characterized for physicochemical properties by multiple morphological and structural methods (e.g., SEM, TEM, and XRD). These nanoparticles were tested for catalytic upgrading of the pyrolysis bio-oil from pinewood in a continuous packed-bed

97

reactor. Hydroprocessing experiments were performed at a temperature of 350–425°C, 6.89–12.41 MPa (1,000–1,800 psig) hydrogen pressure with a hydrogen flow rate of 500 ml/min at a liquid hourly space velocity of 0.1–1.0 h^{-1}. The results from sulfided catalytic experiments were compared with CoMo carbide nanoparticles. Sulfided CoMo nanoparticles demonstrated higher catalytic activity and resulted in increased hydrocarbon fraction yields. Moreover, the quality of the hydrocarbon fraction, as determined by the acid value, higher heating value, and water content analyses, also improved.

Keywords: Pinewood, Pyrolysis, Bio-oil, Hydrotreating, Carbon-encapsulated Co–Mo bimetallic carbide and sulfide nanoparticles, Biochar

4.1. Introduction

Fast pyrolysis is a thermochemical technology performed at temperatures from 400°C to 550°C in the absence of oxygen [1], and it is considered as an economical route to convert lignocellulosic biomass to the bio-oil, a potential alternative fuel resource [2]. Raw bio-oils can be further upgraded to liquid fuels via technologies including hydrodeoxygenation (HDO) [3], catalytic cracking [4], supercritical treatment [5], olefination, esterification, and steam reforming [6]. Catalytic cracking can only partially deoxygenate the raw bio-oil, resulting in low liquid yields. Steam reforming produces a low-energy-density gaseous fuel, while supercritical treatment requires high capital cost due to the requirement of high pressure vessels. Olefination and esterification may upgrade the bio-oil to a boiler fuel. HDO is one of the most promising methods for catalytically upgrading the bio-oil to the liquid transportation fuel [7]. Numerous catalysts have been studied for HDO of bio-oils and their oxygen-containing model compounds [8]. These catalysts include Mo-based sulfides, noble metals, base metals, metal phosphides, other metal catalysts, and bifunctional catalysts [9]. Bimetallic CoMo- and NiMo-based sulfide catalysts have been used as industrial hydrotreating catalysts in petroleum refining for decades [10]. They are also used in HDO of the bio-oil with a drawback of adding toxic sulfur to the feed [11]. Therefore, one of the challenges in hydroprocessing of the bio-oil is to design more effective HDO catalysts [1].

Transition metal carbides show characteristics of being wear resistant and chemically inert and having high melting points [12–15]. They are often used in traditional applications such as cutting and grinding tools, bearings, textile-machinery components, and oxidation-resistant gas burners, as well as for new and promising applications such as electronics and optoelectronics [12]. Recently, transition metal carbides have been studied as catalytic materials and demonstrated exceptionally high activity [16–21]. It is reported that the transition-metal carbides, especially tungsten and molybdenum carbides, have excellent noble-metal-like catalytic activity, stability, and selectivity for a wide range of reactions [22], such as ammonia synthesis and decomposition [23], hydrocarbon reforming [24, 25], Fischer–Tropsch synthesis [26], hydrogenation (HYD) [27], hydrotreating [28, 29], hydrodesulfurization (HDS) [30–32], heterogeneous asymmetric reactions [33], and fuel cell electrode catalysts [34]. Supported Co–Mo catalysts have been used in the petroleum refining industry for more than half a century due to their reliable activity and thermal resistance [35]. Currently, one of the challenges for refiners is to develop novel catalysts to produce high-purity fuels [35]. Due to the low cost, the satisfactory reactivity, and the stability [36], carbide catalysts are desirable alternatives for noble-metal catalysts in many catalytic processes [37]. Molybdenum-based carbides have attracted the attention of researchers in hydrogen-involved reactions, such as hydrogenolysis, HYD, dehydrogenation, isomerization, HDS, hydrodenitrogenation (HDN), ammonia synthesis, etc. For example, Mo_2C was found to behave similarly to Ru in CO–H_2 reactions [38]. In fact, carbides of the Group IV–VI metals have been widely examined for their activity in oxidation, HYD/dehydrogenation, isomerization, hydrogenolysis, and CO–H_2 reactions. In many cases, the metal carbides have been found to rival the performance of the less economical Group VIII metals. While the refractory carbides do not show high activity for oxidation reactions (for example, the rate of H_2 oxidation follows the order metal $\gg$ carbide > oxide for Group V and VI metals [39], and the rate of NH_3 oxidation over refractory carbides is lower than that over Group VIII metals [40]), they are as active as the transition metals themselves for HYD and dehydrogenation reactions. In isomerization reactions, water content (WC), Pt, and Ir are unique in their high activity and selectivity [41]. However, it is not necessary for the refractory carbides to be more or even equally active in catalyzing given reactions compared with noble-metal catalysts, because the lower cost of the carbides will offset the loss in catalytic activity.

Most research findings focused on the outstanding functions and new preparation methods of monometallic molybdenum carbides, rather than those of bimetallic molybdenum-based carbides, despite the potential advantages of producing multifunctional catalysts from bimetallic ones [35]. There are several proposed methods for preparing molybdenum carbides. The traditional methods for the synthesis of a metal carbide involved the direct carburization of metals with graphitic carbon at high temperature. There are several approaches to prepare cobalt–molybdenum bimetallic carbide catalysts. The two-stage reaction method, which was first developed by Newsam *et al.* [42], successfully synthesized $Co_6Mo_6C_2$ (Co_3Mo_3C) as well as Co_6Mo_6C were with $CoMoO_4$ as the precursor. Bussell *et al.* [43] synthesized the bulk and alumina-supported Co_3Mo_3C through a temperature-programed nitridation and subsequently the topotactic carburization process. A topotactic transformation process was characterized by internal atomic displacements including loss or gain of atoms or ions, so that the initial and final lattices were in coherence. Xiao *et al.* [44] reported a thermal carburization approach for the preparation of Co–Mo carbides. An oxidic precursor was used in this method and the carburization process was carried out under a C_2H_6/H_2 mixture flow. Liang *et al.* [45, 46] used a carbothermal hydrogen reduction method for the preparation of Co–Mo carbide, nanostructured β-Mo_2C and W_2C [47]. Wang *et al.* [35] prepared Co_3Mo_3C and Co_6Mo_6C via the thermal decomposition of excess hexamethylenetetramine (HMT)-contained mixed-salt precursors, in which HMT acted as both the molybdate ion ligand and the carbon source. This simple synthetic method was demonstrated to be effective to prepare MCM-41-supported Co_3Mo_3C catalyst.

The objective of this study is to prepare and characterize nanostructured Co–Mo carbides over biochar by the carbothermal reduction (CR) method. The activities of these biochar-based catalysts were compared in HDO reactions of the bio-oil to liquid fuels. Nanostructured sulfided Co–Mo particles were also used as catalysts in the HDO process.

4.2. Experimental

4.2.1. *Materials and analyses*

The following chemicals were used in this research effort: cobalt(II) nitrate hexahydrate ACS reagent, $\geq$98%, Sigma-Aldrich; ammonium molybdate $((NH_4)_6Mo_7O_{24} \cdot 4H_2O)$, 99%, Sigma-Aldrich; carbon disulfide, ACS reagent, $\geq$99.9%, Sigma-Aldrich.

Southern yellow pinewood was used as the feedstock to produce bio-oils via fast pyrolysis in an auger reactor. The loblolly pine lumber was first chipped to paper-chip size (8 mm $\times$ 20 mm $\times$ 35 mm) chips in a wood chipper (Carthage Machine Company). The wood chips were air-dried for 3–5 days at ambient temperature to 8–10 wt.% moisture content before grinding into smaller particles. Biochar used in the experiments was obtained from the fast pyrolysis auger reactor from a feed stock of the southern yellow pinewood chips. The char was first boiled in 0.1 M HNO_3 solution overnight to remove any soluble alkali ions, alkaline earth ions, and bio-oil residue, then washed several times with hot deionized (DI) water followed by drying in an oven at 105°C overnight.

The C, H, and N elemental composition of the pinewood and biochar was analyzed with a CE-440 elemental analyzer (Exeter Analytical, Inc., MA, USA). At least three measurements were conducted for every sample. Elemental analysis results (Table 4.1) show the C, H, and N composition in the biochar sample are 48.3 $\pm$ 1.2 wt.%, 1.0 $\pm$ 0.2 wt.%, and 0.4 $\pm$ 0.05 wt.%, respectively.

Mineral analysis was conducted on an ICP spectrophotometer (Optima model 4300 DV, PerkinElmer Instruments, MA, USA). The pinewood and biochar samples were combusted in air, and the ash was extracted with weak acids. The extracted solution was then used for ICP analysis. Mineral analysis results (Table 4.2) demonstrated that there were

Table 4.1.   Elemental analysis of pinewood and biochar (wt.%).

Samples	Carbon (%)	Hydrogen (%)	Nitrogen (%)	Remaining (%)
Pinewood	46.7 $\pm$ 0.5	6.3 $\pm$ 0.2	0.2 $\pm$ 0.05	46.8
Biochar	48.3 $\pm$ 1.2	1.0 $\pm$ 0.2	0.4 $\pm$ 0.05	50.3

Table 4.2. Mineral analysis results of pinewood and biochar.

Samples	Si	Na	Ca	Mg	K
Pinewood (ppm)	850 ± 65	120 ± 10	125 ± 8	91 ± 3	105 ± 5
Biochar (%)	3.1 ± 0.3	0.4 ± 0.1	0.5 ± 0.05	0.4 ± 0.02	0.2 ± 0.05

3.1 wt.% Si, 0.4 wt.% Na, 0.5 wt.% Ca, 0.4 wt.% Mg, and 0.2 wt.% K in raw biochar, while only 1.5 wt.% Si was left in the acid-washed char.

4.2.2. *Preparation of carbon-encapsulated bimetallic Co–Mo carbide nanoparticles over biochar*

Carbon-encapsulated molybdenum carbide nanoparticles were prepared using the incipient wetness impregnation method followed by carburization. An aqueous solution of ammonium heptamolybdate and cobalt nitrate was first prepared with a Co/Mo molar ratio 1:1. This aqueous solution was prepared using the following procedure: 36.8 g of ammonium heptamolybdate was added to 100 ml DI water. The solution was heated to 80°C and stirred until a clear, transparent solution was obtained. The solution was then cooled to 40°C, where 60.6 g of cobalt nitrate were added to the ammonium heptamolybdate solution. The mixture was stirred until a transparent solution obtained. The obtained bimetallic salt solution then was added in a dropwise fashion to a beaker containing 50 g of biochar and stirred continuously. The impregnation step lasted until removal of the solvent at 80°C by evaporation. The mass obtained was further dried at 110°C overnight, and ready for the next carbothermal reduction step. Fifteen grams (15 g) of the dried sample was packed in the middle of a 1-inch OD ceramic tubular reactor. The carrier gas — argon (99.99% purity) — was first introduced into the reactor at a flow rate of 50 ml/min for 30 min. The reactor was heated at a rate of 10°C/min to 1,000°C and kept at 1,000°C for 1 h. The furnace was cooled down by 10°C/min to room temperature under an argon flow. The product stream from the reactor passed through a gas–liquid separator, where the temperature was held at 0°C. Liquid products were collected from the cold condenser. The gas-phase product from the condenser was monitored and analyzed used an online residue gas analyzer (RGA) and an Agilent 6890 gas chromatograph (GC).

Carbon-encapsulated molybdenum carbide nanoparticles were sulfided with a solvent mixture of 5 vol.% carbon disulfide and cyclohexane. Twenty-five milliliters (5 vol.%) of carbon disulfide was added into 500 ml of cyclohexane solvent, and the solvent mixture was pumped through a high-pressure pump system (Teledyne Isco 500HP High Pressure Syringe Pump). Ten grams of carburized Co–Mo/biochar sample was packed in the middle of a 1-inch OD ceramic tubular reactor. The carrier gas — argon (99.99% purity) — was first introduced into the reactor at a flow rate of 50 ml/min for 30 min. The reactor was heated at a rate of 5°C/min to 300°C and kept at 300°C for 10 min, and then CS_2-cyclohexane mixture was pumped into the reaction system with LHSV of 1 h^{-1} for 5 h. The furnace was cooled down by 10°C/min to room temperature under an argon flow.

4.2.3. Preparation of bio-oil from pinewood chips by fast pyrolysis in an auger reactor

An auger-fed pyrolysis reactor produced the required bio-oils. Pyrolysis occurred in a reactor tube, and the pyrolysis vapor passed through a series of condensers where it was condensed to form the bio-oil. Nitrogen was supplied continuously to the reactor heated zone as an inert gas to exclude all oxygen from the pyrolysis reactor. The temperature of the initially formed vapor was about 30°C below the set pyrolysis temperature of 475°C. The pyrolysis vapor exited the heated reactor tube into a first water-cooled condenser where its temperature dropped to about 75–80°C. A second condenser lowered the temperature to 25–35°C, and then aerosol (liquid/vapor) continued to enter the third and the fourth condensers, both of which maintained almost the same temperature as the second condenser (25–35°C). The noncondensable gases produced in this process were collected from the exit of the fourth condenser by using a gas sampler (GAV-200 MK 2 gas sampler kit, SGE Analytical Science) and analyzed by GC. Liquid condensates were collected from all four condensers and analyzed as a whole bio-oil sample. Char was collected in a sealed vessel at the end of the reactor tube and weighed. The noncondensable gases passed through a Ritter gas flow meter to measure the volume and flow rate. All these measurements enabled the calculation of pyrolysis yields and a relatively good mass balance closure. A minimum run time of

30 min was performed for each feedstock to ensure an accurate mass balance calculation and to obtain sufficient bio-oil for various analyses.

Raw bio-oil and the oil-phase products from the hydroprocessing treatments were characterized following ASTM methods. WC in the bio-oil sample was measured by Karl Fisher titration according to ASTM standard E203 with a Cole-Parmer Model C-25800-10 titration apparatus (Thermo Fisher Scientific Inc, MA, USA). Kinematic viscosity at 40°C of the bio-oils was measured according to ASTM D445. The acid value (AV) of the sample was determined by using ASTM D664: 1 g of sample was dissolved in isopropanol/water (v/v = 35:65) solution and then titrated with 0.1 N NaOH to a pH value of 8.5. The AV was then calculated as the required milligram (mg) amounts of NaOH equivalent to 1 g of sample. Higher heating value (HHV) was determined using a Parr 6200 oxygen bomb calorimeter (Parr Instrument Co., Moline, IL, USA) following ASTM D240. Elemental analysis (CHNO) for determination of percent carbon (C), percent hydrogen (H), and percent nitrogen (N) were determined by using the EAI CE-440 elemental analyzer, and percent oxygen (O) was obtained through difference determined by the ASTM D5291 method.

4.2.4. *Catalytic upgrading the bio-oil via hydrotreatment*

The hydrotreating of bio-oil was carried out in a lab-built, continuous flow, fixed-bed reactor (1-inch stainless steel tubular reactor) system. Five (5) grams of the catalyst was loaded into the reactor. The system was first purged with an argon flow for 30 min, followed by prereducing stage with a hydrogen flow at 400°C for 2 h. Before feeding the bio-oil, the pressure of the reactor was gradually increased to 6.89 MPa (1,000 psi). In the meantime, the reaction temperature was adjusted to 300–400°C. The bio-oil was pumped into the catalyst tube with a high-pressure pump system (Teledyne Isco 500HP). The hydrogen flow rate was controlled with a mass flow controller (MFC, Brooks Instruments), and the reaction system pressure was controlled with a back-pressure regulator. The reaction was operated under the following conditions: 350–425°C, a liquid hourly space velocity (LHSV) of 0.1–1.0 h^{-1} and a pressure of 6.89–12.41 MPa (1,000–1,800 psig).

The vent gas flow rate was measured with an Agilent gas flow meter. The analysis of gas-phase products was carried out with an online Agilent 7890 GC equipped with two thermal conductivity detectors (TCDs) and a flame ionization detector (FID). Helium and nitrogen were used as the carrier gases. Liquid products were collected using a condenser that was kept at 0°C. Products exiting from the packed-bed reactor were cooled in the condenser, and the liquid products were collected in a sampling bottle at 2 h intervals. The collected liquid products were centrifuged for 1 h to separate the aqueous phase and the oil phase. Oil samples were analyzed using an Agilent 7683B Series Injector coupled to an Agilent 6890 Series gas chromatograph system and a 5973 Mass Selective Detector, i.e., a quadrupole-type GC-MS system, as well as a FID detector. An Agilent DB-WAXetr (50 m × 0.32 mm I.D., 1.0 μm) capillary column was used. A constant column flow of 1 ml/min (24 cm^3/s) helium was applied. The injector was kept at 250°C. Samples were injected (1 μl) with a split ratio of 100:1. The temperature-programed separation started at 40°C for 5 min, and then the temperature was increased at a rate of 10°C/min to 250°C for 10 min. The FID detector worked at a temperature of 250°C with helium makeup gas at 30 ml/min. For the MS, the transfer line and EI source temperatures were 250°C and 200°C, respectively. Quadrupole conditions involved an electron energy of 70 eV and an emission current of 150 μA.

4.2.5. *Detailed hydrocarbon analysis (DHA) of liquid hydrocarbons*

Detailed hydrocarbon analysis (DHA) was performed over a Perkin Elmer Clarus 680 GC with a FID detector using the PIANO method. PIANO describes the method for determining the amount of paraffins (P), isoparaffins (I), aromatics (A), naphthenes (N), and olefins (O) within a sample. A liquid sample with a volume of 1 μl was injected into a 100-m GC column with about 200:1 split ratio. A flame ionization detector and retention time library was used to identify compounds with a carbon number up to 14. This method is based on the ASTM test method D 5134-92, but uses a 100-m capillary column instead of a 50-m column. The PIANO method is mainly used for gasoline-type samples, which is why it is limited to compounds with carbon numbers less than 14. Any C_{15} compound or heavier were reported as unknown compounds. The initial temperature of the GC injector was set at 200°C and held at this temperature for 43.15 min. The sample injector of the GC was heated to 450°C at 100°C/min and held at this temperature throughout the end of the test. A DHA analytical column (100 m × 0.25 μm ID) was used to separate sample components. Hydrogen was used as a carrier gas with a flow rate of 100 ml/min. The initial oven temperature was held at 35°C for 5 min, heated to 50°C at 10°C /min and held for 21.5 min. Then, the oven temperature was ramped to 150°C with a heating rate of 3°C/min and kept at 150°C for 4.67 min. The FID detector temperature was 250°C with a hydrogen flow rate of 42 ml/min and an air flow rate of 450 ml/min.

4.2.6. *Thermogravimetric analysis (TGA) experiments*

Thermogravimetric analysis of the samples was carried out in a TGA (Shimadzu TGA-50H) through isothermal analyses. The system was capable of quantitatively measuring the change in mass of a sample as a function of temperature up to 1,500°C. The change in mass was then related to the changes taking place in the sample during decomposition. For each sample prepared, argon (99.99% purity, 50 ml/min) was flown through the TGA at 50 ml/min as the temperature was ramped at 10°C/min. Each sample was repeated for at least three times.

4.2.7. *Temperature-programed decomposition (TPD)*

Temperature-programed decomposition (TPD) experiments involved heating the sample in an argon purging gas (20 ml/min) at a programed heating rate to induce thermal decomposition of metal-promoted samples. TPD experiments were carried out using an Autochem 2920 (Micromeritics, GA, USA). One gram of the sample was used in each run. Volatile species from the TPD process were measured with an online mass spectrometer.

4.2.8. *Characterization*

X-ray powder diffraction (XRD) patterns of the samples were obtained using a Rigaku Ultima III X-ray Diffraction System (Tokyo, Japan) operated at 40 kV and 44 mA using Cu-K radiation with a wavelength of 1.5406 Å, from 20° to 80° at a scan rate of 0.02°s^{-1}. The Jade powder diffraction analysis software from Materials Data, Inc (CA, USA) was used for both qualitative and quantitative analysis of polycrystalline powder materials. The morphology of the samples was investigated with a scanning electron microscope (SEM). All samples were precoated with 10 nm Pt before being introduced into the vacuum chamber. The system was operated with accelerating voltage of 10 kV. The sample particle sizes were examined with a JEOL JEM-100CX II transmission electron microscope (TEM, Tokyo, Japan) operated at accelerating voltage of 200 kV. All samples were sonicated in ethanol solution for 1 min before transferred to copper grids.

4.3. Results and discussion

4.3.1. *TPD*

Temperature-programed thermal treatment experiments followed by mass spectrometry were conducted to detect the gases produced during the thermal activation of biochar produced by pinewood fast pyrolysis and Co- and Mo-promoted biochar samples. The surface functionalities chemically bound to chars decompose upon heating by releasing gaseous compounds at different temperatures. The evolution of five species in the gas phase, namely H_2, CH_4, H_2O, CO, and CO_2, was monitored. A typical profile is shown in Figure 4.1. The interpretation of the thermal desorption profiles provided useful information on the type of species desorbed from the carbon surface and on the nature of interactions of the gaseous species and carbon. The amounts of the different surface groups can be estimated from the TPD spectra.

While being heated, the surface functionalities chemically bound to carbons decompose, releasing various gaseous compounds at different temperatures. The dominant gases evolved during thermal desorption are oxides. Figure 4.1(a) shows typical TPD curves recording evolution of five gaseous species (i.e., H_2, CH_4, H_2O, CO_2, and CO) during thermal activation of biochar samples. Oxygen-containing functional groups decomposed mostly as CO, CO_2, and H_2O. Surface complexes

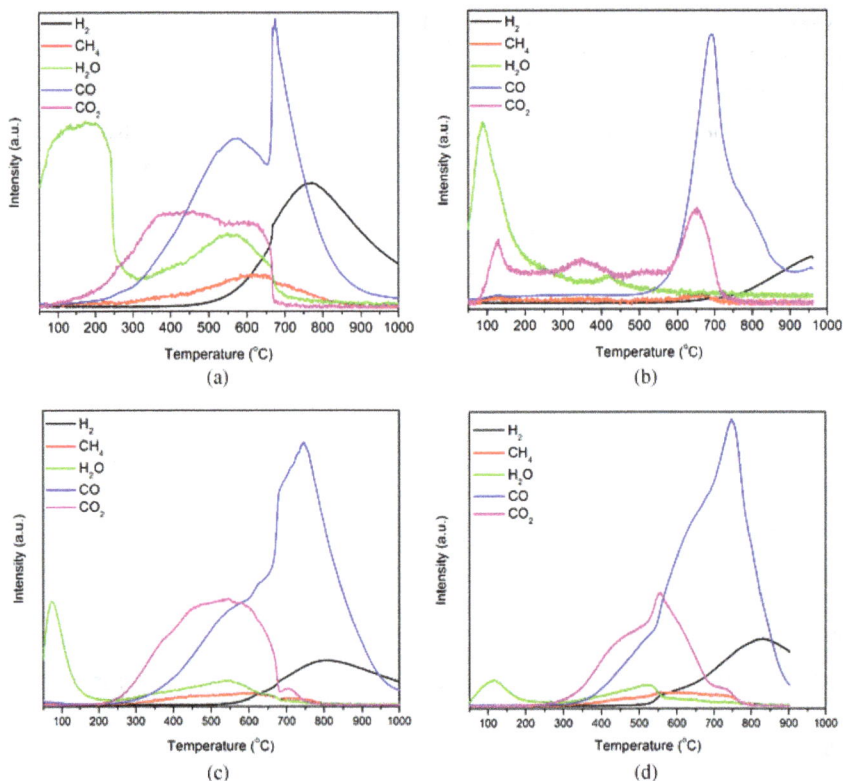

Figure 4.1. Hydrogen, methane, water, carbon monoxide, and carbon dioxide evolution for the temperature-programed thermal treatment of the biochar (a) and the different metal-promoted biochar at a rate of 10°C/min in argon atmosphere, (b) Co-biochar, (c) Mo-biochar, and (d) CoMo-biochar.

yielding CO_2 groups decomposed at two different temperatures, corresponding to two types of functional groups. Carboxylic acids (—COOH → CO_2) were responsible for the low-temperature peak (100–400°C), and the high-temperature CO_2 evolution was attributed to carboxylic anhydrides (350–630°C) and lactones (190–630°C). The CO-yielding groups were also represented by two distinct groups of peaks, corresponding to phenols and quinine (600–1,000°C), and ether structures (~700°C). The peak of water release at low temperatures was generally attributed to the removal of physisorbed water (room temperature to 200°C). The peak at high temperatures (~530°C) was assigned either to the evolution of water hydrogen bonded to oxygen complexes, to condensation of phenolic groups, or to dehydration reactions of neighboring carboxylic groups to give carboxyl anhydrides. The methane peak at 620°C originates from decomposition of the methyl group attached to aromatic units.

Hydrogen was present on a carbon surface as chemisorbed water, as surface functionalities (e.g., carboxylic acids, phenolic groups, and amines), or was bonded directly to carbon atoms as a part of aromatic or aliphatic structures. The carbon–hydrogen bond is very stable but breaks on heating at about 1,000°C. Nevertheless, the complete desorption of hydrogen does not happen at temperatures below 1,200°C. Heat treatment in an inert atmosphere eliminates part of the hydrogen via surface reduction.

Since the temperature region is higher than 600°C, the breakdown of the wood structure is completed and char formed. The char still contains some oxygen-containing structures, mainly aromatic heterocyclic structures, which decomposed with the formation of carbon monoxide and methane.

Figure 4.1(b) shows the evolution curves of H_2 (*m/z* 2), CH_4 (*m/z* 16), H_2O (*m/z* 18) and CO_2 (*m/z* 44) during decomposition of Co-biochar. TPD curves were significantly different compared to those of the unpromoted char. The CO peak of 600°C corresponding to phenols was not noticeably changed, but the CO peak (700°C) assigned to ether disappeared. This implied that doped nickel ions promoted the hydrolysis of the ether. There was a sharp CO_2 peak at 430°C that was assigned to the carbothermal reduction of nickel oxide. Hydrogen evolution was observed when the temperature was above 900°C, probably due to graphitization of the char material promoted by nickel metal. Oya *et al.* [48] investigated the catalysis of carbon crystallization by transition metals and proposed a dissolution/precipitation mechanism. In the present work, this mechanism explains the cobalt-catalyzed biochar carbonization results. Cobalt oxide dissolved in the char matrix may first be reduced by surface functional groups of the char. The reduced metallic cobalt then reacts with amorphous carbon to form cobalt carbides.

$$CoO + \text{active functional groups} \rightarrow Co + CO_2 + CO + H_2O$$
$$Co + \text{amorphous carbon} \rightarrow Co_3C$$

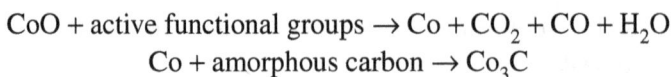

Figure 4.1(c) shows the TPD profiles of evolved gas species during temperature-programed thermal treatment of the molybdenum-promoted biochar sample. The TPD curves were almost identical to those of the unpromoted char when the temperature was below 760°C. There was a large CO peak that appeared at 760°C. This CO peak was attributed to the carbothermal reduction of MoO_3 on the char surface. The carbothermal reduction reaction included two steps. First, MoO_3 was reduced to metal molybdenum by active carbon atoms over the char surface.

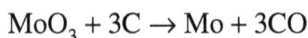

$$MoO_3 + 3C \rightarrow Mo + 3CO$$

Then, the reduced molybdenum reacted with active carbon atoms to form Mo_2C.

$$2Mo + C \rightarrow Mo_2C$$

TPD spectra of Co–Mo over biochars are shown in Figure 4.1(d). The TPD results showed that the carbothermal reduction reaction proceeded in several stages over the temperature range studied. There was some CO_2 formation at about 200–525°C. It is suggested that CoO was reduced to cobalt metal in this region. The CO_2 peak between 525°C and 700°C was attributed to the reduction of MoO_3 to MoO_2 by surface carbon containing functional groups. Part of CO_2, CO, and CH_4 between 525°C and 750°C may be produced by the catalytic decomposition of CH_xO_y in biochars by metallic cobalt through the reaction of CH_xO_y $* \rightarrow CO + CO_2 + CH$. The significant CO peak at 740°C was mainly due to the carbothermal reduction reaction, and β-Mo_2C was formed by the reaction between MoO_2 and carbon species from reactive carbon atoms or groups on carbon material. A small water desorption peat at 115°C was attributed to physically adsorbed moisture on the sample. The wide water peak 300–850°C was assigned either to the evolution of water hydrogen-bonded to oxygen complexes, to condensation of any phenolic groups, or to dehydration reactions of neighboring carboxylic groups to give carboxyl anhydrides on biochars. The hydrogen peak at 826°C was assigned the thermal decomposition of CH_x ($x = 1$–3) groups bonded directly to carbon atoms as a part of aromatic or aliphatic structures.

4.3.2. *TGA*

Thermogravimetric analysis (TGA) and derivative thermogravimetric analysis (DTG) data demonstrate a continuous mass loss associated with increasing of temperature (Figure 4.2), which was attributed to the breaking of chemical linkages and removal of volatile products from the biochars. There were six possible steps of mass loss occurring in the biochar with significant mass loss phases at 190–350°C, 350–700°C, 700–900°C (Figure 4.2(a)), mainly due to decomposition of oxygen-containing functional groups. The carbonization process of the biochar was completed around 900°C, since all oxygen-containing groups were eliminated from biochar at this temperature, which was also confirmed from TPD results.

There were three significant mass loss steps that were shown in Figure 4.2(b). The first and the highest mass loss of the cobalt-promoted biochar occurred in the temperature range from 71°C to 385°C with a peak temperature of 279°C. This mass loss stage was very complex: first, cobalt promoted obviously the decomposition of the biochar; second, the thermal decomposition of cobalt nitrate contributed to the mass loss. The reduction of cobalt oxide from the decomposition of cobalt nitrate also should be included. There was about 33.5% mass loss in this step. The second loss of mass was found around 385–700°C and corresponded to the decomposition of functional groups in the biochar structures. The mass loss in this zone was 10%. These results indicated that oxygen functional groups started to release in this temperature zone and led to initial carbonization of the carbonaceous biochar. After 700°C until 1,000°C, the mass decreased gradually.

For biochar promoted by ammonium heptamolybdate, a continuous gradual weight loss between 130°C and 320°C corresponded to the decomposition of ammonium heptamolybdate (Figure 4.2(c)). At around 320–385°C, the shoulder peak was attributed to the partially reduction of Mo^{3+} in ammonium heptamolybdate by active carbon species in biochar. The DTG curve showed that there was a mass loss peak at 430°C, indicating biochar decomposition. From 585°C to 730°C, around 4.38% of the mass was lost corresponding to the reduction of Mo oxides to metallic Mo. The sharp peak at 880°C was attributed to the formation of Mo_2C due to the reaction between remaining carbon and molybdenum, while biochar was catalytically graphitized by molybdenum carbide.

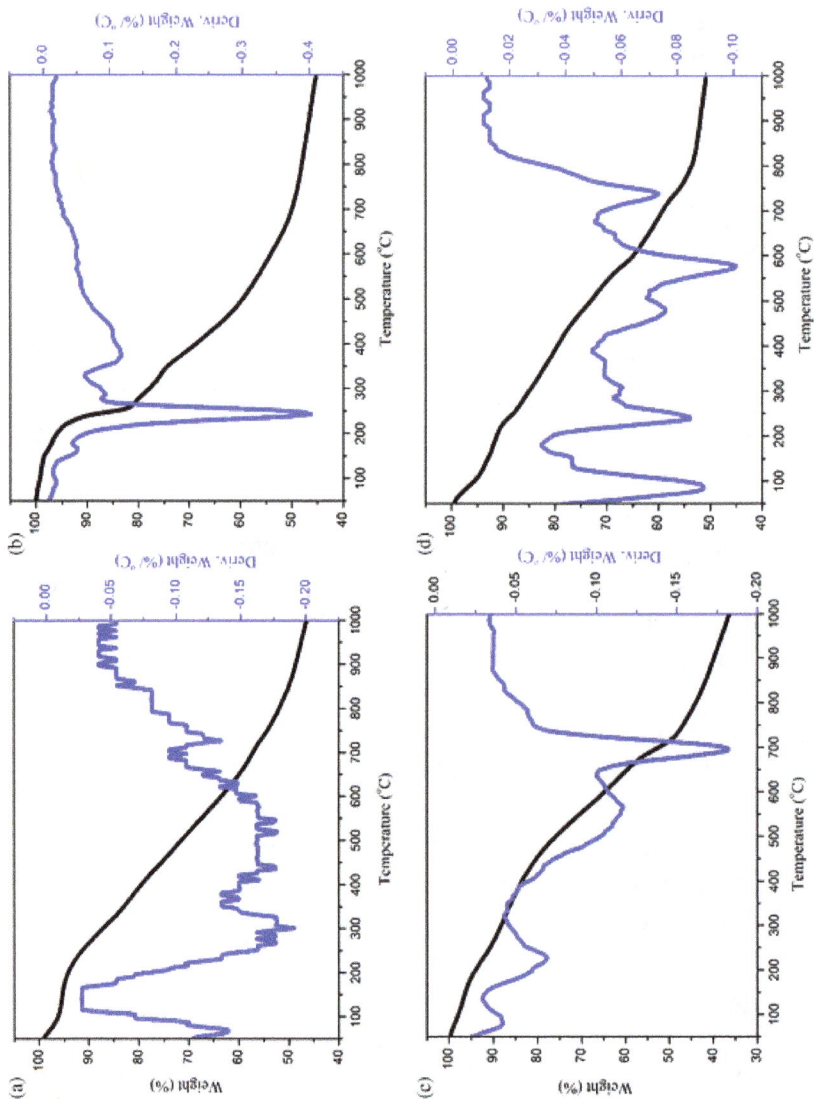

Figure 4.2. TGA and DTG curves of the raw biochar and the metal-biochar samples heated at a rate of 10°C/min in argon atmosphere: (a) raw biochar, (b) Co-biochar, (c) Mo-biochar, and (d) CoMo-biochar.

TGA/DTG curves of CoMo-biochar sample are plotted in Figure 4.2(d). The catalytic decomposition process could be divided into five stages. The initial mass loss (1) corresponded to the loss of physically adsorbed water. This loss occurred between 50°C and 160°C with a tiny peak temperature of 112.5°C. The second mass loss (2) was observed around 160–317°C and corresponded to the depolymerization of biochar and the decomposition of cobalt nitrate. The maximum mass loss peak was centered at 237.5°C. Most of the oxygen-containing groups in the alkyl side chain of the lignin basic units were catalytic decomposed in this stage. The mass decreased rapidly because of the breakage of large amounts of ether bonds and C–C bonds connected on the phenyl propane units. And then a lot of small-molecule gases and macromolecular condensable volatiles were generated. The third stage (3) is the decomposition of biochar structures, which was followed by (4) reduction of cobalt oxide by active carbon species in biochar, and (5) carbonization of biochar. Cobalt has been known to be one of the efficient catalysts for the graphitization of amorphous carbons.

4.3.3. *Characterization*

4.3.3.1. *XRD*

XRD pattern of biochar and Co- or Mo-promoted biochar samples thermally treated at 1,000°C were plotted in Figure 4.3(a). XRD pattern of biochar did not show any diffraction peaks.

For the Co-biochar sample, the strong diffraction peaks at 2θ values of 44.35°, 51.65°, and 75.95° corresponded to (111), (200), and (220), and crystal planes indicated formation of face-centered cubic (FCC) crystalline cobalt (JCPDS 15–0806).

The Mo-biochar showed the signal of β-Mo_2C hexagonal phase (JCPDS 35-0787), where 2θ peaks located at 34.3° (100), 39.6° (101), 37.8° (002), 52.3° (102), 34.4° (100), 61.5° (110), 69.8° (103), and 74.5° (112). Additionally, the average crystallite size determined by (101) peak using Scherrer formula, was 21.0 nm.

XRD patterns of CoMo-biochar showed diffraction peaks of β-Mo_2C, which were observed after carbon thermal reduction at 1,000°C under the argon flow.

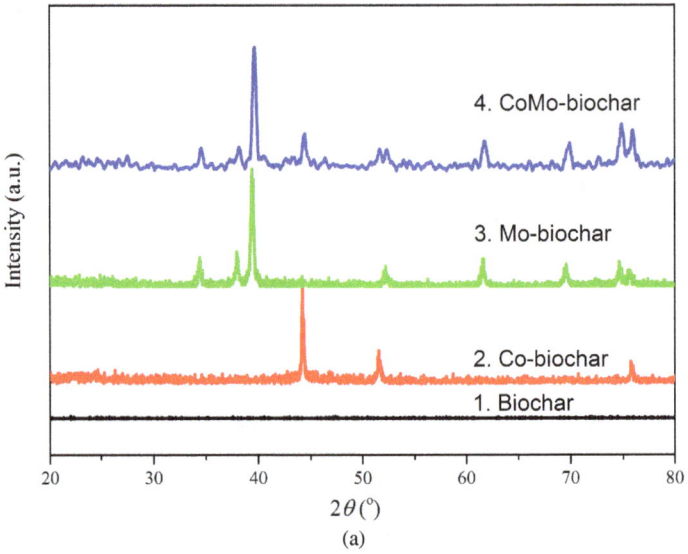

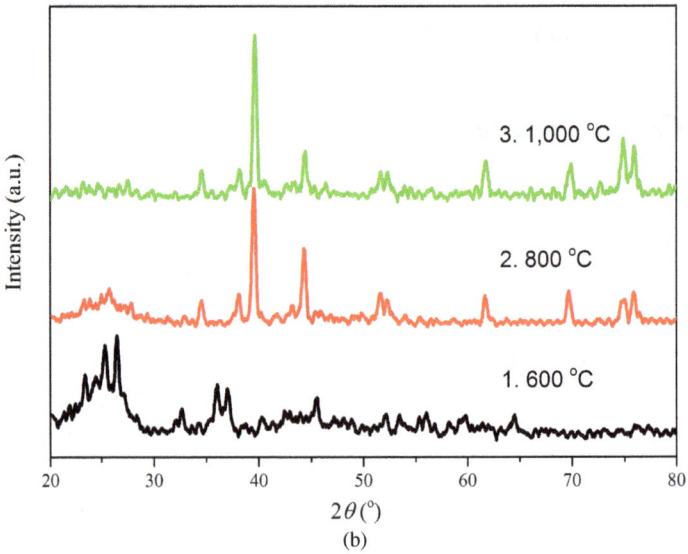

Figure 4.3. (a) XRD patterns collected on the as prepared (1) biochar, (2) Co-biochar, (3) Mo-biochar, and (4) CoMo-biochar treated at 1,000°C. (b) CoMo-biochar treated at (1) 600°C, (2) 800°C, and (3) 1,000°C.

The Co–Mo bimetallic catalysts supported over biochars were thermal treated at 600°C, 800°C, and 1,000°C, respectively. Their XRD patterns are shown in Figure 4.3(b). The XRD pattern of the 600°C carburized Co–Mo showed XRD peaks at 2θ values of 27.3°, 28.2°, and 37.01°. These peaks can be ascribed to the MoO_2; this result indicated that MoO_3 was reduced to MoO_2 by reduction agents of carbon or surface carbon species at 600°C. As the carburizing temperature increased to 800°C, typical diffraction peaks of β-Mo_2C showed clearly at 34.4°, 37.8°, 39.4°, 52.0°, 61.6°, 69.5°, 74.6°, and 75.7°, and there were also several peaks with $2\theta =$ 32.80°, 35.84°, 40.50°, 43.08°, 47.10°, 65.74°, and 73.72° that matched the face-centered cubic Co_6Mo_6C structure. These results confirmed that a small quantity of Co_6Mo_6C has been prepared by this heat treatment process. No peaks associated with molybdenum oxides were observed for the 800°C sample. This demonstrated that supported molybdenum carbide catalysts can be fabricated with nanocarbons as the carbon source. Very broad peaks of Mo_2C were observed over the catalyst with nanocarbon as the support, indicating that molybdenum carbide was highly dispersed over biochars and the particle size was much smaller. As the carburizing temperature further increased to 1,000°C, the β-Mo_2C diffraction peaks became more intense, suggesting that more β-Mo_2C was formed in the bimetallic catalyst. The sharp peaks of Mo_2C over biochars showed that the size of the carbide particles was relatively large at 1,000°C.

4.3.3.2. *SEM*

The SEM image of the raw biochar is shown in Figure 4.4(a). The surface of the biochar was smooth and clean. There is no dust or devolatilization coking residue on the char surface. Figure 4.4(b) shows the morphology of cobalt promoted biochar after thermal treatment at 1,000°C. The results showed nanosized cobalt particles uniformly distributed on the surface of the biochar with particle sizes between 100 nm and 300 nm. From SEM pictures, the observed nanoparticles were found on the surface of the wood char. At elevated temperatures, cobalt oxide over the char surface was first reduced to cobalt metal particles, and then these nanoparticles may merge and agglomerate to large size under high temperature.

Figure 4.4(c) shows the morphology of molybdenum promoted biochar after thermal treatment at 1,000°C. It was observed that the char surface was covered with a layer of small bar-shaped nanoparticles. These particles had

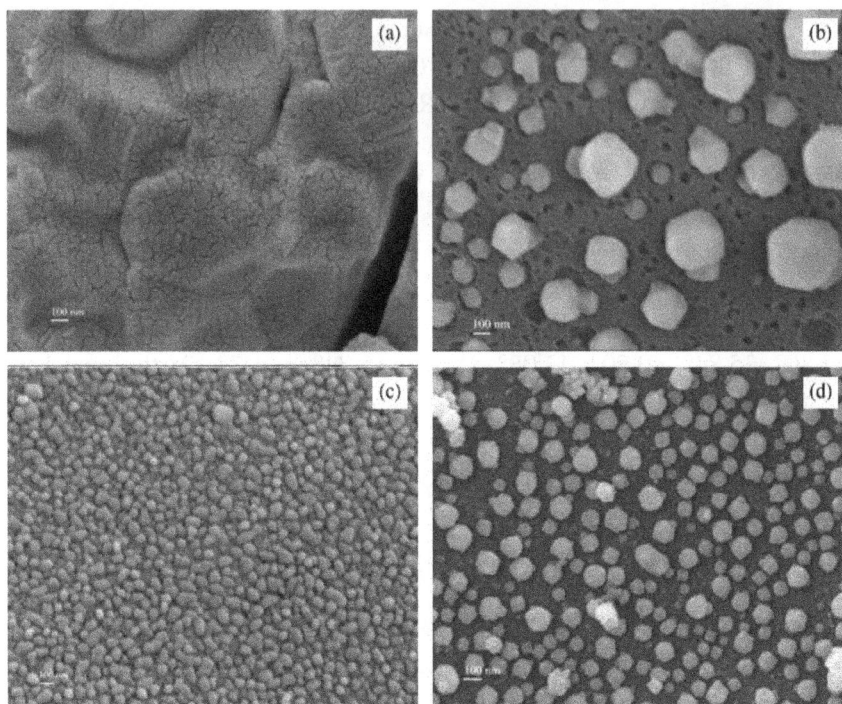

Figure 4.4. SEM images of the biochar (a) and biochar samples doped with Co (b), Mo (c), and CoMo (d) after heat treatment at 1,000°C under an argon flow for 1 h.

sizes between 20 nm and 200 nm. XRD results proved these nanobars were molybdenum carbide; these molybdenum carbides were mostly carbon-encapsulated nanoparticles. Like the cobalt-promoted char at elevated temperatures, molybdenum oxide on the char surface was first reduced to molybdenum metal particles at 760°C, and molybdenum metal continued to react with carbon to form molybdenum carbide. Molybdenum carbide is an excellent catalyst for the chemical vapor deposition (CVD) process to produce nanocarbons. Under high temperature (>800°C), the volatiles desorbed from the wood char were catalytically decomposed to elemental carbon and precipitated onto the Mo_2C particle surfaces. Carbon-encapsulated molybdenum carbide nanoparticles were formed on the biochar surface.

Figure 4.4(d) shows SEM image of CoMo-biochar thermally treated at 1,000°C. The morphologies of the decomposed CoMo-biochar sample showed that Co–Mo carbide particles were uniformly distributed over the biochar surface with particle sizes of 20–100 nm.

4.3.3.3. *High-resolution transmission electron*
microscopy (HRTEM)

Figure 4.5 shows high-resolution transmission electron microscopy (HRTEM) images of the thermal treated CoMo-biochar sample. It was found that Mo_2C nanoparticles with sizes ranging from 5 nm to 20 nm were encapsulated in 1–15 layers of graphene structure (Figures 4.5(a) and 4.5(b)). Figure 4.5(c) shows the HRTEM image of β-Mo_2C with the d-spacing values of 0.227 nm for the (101) crystallographic planes. There were 1–5 layers of graphene over (101) plane. While Figure 4.5(d) shows the HRTEM image of β-Mo_2C with the d-spacing values of 0.259 nm for the (110) planes. There were more than 10 layers of graphene over (110) plane.

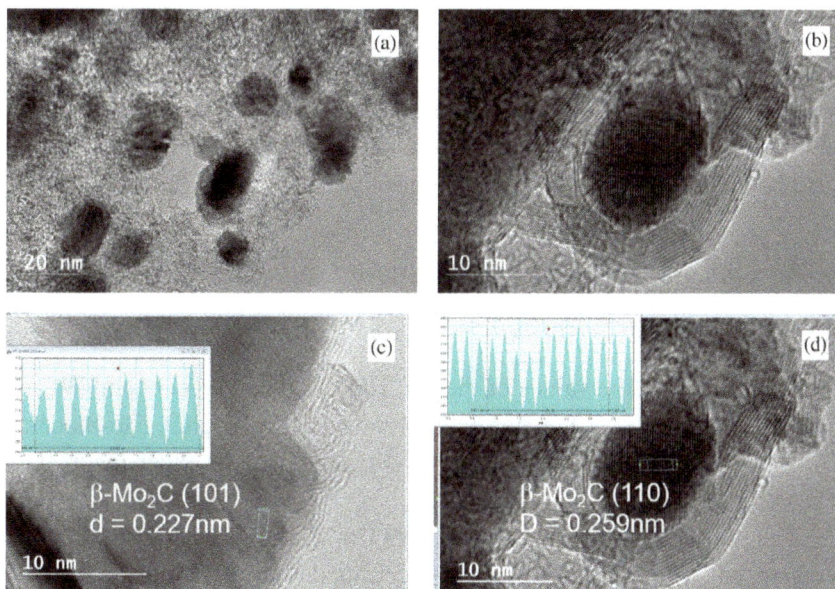

Figure 4.5. HRTEM images of the biochar samples prepared with CoMo after heat treatment at 1,000°C under an argon flow for 1 h. (a) and (b) Mo_2C nanoparticles with sizes ranging from 5 nm to 20 nm encapsulated in 1–15 layers of graphene structure, (c) β-Mo_2C with the d-spacing values of 0.227 nm for the (101) crystallographic planes, and (d) β-Mo_2C with the d-spacing values of 0.259 nm for the (110) planes.

Table 4.3. Characteristics of the bio-oil produced from fast pyrolysis of loblolly pinewood.

Bio-oil properties	Value
Water content (wt.%)	28.36
Viscosity @ 40°C (cSt)	12.59
Density (g/ml)	1.19
pH	2.40
Acid value (AV) (mg KOH/g)	90.23
Higher heating value (HHV) (MJ/kg)	16.25
Carbon (wt.%)	42.05
Hydrogen (wt.%)	7.86
Nitrogen (wt.%)	0.05
Oxygen (wt.%)	50.04

4.3.4. *Bio-oil production from pinewood*

Bio-oil was produced by the fast pyrolysis process at a temperature of 400–450°C under nitrogen atmosphere using an auger-fed pyrolysis reactor. The characteristics of the bio-oil are listed in Table 4.3.

4.3.5. *Catalytic upgrading of the bio-oil to liquid hydrocarbons*

4.3.5.1. *Catalytic performance of biochar-based catalysts*

Table 4.4 presents the values of the AV, HHV, oxygen percent and WC of bio-oils upgraded over biochar-based catalysts (Co-biochar, Mo-biochar, CoMo-biochar, and sulfided CoMo-biochar). Each experiment was repeated three times.

The AV for the upgraded bio-oil over Co-biochar, Mo-biochar, CoMo-biochar, and sulfided CoMo-biochar declined from 95.9 mg KOH/g of raw bio-oil to 42.1 mg KOH/g, 19.9 mg KOH/g, 5.6 mg KOH/g, and 4.3 mg KOH/g, respectively. Among these catalysts, AV of the products from CoMo-biochar and sulfided CoMo-biochar catalysts was the lowest comparing to the Co-biochar and Mo-biochar catalysts.

The HHV of the hydrotreated bio-oil samples over Co-biochar, Mo-biochar, CoMo-biochar, and sulfided CoMo-biochar catalysts

Table 4.4. Catalytic performance of biochar-based catalysts on bio-oil hydroprocessing.

Catalyst	AV (mg KOH/g)	HHV (MJ/kg)	Oxygen content (wt.%)	WC (wt.%)
Raw bio-oil	90.23	16.25	50.04	28.36
Co-biochar	42.1	30.2	16.9	7.2
Mo-biochar	19.9	35.8	12.7	4.9
CoMo-biochar	5.6	39.5	5.2	2.6
Sulfided CoMo-biochar	4.3	40.6	3.9	1.9

increased from 16.25 MJ/kg of raw bio-oil to 30.2 MJ/kg, 35.8 MJ/kg, 39.5 MJ/kg, and 40.6 MJ/kg, respectively. CoMo-biochar and sulfide CoMo-biochar catalysts showed the best activity on improving the HHV of upgraded products.

The oxygen content of the hydrotreated bio-oil samples over Co-biochar, Mo-biochar, CoMo-biochar, and sulfided CoMo-biochar catalysts decreased significantly from 50.04% of raw bio-oil to 16.9%, 12.7%, 5.2%, and 3.9%, respectively. Among these catalysts, CoMo-biochar and sulfide CoMo-biochar demonstrated the lowest oxygen content.

The WC percentages of the hydrotreated bio-oil samples over Co-biochar, Mo-biochar, CoMo-biochar, and sulfided CoMo-biochar catalysts decreased from 28.36% of raw bio-oil to 7.2%, 4.9%, 2.6% and 1.9%, respectively.

The overall yields (OYs), aqueous fraction (AF), organic fraction (OF), gas products, and solid char for hydroprocessing of the bio-oils with biochar-based catalysts are given in Table 4.5.

The OYs over Co-biochar, Mo-biochar, CoMo-biochar, and sulfided CoMo-biochar catalysts were 62.6%, 67.9%, 80.7%, and 82.5%, respectively. The AF yields over Co-biochar, Mo-biochar, CoMo-biochar, and sulfided CoMo-biochar catalysts were 50.7%, 54.2%, 56.9%, and 58.7%, respectively. The OF yields over Co-biochar, Mo-biochar, CoMo-biochar, and sulfided CoMo-biochar catalysts were 13.2%, 17.5%, 25.4%, and 26.1% respectively. Among all catalysts, CoMo-biochar and sulfided CoMo-biochar catalysts showed higher OY, AF, and OF, which indicated a higher catalytic performance. The high activity of the sulfided CoMo-biochar catalyst might be due to the formation of Co–Mo–S phase, which was believed to be the active phase and formed by the insertion of Co

Table 4.5. Performance of biochar-based catalysts on upgrading bio-oil.

Catalyst	OY (wt.%)	AF yield (wt.%)	OF yield (wt.%)	Gas (wt.%)	Char (wt.%)
Co-biochar	62.6	50.7	13.2	8.3	—
Mo-biochar	67.9	54.2	17.5	11.9	—
CoMo-biochar	80.7	56.9	25.4	9.1	8.4
Sulfided CoMo-biochar	82.5	58.7	26.1	9.8	5.3

Table 4.6. Distribution of gaseous products from hydroprocessing over biochar-based catalysts.

Gas (%)	Co-biochar	Mo-biochar	CoMo-biochar	Sulfided CoMo-biochar
H_2	81.2	76.9	69.9	65.3
CH_4	1.0	1.8	2.7	3.0
CO	0.6	0.5	0.6	0.5
CO_2	5.2	6.9	7.3	8.4
C_2H_4	12.0	13.9	19.5	22.8

atoms to the interlayer of MoS_2 planes. The gas and solid char yields are also shown in Table 4.5.

Table 4.6 displays the gaseous product distribution during hydroprocessing of the bio-oil. The vent gaseous products from hydroprocessing over sulfided CoMo-biochar catalyst was composed of 65.3% hydrogen, 3.0% methane, 0.5% carbon monoxide, 8.4% carbon dioxide, and 22.8% ethane. Lower hydrogen concentration in the vent gas means higher hydrogen consumption and higher deoxygenation activity of the catalyst.

4.3.5.2. *Testing of different process conditions*

Sulfided CoMo-biochar catalyst demonstrated a higher activity on hydroprocessing tests of the bio-oil. It was selected and examined further as the hydroprocessing catalyst at various process conditions: temperature (T), pressure (P), and liquid hourly space velocity (LHSV). The hydroprocessing was performed under the temperature of 350–425°C, and pressures of 1,000 psig, 1,500 psig, and 1,800 psig, and LHSVs of 0.1 h^{-1}, 0.3 h^{-1}, 0.5 h^{-1}, and 1 h^{-1}.

4.3.5.2.1. Effect of temperature

The temperatures studied were 350°C, 375°C, 400°C, and 425°C, with the remaining conditions maintained constant. Five grams of the catalyst were loaded into the reactor, and the bio-oil and hydrogen were fed into the reactor. The operating pressure was maintained at 1,500 psig, while the LHSV was maintained at 0.3 h^{-1}.

The properties of upgraded bio-oil products from the hydroprocessing treatments are listed in Table 4.7. The AV values of the treated bio-oils decreased with increasing of temperature, from 16.3 mg of KOH/g at 350°C to 1.0 mg of KOH/g at 425°C. HHV of the organic products increased with increasing of temperature, from 27.9 MJ/kg at 350°C to 43.8 MJ/kg at 425°C. The properties of the OF were also compared to optimize reaction temperature. It was observed that OF values increased with increasing of temperature. The WC percentages of the organic fractions were 8.5%, 0.7%, 0.1%, and 0.0%, respectively.

The vent gas composition under different reaction temperatures was analyzed as shown in Table 4.8. Hydrogen percentage was decreasing with increasing of temperature, indicating that more hydrogen was consumed at a high reaction temperature. Light hydrocarbons (CH_4 and C_2H_6) increased with increasing of temperature.

4.3.5.2.2. Influence of pressure on the catalyst performance

Experiments under pressures of 1,000 psig, 1,500 psig, and 1,800 psig were conducted. Temperature was held constant at 400°C. The liquid

Table 4.7. Effect of temperature on properties of the oil-phase products and yields of OF and AF.

Temperature (°C)	AV (mg of KOH/g)	HHV (MJ/kg)	C (%)	H (%)	O (%)	WC (%)	OY (wt.%)	AF (wt.%)	OF (wt.%)
350	16.3	27.9	67.3	8.9	23.8	8.5	68.5	53.0	18.2
375	3.1	40.5	82.5	11.6	5.9	0.7	75.3	57.6	21.0
400	1.8	42.0	86.8	12.3	0.8	0.1	86.2	60.5	25.0
425	1.0	43.8	87.1	12.9	0.0	0.0	88.3	61.3	26.4

Note: Hydroprocessing conditions: five grams of sulfided CoMo-biochar catalyst, 1,500 psig, LHSV of 0.3 h^{-1} and hydrogen flow rate of 500 ml/min.

Table 4.8. Vent gas composition under different reaction temperatures.

Temp (°C)	H_2 (%)	CH_4 (%)	CO (%)	CO_2 (%)	C_2H_6 (%)
350	91.9	0.7	0.8	3.9	2.7
375	81.7	1.3	0.6	6.4	10.1
400	68.8	3.0	0.6	8.9	18.7
425	66.3	8.4	1.4	10.3	13.6

Note: Hydroprocessing conditions: five grams of sulfided CoMo-biochar catalyst, 1,500 psig, LHSV of 0.3 h^{-1} and hydrogen flow rate of 500 ml/min.

space velocity was maintained at 0.3 h^{-1}. The effects of pressure on organic products from hydroprocessing treatments are listed in Table 4.9.

An increase in total pressure shifted the equilibrium towards the product side of the reaction and increased the bio-oil conversion. With increasing of pressure, the AV of organic products decreased from 3.3 mg KOH/g at 1,000 psig to 1.8 mg KOH/g and 0.5 mg KOH/g at 1,500 psig and 1,800 psig, respectively, while HHV increased with increasing of pressure, i.e., 27.9 MJ/kg, 42.0 MJ/kg, and 44.1 MJ/kg for the organic products of 1,000 psig, 1,500 psig, and 1,800 psig, respectively. No significant difference in WC (0.5%, 0.1%, and 0.0% for 1,000 psig, 1,500 psig, and 1,800 psig) was observed for the organic products. OY and OF increased with increasing of pressure.

The vent gas composition under different pressures was analyzed (Table 4.10). Hydrogen percentage was decreasing with increasing of pressure, meaning that more hydrogen was consumed at a high pressure. Light hydrocarbons (CH_4 and C_2H_6) increased with increasing of pressure.

4.3.5.2.3. Impact of space velocity on the catalyst performance

Another set of experiments were completed by varying the LHSV. The following LHSVs were used: 0.1 h^{-1}, 0.3 h^{-1}, 0.5 h^{-1}, and 1.0 h^{-1}. The temperature and pressure were held constant at 400°C and 1,500 psig, respectively. The effect of LHSV on properties of the oil-phase products and yields of OF and AF is shown in Table 4.11. As the LHSV was increased from 0.1 h^{-1} to 1.0 h^{-1}, the AV increased from 0.3 mg KOH/g to 21.6 mg KOH/g, while the HHV decreased from 43.9 MJ/kg to 33.6 MJ/kg.

Table 4.9.   Effect of pressure on properties of the oil-phase products and yields of OF and AF.

Pressure (psig)	AV (mg of KOH/g)	HHV (MJ/kg)	C (%)	H (%)	O (%)	WC (%)	OF (wt.%)	AF (wt.%)	OF (wt.%)
1,000	3.3	27.9	81.3	11.8	23.8	0.5	74.0	54.7	21.1
1,500	1.8	42.0	86.8	12.3	0.8	0.1	86.2	60.5	25.0
1,800	0.5	44.1	87.0	13.0	0.0	0.0	89.5	61.3	25.9

Note: Hydroprocessing conditions: five grams of sulfided CoMo-biochar catalyst, 400°C, LHSV of 0.3 h^{-1} and hydrogen flow rate of 500 ml/min.

Table 4.10.   Vent gas composition under different pressure.

Pressure (psig)	H_2 (%)	CH_4 (%)	CO (%)	CO_2 (%)	C_2H_6 (%)
1,000	76.0	4.1	0.4	10.2	9.4
1,500	68.8	3.0	0.6	8.9	18.7
1,800	65.6	3.7	1.0	7.2	22.4

Note: Hydroprocessing conditions: five grams of sulfided CoMo-biochar catalyst, 400°C, LHSV of 0.3 h^{-1} and hydrogen flow rate of 500 ml/min.

Table 4.11.   Effect of LHSV on properties of the oil-phase products and yields of OF and AF.

LHSV (h^{-1})	AV (mg of KOH/g)	HHV (MJ/kg)	C (%)	H (%)	O (%)	WC (%)	OY (wt.%)	AF (wt.%)	OF (wt.%)
0.1	0.3	43.9	87.2	13.9	0.0	0.0	88.5	61.2	26.7
0.3	1.8	42.0	86.8	12.3	0.8	0.1	86.2	60.5	25.0
0.5	6.9	40.2	80.2	11.0	6.7	3.2	79.6	58.6	23.8
1.0	21.6	33.6	77.5	9.9	10.5	16.2	68.3	55.0	17.9

Note: Hydroprocessing conditions: five grams of sulfided CoMo-biochar catalyst, temperature of 400°C, 1,500 psig, and hydrogen flow rate of 500 ml/min.

The vent gas composition under different LHSVs was analyzed (Table 4.12). Hydrogen percentage was increasing with increasing of LHSV, and light hydrocarbons (CH_4 and C_2H_6) also decreased with increasing of LHSV.

Table 4.12. Vent gas composition under different LHSVs.

LHSV (h^{-1})	H$_2$ (%)	CH$_4$ (%)	CO (%)	CO$_2$ (%)	C$_2$H$_6$ (%)
0.1	65.7	5.3	0.7	6.5	21.8
0.3	68.8	3.0	0.6	8.9	18.7
0.5	75.9	3.8	0.5	8.5	11.2
1.0	83.8	1.7	0.9	9.2	4.4

Note: Hydroprocessing conditions: five grams of sulfided CoMo-biochar catalyst, temperature of 400°C, 1,500 psig, and hydrogen flow rate of 500 ml/min.

Table 4.13. Typical product distribution of the C$_5$+ liquid fraction for the sulfide CoMo-biochar catalyst by group type and carbon number (in mol %).

	Paraffins	I-paraffins	Olefins	Napthenes	Aromatics	Unknowns	Total
C$_5$	0.3	0.1	0.1	0	0	0	0.5
C$_6$	1	0.2	1.7	0.2	0.1	0	3.2
C$_7$	2.5	0.3	2.3	1.7	0.5	1.2	8.5
C$_8$	2	0.6	1.26	2.5	1.2	9.7	17.26
C$_9$	2.1	7.5	3.2	4.8	1.8	3.5	22.9
C$_{10}$	2.5	4.2	0.9	1.9	7.8	1.2	18.5
C$_{11}$	1.4	1.9	0.3	0.5	3.1	0.7	7.9
C$_{12}$	0.7	0.2	0.1	0.7	1	3	5.7
C$_{13}$	0.3	0.1	0.1	0.1	0.2	2.8	3.6
Total	12.8	15.1	9.96	12.4	15.7	22.1	65.96

4.3.6. *Analysis of liquid fuel products: DHA analysis*

After the hydroprocessing reaction, liquid hydrocarbon samples were collected from the condenser. Table 4.13 shows the DHA of mixed liquid hydrocarbons obtained with sulfided CoMo-biochar from the most effective process conditions (temperature of 400°C, pressure of 1,500 psig, LHSV of 0.3 h^{-1}, and hydrogen flow rate of 500 ml/min). The DHA of the upgraded products predominately contained paraffins, isoparaffins, naphthenes, aromatics, and olefins. The characterization of organic products of the bio-oil hydroprocessed over sulfide CoMo-biochar catalyst was performed by GC using the PIANO analysis. Liquid oil products were quantitatively reported in the volatile range up to molecules with 13 carbons with the five families of hydrocarbon groups (paraffins, isoparaffins, aromatics, naphthenes, and olefins).

4.4. Conclusions

Nanostructured Co–Mo carbide nanoparticles were successfully produced using an easy synthesis method by carbothermal reduction of CoMo-promoted pine biochar at 1,000°C. The resulting carbides were characterized with XRD, TEM, field emission scanning electron microscopy (FE-SEM), and temperature-programed desorption–mass spectroscopy (TPD-MS). These nanoparticles were tested for catalytic upgrading of the bio-oil in a continuous packed-bed reactor. Hydroprocessing experiments were performed at a temperature of 350–425°C and hydrogen pressure of 1,000–1,800 psig with a hydrogen flow rate of 500 ml/min at a LHSV of 0.1–1.0 h^{-1}. The results from sulfided catalytic experiments were compared with CoMo carbide nanoparticles. Sulfided CoMo nanoparticles demonstrated higher catalytic activity and resulted in increased hydrocarbon fraction yields. Moreover, the quality of the hydrocarbon fraction, as determined by the AV, HHV, and WC analysis, also improved. The hydrocarbon fraction was analyzed by detailed hydrocarbon analysis.

References

1. Wang, H., Male, J., and Wang, Y. (2013). Recent advances in hydrotreating of pyrolysis bio-oil and its oxygen-containing model compounds. *ACS Catalysis*, 3(5), 1047–1070.
2. Serrano-Ruiz, J. C. and Dumesic, J. A. (2011). Catalytic routes for the conversion of biomass into liquid hydrocarbon transportation fuels. *Energy & Environmental Science*, 4(1), 83–99.
3. Huber, G. W., Iborra, S., and Corma, A. (2006). Synthesis of transportation fuels from biomass: chemistry, catalysts, and engineering. *Chemical Reviews*, 106(9), 4044–4098.
4. Elliott, D. C. (2007). Historical developments in hydroprocessing bio-oils. *Energy & Fuels*, 21(3), 1792–1815.
5. Samolada, M. C., Baldauf, W., and Vasalos, I. A. (1998). Production of a bio-gasoline by upgrading biomass flash pyrolysis liquids via hydrogen processing and catalytic cracking. *Fuel*, 77(14), 1667–1675.

6. Vispute, T. P. and Huber, G. W. (2009). Production of hydrogen, alkanes and polyols by aqueous phase processing of wood-derived pyrolysis oils. *Green Chemistry*, 11(9), 1433–1445.

7. Furimsky, E. (2000). Catalytic hydrodeoxygenation. *Applied Catalysis A: General*, 199(2), 147–190.

8. Zhao, H. Y., Li, D., Bui, P., and Oyama, S. T. (2011). Hydrodeoxygenation of guaiacol as model compound for pyrolysis oil on transition metal phosphide hydroprocessing catalysts. *Applied Catalysis A: General*, 391(1–2), 305–310.

9. Laurent, E. and Delmon, B. (1994). Influence of water in the deactivation of a sulfided NiMo/γ-Al$_2$O$_3$ catalyst during hydrodeoxygenation. *Journal of Catalysis*, 146(1), 281–291.

10. Elliott, D. C., Hart, T. R., Neuenschwander, G. G., Rotness, L. J., and Zacher, A. H. (2009). Catalytic hydroprocessing of biomass fast pyrolysis bio-oil to produce hydrocarbon products. *Environmental Progress & Sustainable Energy*, 28(3), 441–449.

11. Wildschut, J., Mahfud, F. H., Venderbosch, R. H., and Heeres, H. J. (2009). Hydrotreatment of fast pyrolysis oil using heterogeneous noble-metal catalysts. *Industrial & Engineering Chemistry Research*, 48(23), 10324–10334.

12. Pierson, H. O. (1996). *Handbook of Refractory Carbides and Nitrides*, Noyes Publications, New Jersey.

13. Storms, E. K. (1967). *The Refractory Metal Carbides*, Academic Press, New York.

14. Campbell, I. E. and Sherwood, E. M. (1967). *High-Temperature Materials and Technology*, John Wiley & Sons, New York.

15. Toth, L. E. (1971). *Transition Metal Carbides and Nitrides*, Academic Press, New York.

16. Oyama, S.T. (1996). *The Chemistry of Transition Metal Carbides and Nitrides*, Blackie Academic and Professional, Glasgow.

17. Levy, R. B. and Boudart, M. (1973). Platinum-like behavior of tungsten carbide in surface catalysis. *Science*, 181, 547–549.

18. Ribeiro, F. H., Boudart, M., Dalla Betta, R. A., and Iglesia, E. (1991). Catalytic reactions of *n*-alkanes on β-W$_2$C and WC: the effect of surface oxygen on reaction pathways. *Journal of Catalysis*, 130, 498–513.

19. Lee, J. S., Yeom, M. H., Park, K. Y., Nam, I. S., Chung, J. S., and Kim, Y. G. (1991). Preparation and benzene hydrogenation activity of supported molybdenum carbide catalysts. *Journal of Catalysis*, 128, 126–136.

20. Leclercq, L., Provost, M., and Leclerq, G. (1989). Catalytic properties of transition metal carbides. *Journal of Catalysis*, 117, 384–395.

21. Oyama, S.T. (1992). Preparation and catalytic properties of transition-metal carbides and nitrides. *Catalysis Today*, 15(2), 179–200.

22. Cheng, J. and Huang, W. (2010). Effect of cobalt (nickel) content on the catalytic performance of molybdenum carbides in dry-methane reforming. *Fuel Processing Technology*, 91, 185–193.
23. Oyama, S. T. (1992). Preparation and catalytic properties of transition metal carbides and nitrides. *Catalysis Today*, 15(2), 179–200.
24. Borowiechi, T. and Golebiowski, A. (1994). Influence of molybdenum and tungsten additives on the properties of nickel steam reforming catalysts. *Catalysis Letters*, 25, 309–313.
25. LaMont, D. C. and Thomson, W. J. (2004). The influence of mass transfer conditions on the stability of molybdenum carbide for dry methane reforming. *Applied Catalysis A: General*, 274, 173–178.
26. Kojima, I., Miyazaki, E., Inoue, Y., and Yasumori, I. (1982). Catalysis by transition metal carbides. *Journal of Catalysis*, 73, 128–135.
27. Woo, H. C., Park, K. Y., Kim, Y. G., Nam, I. S., Chung, J. S., and Lee, J. S. (1991). Mixed alcohol synthesis from carbon-monoxide and dihydrogen over potassium-promoted molybdenum carbide catalysts. *Applied Catalysis*, 75(2), 267–280.
28. Lee, K. S., Abe, H., Reimer, I. A., and Bell, A. T. (1993). Hydrodenitrogenation of quinoline over high-surface-area Mo2N. *Journal of Catalysis*, 139, 34–40.
29. Sajkowski, D. J. and Oyama, S. T. (1996). Catalytic hydrotreating by molybdenum carbide and nitride unsupported Mo_2N and Mo_2C/Al_2O_3. *Applied Catalysis A: General*, 134, 339–345.
30. Diaz, B., Sawhill, S. J., Bale, D. H., Main, R., Phillips, D. C., Korlann, S., Self, R., and Bussell, M. E. (2003). *Catalysis Today*, 86, 191–198.
31. Rodriguez, J. A., Dvorak, J., and Jirsak, T. (2000). Chemistry of SO_2, H_2S, and CH_3SH on carbide-modified Mo(110) and Mo_2C powders: photoemission and XANES studies. *Journal of Physical Chemistry B*, 104, 11515–11520.
32. Yu, C. C., Ramanathan, S., Dhandapani, B., Chen, J. G., and Oyama, S. T. (1997). Bimetallic Nb–Mo carbide hydroprocessing catalysts: synthesis, characterization, and activity studies. *Journal Physical Chemistry B*, 101, 512–518.
33. Zahidi, E. M., Oudghiri-Hassani, H., and McBreen, P. H. (2001). Formation of thermally stable alkylidene layers on a catalytically active surface. *Nature*, 409, 1023–1026.
34. Pang, M., Li, C., Ding, L., Zhang, J., Su, D., Li, W., and Liang, C. (2010). Microwave-assisted preparation of $Mo_2C/CNTs$ nanocomposites as efficient electrocatalyst supports for oxygen reduction reaction. *Industrial & Engineering Chemistry Research*, 49(9), 4169–4174.
35. Wang, X.-H., Zhang, M.-H., Li, W., and Tao, K.-Y. (2008). Synthesis and characterization of cobalt–molybdenum bimetallic carbides catalysts. *Catalysis Today*, 131, 111–117.

36. Lee, J. S., Oyama, S. T., and Boudart, M. (1987). Coking, aging, and regeneration of zeolites: III. Comparison of the deactivation modes of H-mordenite, HZSM-5, and HY during n-Heptane cracking. *Journal of Catalysis*, 106, 242–250.

37. Lausche, A. C., Schaidle, J. A., Schweitzer, N., and Thompson, L. T. (2013). Nanoscale carbide and nitride catalysts. In: *Comprehensive Inorganic Chemistry II, Second Edition*, J. Reedijk and K. Poeppelmeier, eds., Elsevier, Amsterdam, pp. 371–404.

38. Leclerq, L., Imura, K., Yoshida, S., Barbee, T., and Boudart, M. (1978). Synthesis of New Catalytic Materials: metal carbides of the group VI B elements. In: *Preparation of Catalysts II*, B. Delmon *et al.*, eds., Elsevier, Amsterdam, pp. 627–630.

39. Ilchenko, N. I. (1977). Oxidative catalysts over transition metal carbides. *Kinetics and Catalysis*, 18, 153–163.

40. Ilchenko, N. I., Chebotareva, N. P., and Shvidak, N. V. (1976). Oxidation of ammonia on transition metal carbides. *Reaction Kinetics and Catalysis Letters*, 4(3), 343–349.

41. Boudart, M. and Ptak, L. D. (1970). Reactions of neopentane on transition metals. *Journal of Catalysis*, 16(1), 90–96.

42. Newsam, J. M., Jacobson, A. J., McCandlish, L. E., and Polizzotti, R. S. (1988). The structures of the η-carbides Ni6Mo6C, Co6Mo6C, and Co6Mo6C2. *Journal of Solid State Chemistry*, 75(2), 296–304.

43. Korlann, S., Diaz, B., and Bussell, M. E. (2002). Synthesis of bulk and alumina-supported bimetallic carbide and nitride catalysts. *Chemistry of Materials*, 14(10), 4049–4058.

44. Xiao, T.-C., York, A. P. E., Al-Megren, H., Williams, C. V., Wang, H.-T., and Green, M. L. H. (2001). Preparation and characterisation of bimetallic cobalt and molybdenum carbides. *Journal of Catalysis*, 202(1), 100–109.

45. Liang, C., Ying, P., and Li, C. (2002). Nanostructured β-Mo$_2$C prepared by carbothermal hydrogen reduction on ultrahigh surface area carbon material. *Chemistry of Materials*, 14(7), 3148–3151.

46. Liang, C., Ma, W., Feng, Z., and Li, C. (2003). Activated carbon supported bimetallic CoMo carbides synthesized by carbothermal hydrogen reduction. *Carbon*, 41(9), 1833–1839.

47. Liang, C., Tian, F., Li, Z., Feng, Z., Wei, Z., and Li, C. (2003). Preparation and adsorption properties for thiophene of nanostructured W$_2$C on ultrahigh-surface-area carbon materials. *Chemistry of Materials*, 15(25), 4846–4853.

48. Ōya, A. and Ōtani, S. (1981). Influences of particle size of metal on catalytic graphitization of non-graphitizing carbons. *Carbon* 19(5), 391–400.

Chapter 5

Hydroprocessing Catalysts for Algal Biofuels

Changyan Yang*,†, Bo Zhang*,¶, Rui Li‡, and Qi Qiu§,‖

*School of Chemical Engineering and Pharmacy,
Wuhan Institute of Technology, Hubei, China
†Hubei Key Laboratory for Processing and Application of Catalytic
Materials, Huanggang Normal University, Hubei, China
‡Department of Natural Resources and Environmental Design,
North Carolina A&T State University, Greensboro, NC 27411, USA
§College of Chemistry and Environmental Engineering,
Shenzhen University, Guangdong, China
¶Corresponding author: bzhang_wh@foxmail.com
‖qqiu@szu.edu.cn

Abstract

Algae have been considered as a promising biofuel feedstock, which can be converted to the precursor chemicals of drop-in fuels. Due to the high protein content in algal species and the limitations of conversion technologies, these biofuel precursors require further catalytic removal of heteroatoms such as nitrogen and oxygen, being upgraded to biofuels like green diesel and aviation fuel. This chapter reviews the state-of-the-art in hydroprocessing of microalgae-based biofuels, as well as the catalyst development, the effect of process parameters on hydrotreated algal fuels, and the standards suitable for characterization of algal biofuels. Hydroprocessing of algal fuels is a new and challenging task and still underdeveloped. For the long term, an ideal catalyst for this process should possess following characteristics: high activities towards deoxygenation and denitrogenation, strong resistance to poisons, minimized leaching problems and coke formation, and an economically sound preparation process.

Keywords: Algal biofuels, Hydroprocessing, Hydrodenitrogenation, Hydrodeoxygenation, Catalyst development, Process parameters, Standards

5.1. Introduction

Algae were recently considered as a promising third-generation biofuel feedstock due to their superior productivity, high oil content, and environmentally friendly nature [1]. The algal technology for biofuels production has been greatly advanced in the past decade [2]. Currently, there are three approaches that are used mainly for producing algae-based biofuels. The first technique involves first extracting lipids from algal cells, which is followed by transesterification of triglycerides and alcohol into fatty acid alkyl esters (i.e., biodiesel) and glycerol [3] or upgrading (i.e., algal lipid upgrading, also called ALU pathway) [4]. The second technique employs the hydrothermal liquefaction (HTL) process that produces water-insoluble biocrude oil (simply called bio-oil) by using treatments at high pressure (5–20 MPa) and at the temperature range of 250–450°C [5]. Bio-oil produced after the water separation has lower water content and thus higher energy content than that produced directly by the pyrolysis of biomass. The third technique relies on the pyrolysis technology, which thermally degrades biomass at 300–700°C in the absence of oxygen, resulting in the production of bio-oils, solid residues, and gaseous products. The advantages of this technique include short process time, increased process yield, and environmental compatibility [6, 7]. In general, both traditional lipid extraction and pyrolysis require drying wet biomass to around 10% moisture, which limits the options for algae as feedstock and overall process economy [8–10]. Thus, the ALU process and the HTL process are chosen by US Department of Energy (US DOE) as the two most promising approaches [11–13].

Other techniques including gasification (supercritical water or steam) [14, 15] and biological conversion of macroalgal sugars [11] are still in their infancy. The products from these two processes are hydrogen and ethanol, respectively. It will require extensive efforts prior to bringing up more research interests on these two processes. The algal conversion technologies are summarized in Figure 5.1. Usually, the choice of the conversion technology is dependent on the composition of available feedstock. For example, biological conversion is preferred for marine macroalgae with the high carbohydrate content [16], while HTL uses the whole algae regardless of their biochemical makeup [17].

The compositions of some representative algal species are listed in Table 5.1. Usually, due to the high protein content in algal species and the limitations of aforementioned conversion technologies, microalgae-derived fuels require further catalytic processing to remove oxygen, nitrogen, and other heteroatoms. This chapter reviews hydroprocessing catalysts for algal biocrude oil. The rest of this chapter is structured as follows: In Sec. 5.2, an overview of algal biofuels produced via extraction, esterification, HTL, and pyrolysis is presented. Attention is given to the needs of hydrodenitrogenation (HDN) and hydrodeoxygenation (HDO). Section 5.3 provides a thorough presentation of the current development of hydroprocessing of algal biofuels. In Sec. 5.4, the catalyst development for HDN of algal fuels is analyzed. Section 5.5 outlines the standards and protocols that are suitable for characterization of algae-based hydrotreated fuels. Section 5.6 summarizes the effect of process parameters on the hydroprocessing process. Section 5.7 concludes this chapter.

5.2. Overview of microalgae-derived fuels

5.2.1. *Algal oil recovered by extraction and algae-based biodiesel*

Algal lipid extraction has been investigated extensively over two decades [24], and techniques applied included the use of solvents (such as hexane and chloroform), mechanical approaches (like ultrasound and microwave), and/or chemical rupture. Advantages and disadvantages of these techniques have been reviewed by Ehimen *et al.* [25]. Alternatively, the algal lipid upgrading (ALU pathway) was developed by US national laboratories [11]. This process selectively converts algal carbohydrates to ethanol and lipids to a renewable diesel blendstock, being considered as a promising conversion pathway. Because the low selectivity of extraction approaches, crude algal oil (i.e., algal lipids) often contains neutral lipids,

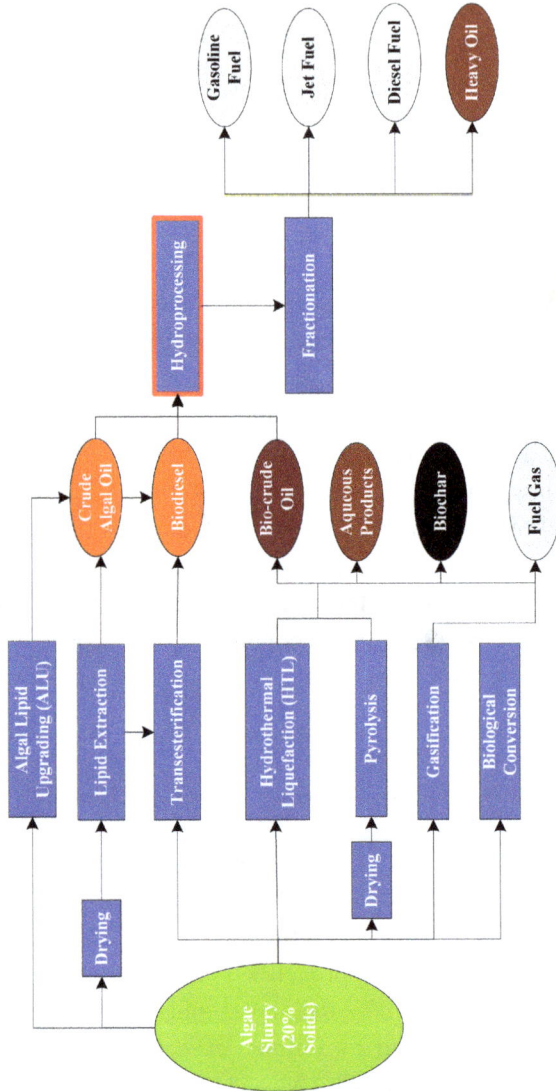

Figure 5.1. Algal conversion technologies.

Table 5.1. Composition of microalgal biomass used for biofuel production.

	Chlorella pyrenoidosa	Chlorella sorokiniana (DOE 1412)	Chlorella sp.	Microcystis sp.	Nannoch loropsis sp.	Nannoch loropsis sp.	Scene desmus	Spirulina platensis
Ultimate/Reference	[18]	[19]	[20]	[21]	[19]	[20]	[22]	[23]
C (%)	49.6	50.2	44.93	42.26	51.9	49.07	50	46.16
H (%)	7	6.8	6.42	6.27	7.5	7.59	7.11	7.14
N (%)	8.2	9.8	6.41	7.88	4.8	6.29	7.25	10.56
S (%)	0.5	0.68	1.57	0.52	0.61	1.42	0.54	0.74
O (%)	25.4	24.3	40.67	43.07	22.4	35.63	30.7	35.44
Proximate								
moisture (%)	10.4	74	4.13	9.59	79.6	5	4.59	4.54
volatile matter (%)	81.2	NR	69.45	70.13	NR	79.69	75.33	79.14
fixed carbon (%)	16.4	NR	16.22	14.14	NR	10.64	12.78	15.24
ash (%)	9.3	2.5	10.2	6.14	7	5.03	7.3	6.56
Component								
protein (%)	NR	44.6	42.7	59.93	14.3	44	36.4	48.36
polysaccharide (%)	NR	10–16	9.42	20.19	NR	21	29.3	30.21
lipid (%)	NR	10.7	2.5	5.22	21.7	30	19.5	13.3
HHV (MJ/kg)	NR	NR	NR	16.2	NR	NR	21.1	20.52

Note: NR: not reported by authors, sp.: species. The moisture content: some of these are "as received" and others have been dried.

polar lipids, chlorophyll a, and undetermined chemicals. For instance, the O, N, S, and P contents in a crude algal oil from *Nannochloropsis salina* were 12.06%, 0.43%, 2,033 ppm, and 246 ppm, respectively [26]. Even after purification, heteroatoms (like N and S) carried in the polar heads of lipids might still exist, deactivating catalysts or shortening their life [27].

In terms of algae-based biodiesel, transesterification of the algal oil extracted from dry biomass has been demonstrated [28]. Meanwhile, studies showed that solvent-based lipid extraction and direct transesterification techniques are inhibited when performed in the presence of a water phase [25, 29]. In order to avoid drying algae and improve the transesterification efficiency, serval methods including acid and base hydrolysis [30], employing alternative solvents [31], and super critical fluids [32, 33], have been developed to process wet algal biomass for oil extraction and/or in situ transesterification. Even though, most of these processes are still not considered economically feasible [34].

Generally, biodiesel, i.e., fatty acid alkyl esters, has a relatively high oxygen content, which makes it less stable, poorer flow property, less efficient than fossil fuels, and not suitable as high-grade fuels [35]. In order to improve the quality, they have been processed via catalytic HDO or deoxygenation, and converted to "Green diesel" that is a mixture of hydrocarbons meeting the American (ASTM) or European (EN) diesel standard [36].

5.2.2. Biocrude oil via hydrothermal liquefaction

Algae are natural wet biomass. Algae harvest requires concentrating the algal cells from below 0.01–0.1 wt.% to 20 wt.% solid content in the slurry. Further drying algae will need more energy and make the process costlier. Hydrothermal liquefaction (HTL), which could directly process wet feedstock with no lipid-content restriction [37], has received increasing attention and been considered as the promising technology for producing algae-derived biofuels.

The HTL of biomass can be done by using the continuous plug flow reactor or the batch reactor. Typically, algal biomass was loaded into a reactor with or without additional water and catalysts, then pressurized with inert gases (e.g., N_2 or He) or reducing gases (e.g., H_2 or CO), and the reactor was heated to a certain temperature (250–374°C) and pressure

(4–22 MPa) for 5–90 min to convert biomass to the biocrude oil [38]. Biocrude oils from algae consist of hydrocarbons and nitrogenated compounds, which might be corefined in an existing fossil refinery to produce energy and chemicals. Since the US Department of Energy (US DOE) added HTL as one of the five major pathways for biomass conversion technologies in 2012 [39], this conversion technology has become the top choice for producing algal biofuels. The development of algal HTL technology has been extensively reviewed by Tian *et al.*, López Barreiro *et al.*, Amin, and Guo *et al.* [37, 40–42].

The use of homogeneous and heterogeneous catalysts in HTL has been investigated and showed positive effects on algal bio-oils. Direct utilization of catalysts in HTL did promote the production of hydrocarbons and H_2/CH_4 from algae [43]. After the HTL reaction, low molecular weight and more polar compounds stay mainly in the aqueous phase, and larger less-polar compounds locate to the oil [44]. However, algae are complex biomasses containing a high amount of protein (N) and other heteroatoms (S, P, K, Na, etc.), which makes it impossible for one-step catalytic HTL to generate desired products. Biocrude oils of algae often have high molecular weight species and high viscosity, containing 5–18% O, 4–8% N, 0.2–1% S, and 3–30 ppm P [13, 38, 45]. Major compounds in algal bio-oil that are identifiable via gas chromatography-mass spectrometry (GC-MS) are heterocyclic nitrogenates (pyrroles, indole, pyridines, pyrazines, imidazoles, and their derivatives) [46], cyclic oxygenates (phenols and phenol derivatives with aliphatic side chains), and cyclic nitrogen and oxygenated compounds (pyrrolidinedione, piperidinedione, and pyrrolizinedione compounds) [47]. In addition, current heterogeneous catalysts for HTL are subject to low efficiency due to the absence of H_2 [48], the presence of supercritical or hot compressed water, and deactivations due to other atoms. It seems that further hydroprocessing of biocrude oils and development of effective catalysts are urgently needed.

5.2.3. *Bio-oil via pyrolysis*

Pyrolysis requires drying feedstock to around 10 wt.% water content and is often not considered as a preferred conversion technology for algae. However, as one of the widely used biomass conversion technologies during last two decades, numerous pyrolysis studies were conducted on algae including *Botryococcus braunii* [49], *Chlorella protothecoides* [50], *Dunaliella tertiolecta* [51], *Spirulina* sp., *Chlorella vulgaris* [52], *Nannochloropsis* sp. [53], residues after lipid extraction [51, 54], and oleaginous algal species [55]. Recent developments of algal pyrolysis research have been reviewed by Marcilla *et al.* and Brennan and Owende [56, 57].

Pyrolysis of algae yields three streams of products (i.e., condensed liquid, gaseous products, and biochar). In most publications, this liquid is called bio-oil. Because a pyrolytic bio-oil normally contains 30–50% water, it will simultaneously form two layers of products: water phase and an oily phase which were called aqueous products (or water solubles) and bio-oil, respectively [23]. The product yields for bio-oil, water solubles, and gases are in ranges of 18–57.9%, 15–30%, and 10–60%, respectively [56]. The problems of algal pyrolytic bio-oils are similar to those of HTL oils and lignocellulosic biomass-based pyrolytic bio-oils. A comparison of properties of HTL and pyrolysis bio-oils is shown in Table 5.2. The high oxygen content in the pyrolytic bio-oil caused low vapor pressure, low heating value, and low thermal stability. In addition, because the high protein content in almost all algal species, the nitrogen content in pyrolytic bio-oil is somewhere between 5% and 13%. Thus, in order to apply algal bio-oils as the transportation fuel, it will require reduction of both nitrogen and oxygen contents.

5.3. Hydroprocessing of algal biofuels

Algae-derived fuels require further catalytic processing to remove oxygen and/or nitrogen. Because sulfur is present in algal biofuels to a low extent, sulfur removal is often not an issue. During a hydrotreating process, sulfur is converted to hydrogen sulfide, and nitrogen is converted to ammonia [60]. These two processes are called hydrodesulfurization (HDS) and HDN, respectively. Due to the thermodynamics limitation of the aliphatic C–N bond hydrogenolysis reaction, HDN from heterocyclic compounds is a more difficult process than sulfur removal [61]. Mechanisms for HDN involve saturating intermediates, elimination, and nucleophilic substitution. An illustration of HDN processes of model nitrogenated chemicals

Table 5.2. Ultimate analysis of HTL and pyrolytic biocrude oils.

Bio-oil source ultimate	HTL			Pyrolysis			High sulfur diesel [22]	US DOE 2022 objective* [13]
	Chlorella sp. [58]	Nannochloropsis sp. [11]	Spirulina sp. [59]	Chlorella sp. [20]	Nannochloropsis sp. [20]	Spirulina platensis [23]		
C (%)	68–72	77.32	68.3	73.2	80.2	67.52	85.9	86
H (%)	8.9–9.4	10.52	8.3	9.61	6.2	9.82	12.98	14
N (%)	6	4.89	6.9	9.25	6.2	10.71	0.57	<0.05
S (%)	0.8	0.68	1.1	0.721	1.59	0.45	0.46	0
O (%)	11.1–16.2	6.52	15.4	7.19	5.81	11.34	0.1	<1
H/C	1.55–1.57	1.63	1.46	1.57	0.928	1.73	1.813	1.95
O/C	0.11–0.18	0.06	0.17	0.07	0.05	0.13	<0.005	<0.009
HHV** (MJ/kg)	32.9–36.1	40.1	32	31.5	37.2	29.3	39.1	

Note: *US DOE 2022 objective is the projected data, which was generated based on the experimental data and used as the input to the modeled projection for 2022.

**HHV: higher heating value.

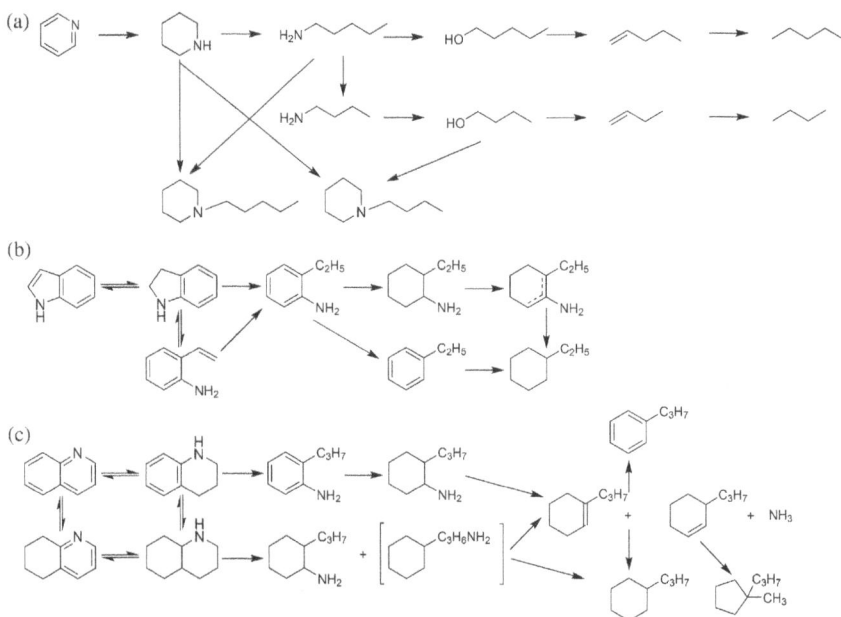

Figure 5.2.   Possible reaction mechanisms of HDN for model chemicals: (a) pyridine/piperidine, (b) quinoline/tetrahydroquinoline, and (c) indole/indoline. (Reprinted with permission from Ref. [1]. Copyright © 2016 Royal Society of Chemistry.)

including pyridine/piperidine, quinoline/tetrahydroquinoline, and indole/indoline is shown in Figure 5.2, which is reillustrated according to Refs. [62, 63]. Comprehensive reviews on HDN processes have been published by Ho and Sánchez-Delgado [63, 64].

Mechanisms for reducing the oxygen content in algae-based fuels include catalytic cracking, HDO, and decarboxylation/decarbonylation. Catalytic cracking does not require hydrogen, but the selectivity for certain hydrocarbons is low. HDO eliminates oxygen as water, while decarboxylation and decarbonylation remove oxygen to form CO_2 and CO, respectively. Three mechanisms might happen simultaneously in a single hydroprocessing reaction. The catalytic cracking process and catalysts including acids (Al_2O_3, $AlCl_3$), alkalines (NaOH, MgO, CaO), zeolites (HZSM-5, HBEA, USY, SAPO5, SAPO11, MCM-41), etc., have been comprehensively reviewed by Zhao *et al.* [65]. Similar to HDN studies, there is a significant amount of HDO research that has been done with model compounds (such as guaiacol, phenol, sorbitol, vanillin, acetic acid, methyl heptonate, cresol, and eugenol) and vegetable oils. The work related to hydrotreating of model compounds has been systematically reviewed by Zhao *et al.* and Arun *et al.* [65, 66].

5.3.1. *Catalytic HDO of extracted algal oil and algae-based biodiesel*

Many hydrotreating studies were conducted to upgrade crude algal oil, while few cases hydrotreated algae-based biodiesel. Because both algal oil and biodiesel normally don't contain too much sulfur and nitrogen, catalytic deoxygenation is an efficient method to upgrade them to green diesel that is also called renewable diesel.

When conducting deoxygenation of algal triglycerides with noble-metal catalysts (Pd/C, Pd/Al$_2$O$_3$, and Pt/US-Y zeolite) at 350°C and 5.5 MPa (800 psi) H$_2$, the Pd/C showed primarily decarbonylation activity, and exhibited stability over 200 h of continuous operation [67]. But these noble-metal catalysts were often considered costly and of low selectivity [68].

Most studies on catalysts based on nonnoble metal nickel (Ni) were done by Dr. J. A. Lercher's group (Germany) [69–71]. A series of supports, ZrO$_2$, TiO$_2$, CeO$_2$, Al$_2$O$_3$, SiO$_2$, H-beta, and HZSM-5, were synthesized or screened for deoxygenation of algal oil that mainly consists of neutral lipids. When using Ni/ZrO$_2$, a 76% yield of total liquid alkanes was attained, which is close to the theoretical yield [69]. The Ni/ZrO$_2$ selectively cleaved C–C and C–O in algal lipids. In another study, the crude algal oil was hydrotreated over with Ni/H-beta (Si/Al = 180) at 260°C and 4 MPa (580 psi) H$_2$, resulting in a 78 wt.% yield of liquid alkanes with the high selectivity towards heptadecane and octadecane [70]. The mechanism of this process was summarized as follows: firstly, double bonds in the alkyl chain were hydrogenated, and then fatty acids and propane were produced through the hydrogenolysis of saturated triglycerides, which was followed by HDO of fatty acids to alkanes. When the particle size of Ni supported on HBEA zeolite was reduced from 5–18 nm to 3 nm, both initial reaction rates and the catalyst stability were enhanced [71].

Another attempt on deoxygenating algal lipids over Ni catalysts was via decarboxylation/decarbonylation mechanisms under lower hydrogen pressures [72]. Four catalysts of $[Ni_{0.67}Al_{0.33}(OH)_2][CO_3]0.17 \cdot mH_2O$ (layered double hydroxide — LDH), Ni/Al_2O_3, Ni/ZrO_2, and Ni/La — CeO_2 were applied to this hydrotreating process. The LDH catalyst was prepared via coprecipitation of $Ni(NO_3)_2$, $Al_2(NO_3)_3$, NaOH, and Na_2CO_3 [73], while others were prepared via excess wetness impregnation with $Ni(NO_3)_2$, followed by calcination and reduction before the use. The hydrotreating of the algal lipids (1.33% in dodecane) was performed at 260°C or 300°C under a H_2 pressure of 4 MPa (580 psi). Although the hydrogen pressure was still high, the Ni–Al layered double hydroxide converted ~50% of algal lipids to hydrocarbons, showing better performance than other catalysts.

Molybdenum catalysts have been widely used for HDS of petroleum products for decades, and the first application of the cobalt molybdate catalyst (HT 400E, Harshaw Chemical Co., Ohio US) on hydrocracking of algal lipids from *Botryococcus bruunii* was reported in 1982 [74]. The hydroprocessing was conducted at 400°C and 20 MPa (3,000 psi) H_2, and upgraded oils can be fractionated into 67% gasoline, 15% jet fuels, 15% diesel, and 3% heavy oil. Only until recently, the sulfided $NiMo/\gamma-Al_2O_3$ was evaluated for hydrotreating algal oil extracted from *Nannochloropsis salina*. This crude algal oil contained neutral lipids (>30 wt.%), polar lipids, and undetermined natural substances. The hydrotreating experiments were performed at 360°C and 3.45 MPa (500 psi) H_2, resulting in a nearly complete conversion (98.7%) of microalgal oil and a 56.2% yield of hydrocarbons with a range of $C_{13}-C_{20}$. After 7-h processing, the catalyst was deactivated due to accumulating oxygenated intermediates [26].

Iron is also an attractive candidate for this kind of conversion due to low cost, rich redox chemistry, and high natural reserves. Kandel *et al.* explored the possibility of using iron nanoparticles supported on mesoporous silica nanomaterials (Fe-MSN) in hydrotreating reactions [75]. A 6-h hydrotreatment (290°C and 3 MPa H_2) gave 67% conversion, and the products were comprised of 16% alcohols, 33% unsaturated hydrocarbons, and 18% saturated hydrocarbons. This Fe–silica catalyst showed high selectivity for HDO over cracking and decarbonylation, which might be due to reverse Mars and van Krevelen mechanism.

With regard to hydrogenation (HYD) of biodiesel, most previous studies were conducted using vegetable oil or other feedstock-based biodiesel [76, 77]. Since the nature of different biodiesels is similar, their conclusions might be applicable to algae-based biodiesel. Recently, a report studied HYD of algae-based biodiesel in dodecane over 5 wt.% Pd/C and 5 wt.% Ni/HY-80 ($SiO_2/Al_2O_3 = 80$). Algal biodiesel was hydrotreated at 300°C and 3 MPa (435 psi) H_2. The performance of Ni/HY-80 was superior to Pd/C catalyst, giving a ~95% yield of hydrocarbons that mainly comprised octadecane, hexadecane, and heptadecane [78].

5.3.2. *Catalytic hydroprocessing of algal HTL biocrude oil*

The developments in the field of catalytic hydroprocessing of cellulosic biomass-derived liquefaction bio-oil between 1980 and 2007 have been documented by Elliott [79]. Hydroprocessing of algae-derived fuels differs from upgrading lignocellulosic biomass-derived oils because of the importance of both deoxygenation and denitrogenation. Thus, an algal bio-oil upgrading process needs to fulfill following purposes: oxygen and nitrogen removal, molecular weight reduction, minimizing hydrogen consumption, and avoiding saturation of the aromatic rings. Recent studies on hydroprocessing of algal HTL bio-oils are summarized in Table 5.3.

5.3.2.1. *Hydroprocessing bio-oil with molybdenum-based catalysts*

The work of hydroprocessing of the HTL biocrude oil, led by Pacific Northwest National Laboratory (PNNL), examined all over material balances and upgraded fuel quality. Their hydroprocessing experiments were performed in a bench-scale (412 ml), the two-stage continuous system [80]. The Ru/C and sulfided $CoMo/F–Al_2O_3$ (KF-1001, Akzo Chemicals Inc. [81]), which consisted of catalyst particles with a diameter of 1/16 inch and loadings of 4% Co and 15% Mo, were used for the first (125–170°C and 13.6 MPa) and second (405°C and 13.6 MPa) stages, respectively. They obtained an upgraded oil yield of 80–85% and concluded that hydroprocessing was effective for deoxygenation, denitrogenation, and desulfurization of the bio-oil from *Nannochloropsis* alga [82]. The

Table 5.3. Studies on catalytic hydroprocessing of algal HTL biocrude oil.

Microalgae provider or species	Catalysts	Experimental details	Key results	Ref.
Nannochloropsis	Sulfided CoMo/F-Al$_2$O$_3$ (4% Co and 15% Mo)	Hydroprocessing was conducted in a bench-scale (412 ml), two-stage continuous system. The operation conditions for first and second stages were (125–170°C and 13.6 MPa) and (405°C and 13.6 MPa), respectively.	They obtained an upgraded oil yield of 80–85% The products in the upgraded bio-oil that had a carbon number range of C$_6$–C$_{32}$ fell primarily in the diesel range (C$_{14}$–C$_{18}$).	[82]
Chlorella	Sulfided CoMo/F-Al$_2$O$_3$	Same as as the above.	The oxygen, nitrogen, and sulfur contents in algal bio-oils were reduced to 2.2%, <0.05%, and <50 ppm, respectively.	[46]
Chlorella grown heterotrophically	Sulfided CoMo/F-Al$_2$O$_3$	The bio-oil was upgraded at 400°C and 10.3 MPa (1500 psi) H$_2$ in a continuous system.	Compared with the phototrophic culture, this alga produced twice amount of bio-oil and upgraded oil.	[83]
Chlorella	Sulfided NiMo/Al$_2$O$_3$ or CoMo/Al$_2$O$_3$	The hydroprocessing was operated at 350°C or 405°C under 6–6.6 MPa of initial H$_2$ pressure in a 500-ml Parr reactor.	The upgraded oil yield was between 41 and 94.8%. The treated oil contained alkane hydrocarbons ranging from C$_9$ to C$_{26}$, and can be fractionated into 25% gasoline, 50% diesel, and 25% heavy fuel oil.	[58]

Scenedesmus sp.	Pt/C, Ru/C, Ni/C, and Co/C	The experiments were carried out at 350°C under 6.9 MPa of initial H_2 pressure for 4 h in a 450-ml Parr reactor.	Ru/C and Pt/C had the best efficiency in HYD, and enhanced the production of octadecane and hexadecane.	[84]
Spirulina platensis, Nannochloropsis sp., and a mixture of *Chlorella sorokiniana, Chlorella minutissima,* and *Scenedesmus bijuga*	Ru/C and sulfided CoMo/Al$_2$O$_3$	HDO was performed at 350°C under 5.17 MPa of H_2 pressure for 4 h in a batch reactor.	HDO reduced nitrogen heteroatoms in biocrude oil to 2.4–3.1%.	[85]
Nannochloropsis sp.	Pt/C, Pt/Al$_2$O$_3$, Pd/C, Pd/Al$_2$O$_3$, Ru/C, Ru/Al$_2$O$_3$, and NiMo/Al$_2$O$_3$	Hydrotreatment was done at 400°C under 5 MPa of H_2 pressure for 1 h in a microreactor.	The highest yield of upgraded oil gained via NiMo/Al$_2$O$_3$ catalyzed the reaction, which gave the lowest N content of 2.05%. The use of Pt/C or Pt/Al$_2$O$_3$ yielded the oil with the lowest O contents (1.6–2.0%).	[86]
Nannochloropsis oceanica	Sulfided CoMo/Al$_2$O$_3$	The two-stage upgrading was performed in the 500-ml reactor at 300–400°C under 4 MPa H_2 for 2 h.	The two-stage upgrading reduced the N and O contents to 1.95 and 0.72%, respectively. The heteroatoms in phenols, pyrroles, and indoles were difficult to remove.	[87]

(Continued)

Table 5.3. *(Continued)*

Microalgae provider or species	Catalysts	Experimental details	Key results	Ref.
Nannochloropsis sp.	Pt/C in the presence of water	The hydrotreating experiments were performed by adding a certain amount of water in HTL bio-oils of *Nannochloropsis* sp., which were followed by treatments in a 4-ml minireactor at 400°C and 3.4 MPa H_2 for 1–4 h.	Pt/C resulted in an oil yield of 77% and 82% carbon recovery. However, the N and O contents in treated oils were still in ranges of 1.99–3.98 and 3.08–6.97%, respectively.	[88]
Nannochloropsis sp.	Pd/C	Same as the above.	The use of Pd/C produced oils with 44 MJ/Kg HHV and a yield of 79%. The most abundant alkane in the treated oil was pentadecane (C_{15}) coexisting with others ranging from C_8 to C_{32}.	[89]
Nannochloropsis sp.	Pt/C, Mo_2C, and HZSM-5	HTL oil upgrading was carried out in a stainless-steel mini batch reactor with 0.5 g of crude bio-oil, the desired amount of catalyst, and 0.4 ml water. Factors of temperature (330–530°C), time (2–6 h), catalyst types, and catalyst loading (5–20 wt.%) were varied.	The reaction temperature was the most influential factor. The most abundant alkane in the treated oils was pentadecane (C_{15}), and others alkanes ranging from C_{10} to C_{31} are also present.	[90]

Organism	Catalyst	Method	Results	Ref.
Chlorella pyrenoidosa	$Pt/\gamma\text{-}Al_2O_3$	HTL oil upgrading was done in supercritical water (400°C) for 1 h, and H_2 or formic acid was used as the source of electrons.	Under supercritical water conditions, reactions caused an oil yield 60–70%. GC-MS showed the treated oil contained a series of *n*-alkanes starting at C_{11}.	[91]
Chlorella pyrenoidosa	Pt/C, Pd/C, Ru/C, sulfided Pt/C, $Pt/C(CO)$, $Pt/C(n\text{-}C_6H_{14})$, Mo_2C, sulfided MoS_2, Al, sulfided $CoMo/\gamma\text{-}Al_2O_3$, $Ni/SiO_2\text{-}Al_2O_3$, HZSM-5, activated carbon, and Al/Ni	The hydrotreatment was done at 400°C and 6 MPa H_2 in a 58-ml reactor filled with 3 g bio-oil, 0.3 g catalyst, and 1.5 ml water.	The process showed upgraded oil yields of 53.1–77.2%. When using Ru/C with Raney Ni as the catalysts, the upgraded oil flows freely, and has 97 wt.% of the material boiling below 400°C and a heating value of 45 MJ/kg.	[92]
Chlorella pyrenoidosa	A mixture of Ru/C with one of the abovementioned catalysts	The hydrotreatment was performed at 400°C for 4 h in a batch reactor. For each run, 3 g bio-oil, 0.3 g catalyst (0.15 Ru/C and 0.15 g other catalysts), 1.5 ml water, and 6 MPa H_2 were loaded into the reactor.	Ru/C and Mo_2C produced the highest oil yield of 77.2% and energy recovery. The treated bio-oil contained straight-chain alkanes ranged from C_{10} to C_{25}, with pentadecane (C_{15}), hexadecane (C_{16}), and heptadecane (C_{17}) as the three most abundant hydrocarbons.	[18]

oxygen, nitrogen, and sulfur contents in algal bio-oils were reduced to 1–2%, <0.5%, and <50 ppm, respectively. The products in the upgraded bio-oil fell primarily in the diesel range. The similar results were also confirmed for *Chlorella* alga, as the oxygen, nitrogen, and sulfur contents in algal bio-oils were reduced to 2.2%, <0.05%, and <50 ppm, respectively [46]. Further, the same process was applied to *Chlorella* grown heterotrophically, which had a lipid content of 57–64% and low nitrogen content of 0.5% [83]. After a hydrotreatment, the oxygen, nitrogen, and sulfur contents of this algal biofuel were reduced to 1.7%, <0.05%, and 18 ppm, respectively.

Concerning both the yield of treated oil and the effects of HDO, HDN, and HDS, the results obtained by researchers at the PNNL are remarkably better than those of other studies that are reviewed in following sections. The possible reasons are threefold: high availability of H_2; the reaction by-products (such as NH_3, H_2O, H_2S, and coke) were removed immediately from the continuous process, minimalizing their windows for reacting with hydrocarbons to form undesired products; and the two-stage upgrading.

A similar work was done by scientists at the University of Leeds and the University of Illinois [58], who hydroprocessed biocrude oil from hydrothermal liquefaction of *Chlorella* in a 500-ml Parr reactor. The biocrude was mixed with sulfided NiMo/Al_2O_3 or CoMo/Al_2O_3 catalysts at a ratio of 25 g:5 g. The hydroprocessing was operated at 350°C or 405°C under 6–6.6 MPa of initial H_2 pressure. Both noncatalytic and catalytic hydroprocessing reduced nitrogen and oxygen contents in the upgraded oil, giving an oil yield between 41% and 94.8%. The treated oil can be fractionated into 25% gasoline, 50% diesel, and 25% heavy fuel oil. However, the lowest N content reached in this study was 2.4% by using NiMo/Al_2O_3 at 405°C, so the catalytic function of catalysts toward HDN needs to be further improved. The authors also pointed out the differences between their work and the PNNL study: (1) a solvent was used to recover biocrude, resulting in higher O and N contents in the biocrude (11–16% O and 6% N) compared to PNNL bio-oil (5–8% O and 4–5% N) and (2) hydroprocessing was conducted in the batch mode.

Researchers have compared the catalytic effects of noble metals (Pt, Pd, and Ru etc.) with transition bimetallic catalysts (NiMo and CoMo) using various species of algae [84–86]. Noble metals showed higher HDO activities than transition metals, but the noble metals were often deactivated within a short period of time. When comparing the HDN activities, Costanzo *et al.* [85] found that the N content in upgraded oils was higher than 2.4% of total oil even if the amount of Ru/C was increased to 30% of the total loading, and Patel *et al.* [86] reported that NiMo/Al$_2$O$_3$ catalyzed reaction resulted in the highest yield of upgraded oil with the lowest N content of 2.05%.

Zhao *et al.* studied two-stage upgrading of the biocrude oil of *Nannochloropsis oceanica* over sulfided CoMo/Al$_2$O$_3$ [87]. The two-stage upgrading was performed in the batch reactors, and reduced the N and O contents of the upgraded oils to 1.95% and 0.72%, respectively. Chemicals including phenols, pyrroles, and indoles were detected in the final oil products.

5.3.2.2. *Hydrotreating bio-oil in the presence of water*

A series of studies on hydrotreating of bio-oil in supercritical water were conducted by Savage [89] and his collaborators at the Henan Polytechnic University. The motivation of their research is from a process engineering perspective, to take advantages of hydroprocessing HTL bio-oil in the same environment as HTL. Their initial hydrotreating experiments were performed by adding a certain amount of water in HTL bio-oils of *Nannochloropsis* sp., which were followed by treatments in a 4-ml mini-reactor at 400°C and 3.4 MPa H$_2$ for 1–4 h by using Pt/C and Pd/C catalysts [88]. The use of Pd/C produced an oil with 44 MJ/kg HHV and a yield of 79%, while Pt/C resulted in an oil yield of 77% and 82% carbon recovery. However, the N and O contents in treated oils were still in ranges of 1.99–3.98 and 3.08–6.97%, respectively. Further, they compared the functions of three catalysts: Pt/C, Mo$_2$C, and HZSM-5 and concluded that reaction temperature was the most influential factor [90]. Among the catalysts studied, applying Mo$_2$C at 530°C for 2 h showed the best deoxygenation performance, and using Pt/C at 530°C for 6 h resulted in the lowest N content of 1.5% in the treated oil.

Their recent research subject changed to HTL oils of *Chlorella pyrenoidosa*, which has a lower O content of 2.1–7.8% but higher N content of 7.8–8.0%. Treatment of this bio-oil was done with Pt/γ-Al$_2$O$_3$, and H$_2$ or formic acid was used as the source of protons. Under supercritical water conditions (400°C), reactions caused an oil yield 60–70% [91]. Although both deoxygenation and denitrogenation functions of Pt/γ-Al$_2$O$_3$ were not effective, this research indicated that 0.025 g/cm^3 water density is the optimal condition for hydrotreating bio-oils in supercritical water.

Later on, 15 HYD catalysts including Pt/C, Pd/C, Ru/C, sulfided Pt/C, Pt/C(CO), Pt/C(n-C$_6$H$_{14}$), Mo$_2$C, MoS$_2$, Al, sulfided CoMo/γ-Al$_2$O$_3$, Ni/SiO$_2$–Al$_2$O$_3$, HZSM-5, activated carbon, and Al/Ni were tested for upgrading HTL oils of *C. pyrenoidosa* [92]. The hydrotreatment was done at 400°C in a 58 ml reactor filled with 3 g bio-oil, 0.3 g catalyst, 1.5 ml water, and 6 MPa H$_2$. The process showed upgraded oil yields of 53.1–77.2%. Ru/C gave the best result for deoxygenation and Al/Ni (Raney nickel) was shown to be a suitable catalyst for denitrogenation. Catalysts of Co–Mo/Al$_2$O$_3$, Mo$_2$C, and MoS$_2$ performed poorly for deoxygenation in the presence of water, but remained high denitrogenation activity that is comparable to that of the noble-metal catalysts.

Furthermore, a mixture of Ru/C with one of the abovementioned catalysts was used. In respect of deoxygenation, denitrogenation, and desulfurization, Ru/C and Mo$_2$C, Ru/C and Pt/γ-Al$_2$O$_3$, or Ru/C and Pt/C showed the best results, giving the O, N, and S contents of 0.1 wt.%, 1.8 wt.%, and 0.065 wt.%, respectively [18]. Ru/C and Mo$_2$C produced the highest oil yield of 77.2% and energy recovery. Although these experiments were performed in an engineering way, the results revealed some insights into hydrotreating reactions, indicating that catalytic synergy in bimetallic catalysts is worth further research. However, the N content of upgraded oils could not be reduced to less than 1.5% under supercritical water conditions according to their reports. Therefore, both the hydrotreating process and the catalyst will require further improvements.

5.3.3. *Catalytic hydrotreating of algal pyrolytic bio-oil*

As a separate unit operation, the study of hydrotreating always followed the waves of developments of algal research or conversion technologies. Even though there are an increased number of algal pyrolysis studies since 2009 to date, only few articles reported hydrotreating of algal pyrolytic bio-oil. Zhong *et al.* [93] studied hydrotreating of fast pyrolysis oil from *Chlorella* over a Ni–Co–Pd/γ-Al$_2$O$_3$ catalyst. Hydroprocessing at 300°C and 2 MPa H$_2$ resulted in a refined oil yield of 89.6% and an 80.4% reduction of the oxygen content. The nitrogen content was reduced from 6.48% to 2.45%.

In another study, bimetallic Ni–Cu/ZrO$_2$ catalysts with various Cu/Ni ratios (0.14–1.00 w/w) were used for HDO of pyrolytic bio-oils of *Chlorella* sp. and *Nannochloropsis* sp. at 350°C and 2 MPa H$_2$ [20]. Comparing with Ni/ZrO$_2$ and sulfided NiMo/Al$_2$O$_3$, the addition of copper could facilitate the reduction of nickel oxide and limit sintering and coking, showing a higher HDO efficiency of 82%. But the denitrogenation activities of catalysts were not compared.

Nam *et al.* [94] studied catalytic upgrading of the pyrolytic bio-oil of *Nannochloropsis oceanica*. The bio-oil was first fractionated by distillation, and then hydrotreated over Pd/C at 130–250°C under 4.1–8.3 MPa H$_2$ for 4 h in a 50-ml batch reactor. They concluded that the distillation step prior to catalytic upgrading led to a better quality of upgraded bio-oil, but this Pd catalyst could not effectively remove nitrogen compounds from the bio-oil.

5.4. Catalyst development for HDN of algal bio-oil

5.4.1. *Catalyst development*

The crude algal oil produced from the ALU pathway has low S and N contents, so the traditional HDS catalysts might be effective enough to remove oxygen and improve its quality. A detailed review of the catalyst design strategies for HDO can be found at Arun *et al.* [66]. For upgrading algal HTL and pyrolytic bio-oils, the catalysts have to be bifunctional, possessing both HDO and HDN activities. Until now, a limited number of catalysts have been investigated, and the results showed both promising possibilities and significant problems.

Most HYD catalysts could denitrogenate the algal fuels to some extent, but the nitrogen content left in the hydrotreated oil was often between 1% and 4%. The residual nitrogen-containing compounds are in forms of pyrrole [93], amides (like *N,N*-dimethylhexanamide, palmitamide, benzenamine), nitriles, quinolone [88], and indole [90]. Although ASTM and EN standards do not regulate the minimal nitrogen content of current transportation fuels, US DOE's goal for the nitrogen content in upgraded algal fuels is less than 0.05% (500 ppm) [13]. In order to meet this goal, further development of highly selective catalysts for the C–N bond breakage is needed.

Meanwhile, a significant number of HDN studies have been done with model chemicals. Because the algal fuels are complex mixtures, the most active catalyst for HDN of model compounds might not show the highest catalytic activity for upgrading algal bio-oils [92]. Conversely, the use of model compounds is a logical way to investigate the possible mechanism of a catalyst that showed a high performance in hydroprocessing of algal fuels. In this section, the HDN studies on algal bio-oils and model chemicals are discussed together, which would give us new insights into the catalyst development strategy.

5.4.1.1. *Molybdenum-based catalysts*

The most active catalyst of sulfided $CoMo/Al_2O_3$ was identified for the HDN and HDO of algal bio-oils by researchers at the PNNL [38]. The hydroprocessing was conducted under relatively severe conditions (405°C and 13.6 MPa H_2) in a continuous reactor. As the traditional HDS catalysts for petroleum, $CoMo/Al_2O_3$ and $NiMo/Al_2O_3$, specifically sulfided form, have been widely studied for HDS and HDN of model chemicals. These catalysts exhibited higher HDS activity than HDN activity in competitive reactions between thiophene and pyridine [95], and the presence of Co or Ni accelerated mainly the HDS reaction [96]. Because algal biofuels have low sulfur contents, the use of sulfided catalysts will require the addition of external sulfur sources, e.g., hydrogen sulfide that was able to enhance denitrogenation and inhibit HYD [97]. If the sulfur source is not available, it will cause sulfur leaching, the reduction of catalyst performance, and contamination of upgraded oils. In addition, sulfided catalysts have a poor hydrostability in the presence of water.

Unpromoted Mo [96], Mo sulfides [98], nitrides [99], carbides, and phosphides [100] were proved to be more active in the HDN reactions of model compounds than sulfided CoMo and NiMo.

The Mo nitrides (Mo_2N) showed as much as 5–10 times more activity for pyridine HDN than the sulfided $Co–Mo/Al_2O_3$ and MoS_2/Al_2O_3

catalysts [101], and the selectivity for C–N bond hydrogenolysis over C–C bond was higher for the nitride catalysts [102]. The bulk phase was predominantly γ-Mo$_2$N with the surface consisting of either β-Mo$_{16}$N$_7$ or mixtures of Mo and β-Mo$_{16}$N$_7$. The most active sites were located at the perimeters of raft-like domains, while lower activity sites were associated with the γ-Mo$_2$N crystallite [103].

The Mo carbides (Mo$_2$C) were proven to have the similar catalytic properties for pyridine HDN to Mo nitrides [104]. HDN of pyridine over Mo carbides and Mo nitrides produced mostly cyclopentane and pentane, respectively. The selectivity difference between Mo carbides and Mo nitrides might be due to differing bonding geometries for pyridine on the Mo carbides and nitrides.

When MoP was tested for the catalytic activity in the HDN reaction of *o*-propylaniline, the intrinsic HDN activity of the surface Mo atoms was about six times higher than that of Mo edge atoms in MoS$_2$/Al$_2$O$_3$ [105].

Further modification of Mo$_2$C, Mo$_2$N, and MoP could improve their activities. Doping Mo$_2$C with platinum (Mo$_2$C–Pt) resulted in a higher HDN efficiency than Mo$_2$C [106]; nickel-promoted Mo nitrides (NiMoN$_x$— Ni 5 wt.%, MoN$_x$ 15 wt.%, supported on γ-Al$_2$O$_3$) were more active than Mo$_2$N [107]; while addition of TiO$_2$ to MoP/MCM-41 enhanced the C–N bond cleavage but inhibited the dehydrogenation function [108].

Generally, the HDN activity of sulfided catalysts or Mo sulfides in the presence of sulfur sources is always superior to other Mo compounds, because sulfur participates in the reactions. However, due to the high N content of algal fuels, Mo nitrides and carbides are more of interest to hydroprocessing of algae-derived biofuels. Since the catalytic performance of Mo nitrides and carbides relies on the surface and crystal structure [109], future research attention should be given to controlling crystal structure, surface modification, and selecting suitable promotors and supports.

5.4.1.2. *Noble-metal catalysts*

Noble-metal catalysts (Ru, Pd, Rh, Ir, and Pt supported on carbon), specifically Ir and Pt sulfides, exhibited higher pyridine HDN activity than the sulfided molybdenum-based catalysts and can be used under milder conditions with high activity [110]. When hydrotreating algal HTL biocrude, Pt/C and a mixture of Ru/C and Pt/γ-Al$_2$O$_3$ showed good HDN activities, reducing the N content in hydrotreated oil to 1.5–1.8 wt.% [18, 89].

Pt and Ru were often used to dope other metal catalysts, such as tungsten carbides [111], Mo carbides [106], Fe [112], and CoMo [113]. The HDN activity over the Pt- or Ru-promoted catalysts was highly dependent on the amount of metallic sites introduced by them [111].

Noble-metal catalysts are most reactive metals for the C–N bond cleavage. Obviously, the cost of noble metals will be a barrier for the process development. However, the use of these metals as promoters with common hydrotreating catalysts might improve their HDN activity, as well as undergoing minimal HYD reactions [114].

5.4.1.3. *Other transition metal catalysts*

Nickel is an attractive metal in hydrotreating because of its high activity and low cost [115]. Ni has been used to promote many other metallic catalysts (like Mo and Mo$_2$N), and test results suggested that Ni–Mo species enhanced the HYD of model chemicals like pyridine [116]. Catalytic hydroprocessing of algal bio-oil with Raney® Ni led to the lowest N content of 1.6 wt.% in upgraded oil [92]. In addition, Ni phosphides exhibited high conversion of HDS and HDN model compounds (99% and 100%, respectively) [117].

As a relatively new HYD catalyst, iron (Fe) is considered as a low cost, environmentally friendly, and sustainable material. However, the use of Fe as the only active metal for HDN of algal oil resulted in lower conversion and production of a significant amount of alcohols [75]. Instead, Fe-promoted Mo, tungsten (W) [118], and vanadium (V) [119] sulfides were reported to give an unusual high HDN effect.

Tungsten carbides and phosphides were often used as a HDN catalyst. One study compared HDN of carbazole over W$_2$C with Mo$_2$C. The results indicated that W$_2$C possessed higher HYD activity but lower total activity [120]. Tungsten phosphides were more extensively studied for their HDN behavior [121]. Bulk WP and WP/SiO$_2$ were found to be more active in HDN than W$_2$C, W$_2$N, WS$_2$, and Ni–Mo–S/Al$_2$O$_3$ catalysts [122].

Transition metal phosphides, such as WP, Ni_2P, CoP, MoP, and Fe_2P, emerged recently as an attractive group of hydroprocessing catalysts, which have an excellent activity for HDS and HDN [123, 124]. The study showed that their catalytic activities for dibenzothiophene HDS and quinoline HDN followed the order: Ni_2P > WP > MoP > CoP > Fe_2P. The crystal structure of metal phosphides is built with blocks of trigonal prisms, which can well accommodate the large phosphorus atoms, leading to a more isotropic crystal morphology and potentially better exposure of surface metal atoms to liquid-phase reactants [125]. Furthermore, they show good heat and electrical conductivities, and high thermal and chemical stability [126]. However, the deactivation of metal phosphides in the presence of water [127] and deficiency of P [117] was observed.

5.4.1.4. *Effects of supports*

The catalyst support is the vehicle for carrying active constituents and affects the chemical and physical properties of the catalyst, the degree of dispersion of the active components, and the overall stability. Catalyst supports also need to provide the reactants with large surface area and suitable pore size. Currently, Al_2O_3 is most widely used as the support in traditional hydroprocessing catalysts and studied for its effects on HDN reactions. For instance, the alumina-supported molybdenum nitride (Mo_2C/Al_2O_3) catalyst was extremely active in the HDN of carbazole, comparing with the sulfided and reduced catalysts [128]. The result indicated the C–N hydrogenolysis occurred on partially hydrogenated carbazole, suggesting the possibility of reducing hydrogen consumption. A modification of the alumina support with borate ions could increase the amount of acidic centers of NiMo catalyst, leading to an increase in the resistance to coking [129]. However, it is generally accepted that the Al_2O_3 was not stable in the presence of a large amount of water and its acid sites could result in carbon deposition [130].

Compared to the Al_2O_3 support, carbon supports showed better ability in water resistance and anticoking. Hydroprocessing of pyridine over carbon-supported NiMo sulfide formed less undesired products than Al-supported CoMo and NiMo catalysts [96]. When the activated mesoporous carbon black support was employed to support Mo carbides, the β-Mo_2C hexagonal compact crystallographic phase was obtained as the unique active phase, improving HDN of indole [131]. The carbon-supported catalysts also showed high resistance to poisoning [132], and high HDN activity/selectivity [133].

SiO_2 and molecule sieves (e.g., MCM-41 [108], SBA-15 [77, 134]) are of interest because of their moderate acidity, high surface area, large pore size, and highly ordered structures. For example, the SiO_2-supported Mo or W phosphides showed superior HDN but lower HDS activity compared to the sulfides [135]. Silica–alumina [136, 137], TiO_2 [138], ZrO_2 [139], and CeO_2–ZrO_2 [140] were also tested as supports in hydrotreating catalysts.

To summarize this subsection, a good HDN catalyst support should improve overall thermal and chemical stability and the dispersity of active components, enhance surface chemistry (for example, the HDN performance was related to the Brønsted acidity of some catalysts [137, 141]), and promote the formation of the highly active crystal structure.

5.4.2. *Catalyst suppliers and preparation*

Hydroprocessing of algal fuels is a new and very challenging task. Most studies were conducted by using commercially available catalysts. Table 5.4 gives a list of catalyst suppliers, whose catalysts have been used for hydrotreating algal fuels. The catalysts used for hydroprocessing lignocellulosic bio-oils and their suppliers can be found in Ref. [79].

Instead of using commercial catalysts, many catalysis scientists synthesized new types of catalysts based on their own formula. Following examples are the preparation methodology for several typical catalysts that have been applied to hydrotreating of algal biofuels.

Morgan *et al.* [73] synthesized layered double hydroxides via coprecipitation under low supersaturation conditions. Two aqueous solutions, one containing the 1.5 M metals in forms of $Ni(NO_3)_2$ and $Al_2(NO_3)_3$ and one containing a mixture of NaOH (3 M) and Na_2CO_3 (1 M), were simultaneously added at a rate of 3 ml/min and mixed vigorously while maintaining the pH at 10. The precipitate was aged in the solution overnight at 75°C, isolated by centrifuging, and washed with deionized water until pH reached 7. The resulting solid was dried at 60°C in a vacuum oven. The formulae were determined with elemental analysis.

Table 5.4.   Catalyst suppliers.

Company name	Catalysts used by researchers	Company web address	Ref.
Sigma-Aldrich	Pt/C, Pd/C, Ru/C, Mo_2C, MoS_2, Al, Ni/SiO_2–Al_2O_3, HZSM-5, Al/Ni, and activated carbon	sigmaaldrich.com	[90, 92]
Zeolyst International	Zeolite beta, ZSM-5, zeolite Y	zeolyst.com	[90]
Akzo Chemicals Inc.	KF-1001	akzonobel.com	[81]
Alfa Aesar	CoMo/γ-Al_2O_3	alfa.com	[92]
Johnson Matthey (London, UK)	CoMo/Al_2O_3 and NiMo/Al_2O_3	matthey.com	[58]
Qilu petrochemical catalyst plant (China)	NiMo/Al_2O_3	qpec.cn	[20]

Lee *et al.* [77] prepared mesoporous silica nanoparticles (SBA-15) with sphere-like and necklace-like structures. To synthesize SBA-15, Pluronic 123 was dissolved in HCl solution at room temperature, and then heated to 40°C. Tetraethyl orthosilicate (TEOS) was added as the silica source. The mixture was stirred for 20 h at 40°C and aged at 100°C for 24 h. Formed product was collected via filtration and dried overnight. The surfactant was removed by calcination at 550°C for 5 h. The SBA-15-supported nickel phosphide, Ni_2P/SBA-15, was prepared by wet impregnation of aqueous solutions of nickel nitrate and ammonium phosphate with a Ni/P molar ratio of 1:1.

The most common method for preparing catalysts is impregnation (incipient wetness, pore volume, deposition–precipitation, etc.). A typical example is the preparation of bimetallic Ni–Cu/ZrO_2 catalysts by Guo *et al.* [20]. The support of ZrO_2 was calcined at 500°C for 3 h to remove water and other impurities, and then impregnated with solutions of $Ni(NO_3)_2$·$6H_2O$ and $Cu(NO_3)_2$·$3H_2O$ under strong agitation. The catalysts were ultrasonic treated for 3 h, and then heated under stirring to form a viscous substance, which was dried at 110°C for 10 h and calcined in air at 550°C for 5 h. The weight fractions of Ni and Cu metals in the catalyst were determined by using inductively coupled plasma-optical emission spectrometry (ICP-OES).

5.5. Characterization of algae-based hydrotreated fuels

The ultimate goal of hydroprocessing of algal biofuels is to synthesize drop-in fuels: automobile fuels (gasoline and diesel) and aviation turbine fuels. These fuels are a mixture of different hydrocarbons. The hydrocarbons of gasoline contain typically 4–12 carbon atoms; diesel contains between 12 and 20 carbon atoms per molecule [142]; and the jet fuel has a carbon number distribution between about 8 and 16 [143]. Most hydroprocessing processes successfully upgraded the algal lipids or the biocrude oil to the diesel range, representing 50–85% of total hydrocarbons; while the hydrocracking process was able to convert the algal lipids mainly to gasoline (67% of total hydrocarbons) [74].

Characterization of algae-based hydrotreated fuels should follow the national or international standards. Gasoline (i.e., unleaded petrol) is the fuel derived from petroleum that meets the requirements of the standards ASTM D4814 in the USA [144] and EN 228 in Europe [145], and the diesel needs to meet the standards of ASTM D975 [146] and EN 590 [147]. The EN 590:2013 is applicable to automotive diesel fuel containing up to 7.0% fatty acid methyl esters (FAMEs). Biodiesel is defined in the standard EN 14214 as FAMEs only [148], while US biodiesel is comprised of mono-alkyl esters of long-chain fatty acids derived from animal fats or vegetable oil as shown in the ASTM D6751 [149]. The standard specifications of aviation turbine fuels are available in ASTM D1655 [150] and ASTM D7566 [151]. According to ASTM 7566-11, up to 50% bioderived synthetic fuels can be blended into conventional jet fuel [152].

Besides using the ASTM or EN standards, hydroprocessed algal fuels were characterized using various analytic protocols by researchers. Organic chemicals were identified by using GC-MS. Elemental analyses for carbon, hydrogen, nitrogen, and sulfur contents were determined using a CHNS/O analyzer. The H/C ratio is a useful indicator for the saturation extent of the aromatics in upgraded oils [153]. Alternately, sulfur and other minerals can be measured by ICP-OES. The HHV can be calculated with the DuLong formula according to elemental analyses [154] or directly measured using an oxygen bomb calorimeter.

Because the oil products still contain compounds that cannot be identified by the GC-MS, more detailed analyses may be necessary. Following the ASTM D1319 method, hydrocarbon types in the oil can be determined by fluorescent indicator adsorption [155]. The similar identification can also be done by the carbon-13 NMR [79]. The boiling point distribution of oil samples can be obtained by performing fast simulated distillation analysis according to the ASTM D7169 [156]. Alternately, the ASTM D2887 is applicable to oil products with a boiling point between 55°C and 538°C [157].

Aforementioned methods have been successfully applied to analyzing upgraded algal fuels. Scientists at the National Institute of Standards and Technology (NIST) combined methods of distillation curve method, NMR, and GC-MS to evaluate the properties of an algal-based hydro-treated renewable diesel [158]. Additional test methods including measuring the speed of sound, the cloud point, density, the cetane index, the storage stability, and the oxidation stability were also developed [159]. The combustion of this algal diesel has been tested in a diesel engine along with petroleum diesel [160].

5.6. Effect of process parameters on hydroprocessing

Besides the catalyst, the parameters of a hydroprocessing process include the reactor configuration, reaction temperature, initial H_2 pressure, residence time, etc. These parameters are important to the overall effectiveness of HYD and the product distribution.

5.6.1. *Crude extracted algal oil*

Normally, laboratory experiments were performed at the batch mode in either a tubular reactor or a high-pressure reactor, like autoclaves and Parr reactors [161]. In one case, the algal oil was treated sequentially in a two-reactor system with Pt/C and Pd/US-Y zeolite, respectively, giving a 95% yield of alkanes [67]. The advantages of using two-stage reactors are that each reactor with a different catalyst could be operated under its optimal conditions, thus potentially achieving a higher efficiency of removing heteroatoms.

In terms of reaction time, most experiments were conducted for 6–8 h, and by then the catalysts had deactivated due to coking. Only one report showed that when Pd/C was used as the catalyst, their operation could last for 200 h if the algal oil was charged at the rate of 0.177 ml/min [67]. The long catalyst life in this study might be due to the continuous operation mode, the use of two-stage reactors, and/or the catalyst support of carbon.

Compared with other parameters, the product distribution is more likely determined by the reaction temperature and the amount of initial H_2 (i.e., the severity of reaction conditions). Most hydroprocessing processes, which were conducted between 260°C and 360°C with an initial H_2 pressure of 2–5.5 MPa, successfully upgraded the algal lipids to the diesel range (C_{13}–C_{20}). The major alkanes in treated oil have a carbon number range of C_{15}–C_{18}. When a more severe condition of 400°C and 20 MPa H_2 was applied, the hydrocracking process was able to convert the algal lipids mainly to gasoline (C_4–C_{12}), representing 67% of total hydrocarbons [74].

5.6.2. *Bio-oil*

Some successful hydroprocessing experiments for upgrading biocrude oils were conducted in a two-stage continuous system [80]. Both the continuous operation and the two-stage treatment are important to the quality of the upgraded oil. For example, if the batch reactors were used for the two-stage configuration, the N content of treated oils was still above 2.4% [85].

Hydroprocessing of bio-oils was often performed for less than 24 h, and the life of catalysts has not been studied systematically. The typical treatment temperature was 350–405°C, while the initial H_2 pressure was around 6 MPa. In order to achieve a good performance, a 10 MPa initial H_2 pressure may be necessary [82, 83]. Most studies were able to obtain the upgraded oil with a carbon number distribution mainly in the diesel range (C_{14}–C_{18}), representing 50–85% of total hydrocarbons [162].

5.7. Closing remarks and prospects

The algal biofuel technology has been accelerated during last decade, especially after 2010. Experts from industry, academia, and national laboratories made invaluable contributions to its development from the biology of algae production to fuel conversion, reducing the cost of algae-based biocrude from \$240 to \$7.50 per gallon [163]. However, in order to meet the long term goal of \$3/gasoline gallon equivalent, it still requires a combination of improvements in all key technologies including production, conversion, and processing [46].

From a prospect of the process development, the continuous operation is highly recommended, which might minimize unwanted by-products and the window that by-products react with the oil. It is possible that a dual continuous system is preferred, because each reactor will be operated under different optimal conditions and/or different catalysts. The typical reaction conditions for hydroprocessing of crude algal oil and biocrude oil in a dual reactor are (260–360°C and 3–20 MPa H_2) and (350–405°C and 6–13.6 MPa H_2), respectively. To obtain better HDO and HDN results, a higher initial H_2 pressure (i.e., the availability of H_2) is required.

From the point of view of catalyst development, the traditional HDS catalysts could be efficient enough for deoxygenation of crude algal oils that have low sulfur and nitrogen contents. However, for oils with high S and N, an ideal catalyst for hydroprocessing of algal bio-oils should possess high activities towards both denitrogenation and deoxygenation. According to Pacific Northwest National Laboratory's reports, the pre-sulfided catalyst of CoMo supported on fluorinated γ-Al_2O_3 was suggested to be the best candidate for this process, if a continuous operation can be applied. However, the function of this catalyst was only confirmed in a limited number of applications, and it did not show the same efficiency in the batch reactor. Accordingly, the door is still open to catalysis scientists who are interested in developing effective and cost-effective catalysts.

To date, it is known that bimetallic catalysts could be a promising choice to fulfill the requirement for upgrading algal bio-oils that contain a high amount of nitrogenated chemicals. The active metals of tungsten carbide (WC), molybdenum carbide (Mo_2C), and molybdenum nitride (MoN) are recommended, while the noble/transition metals of Ni, Co, Ru, Pt, Fe, and Cu could be used to modify the active metal. Because the bio-oil contains 5–10% moisture and hydrotreatment will produce water as a by-product, the supporting material of the catalysts should be water resistant. Therefore, the materials, such as carbon, modified Al_2O_3, modified MCM-41, and zeolite Y, are of interest.

For the long term, following issues shall be considered: (1) The study on reaction mechanism using model compounds is essential to reveal the catalysis mechanism. (2) The expected catalyst life is 2 y. Meanwhile, catalysts need to tolerate poisons (such as sulfur, phosphorus, and water) and minimize leaching problems and coke formation. (3) The economics for preparing catalysts are important, so the cost of active metals and regeneration protocols are import factors.

References

1. Yang, C., Li, R., Cui, C., Liu, S., Qiu, Q., Ding, Y., and Wu, Y. (2016). Catalytic hydroprocessing of microalgae-derived biofuels: a review. *Green Chemistry*, 18(13), 3684–3699.
2. Picardo, M., de Medeiros, J., Monteiro, J., Chaloub, R., Giordano, M., and de Queiroz Fernandes Araújo, O. (2013). A methodology for screening of microalgae as a decision making tool for energy and green chemical process applications. *Clean Technologies and Environmental Policy*, 15(2), 275–291.
3. Wahlen, B. D., Willis, R. M., and Seefeldt, L. C. (2011). Biodiesel production by simultaneous extraction and conversion of total lipids from microalgae, cyanobacteria, and wild mixed-cultures. *Bioresource Technology*, 102(3), 2724–2730.
4. Davis, R., Fishman, D., Frank, E., Wigmosta, M., Aden, A., Coleman, A., Pienkos, P., Skaggs, R., Venteris, E., and Wang, M. (2012). *Renewable diesel from algal lipids: an integrated baseline for cost, emissions, and resource potential from a harmonized model*, Argonne National Laboratory, Argonne, IL.
5. Zhang, B., Keitz, M., and Valentas, K. (2008). Thermal effects on hydrothermal biomass liquefaction. *Applied Biochemistry and Biotechnology*, 147(1–3), 143–150.
6. Wan, Y., Chen, P., Zhang, B., Yang, C., Liu, Y., Lin, X., and Ruan, R. (2009). Microwave-assisted pyrolysis of biomass: catalysts to improve product selectivity. *Journal of Analytical and Applied Pyrolysis*, 86(1), 161–167.
7. Zhang, B., Yang, C., Moen, J., Le, Z., Hennessy, K., Wan, Y., Liu, Y., Lei, H., Chen, P., and Ruan, R. (2010). Catalytic conversion of microwave-assisted

pyrolysis vapors. *Energy Sources, Part A: Recovery, Utilization, and Environmental Effects*, 32(18), 1756–1762.

8. Molina Grima, E., Belarbi, E. H., Acién Fernández, F. G., Robles Medina, A., and Chisti, Y. (2003). Recovery of microalgal biomass and metabolites: process options and economics. *Biotechnology Advances*, 20(7–8), 491–515.

9. Lardon, L., Hélias, A., Sialve, B., Steyer, J.-P., and Bernard, O. (2009). Life-cycle assessment of biodiesel production from microalgae. *Environmental Science & Technology*, 43(17), 6475–6481.

10. Yang, C., Wu, J., Deng, Z., Zhang, B., Cui, C., and Ding, Y. (2017). A comparison of energy consumption in hydrothermal liquefaction and pyrolysis of microalgae. *Trends in Renewable Energy*, 3(1), 76–85.

11. Davis, R., Kinchin, C., Markham, J., Tan, E., Laurens, L., Sexton, D., Knorr, D., Schoen, P., and Lukas, J. (2014). Process design and economics for the conversion of algal biomass to biofuels: algal biomass fractionation to lipid-and carbohydrate-derived fuel products (No. NREL/TP-5100-62368). National Renewable Energy Laboratory (NREL), Golden, CO.

12. Laurens, L. M. L., Nagle, N., Davis, R., Sweeney, N., van Wychen, S., Lowell, A., and Pienkos, P. T. (2015). Acid-catalyzed algal biomass pretreatment for integrated lipid and carbohydrate-based biofuels production. *Green Chemistry*, 17(2), 1145–1158.

13. Jones, S. B., Zhu, Y., Snowden-Swan, L. J., Anderson, D., Hallen, R. T., Schmidt, A. J., Albrecht, K., and Elliott, D. C. (2014). Whole algae hydrothermal liquefaction: 2014 state of technology. Pacific Northwest National Laboratory (PNNL), Richland, WA (US).

14. Stucki, S., Vogel, F., Ludwig, C., Haiduc, A. G., and Brandenberger, M. (2009). Catalytic gasification of algae in supercritical water for biofuel production and carbon capture. *Energy & Environmental Science*, 2(5), 535–541.

15. Duman, G., Uddin, M. A., and Yanik, J. (2014). Hydrogen production from algal biomass via steam gasification. *Bioresource Technology*, 166, 24–30.

16. Chen, J., Bai, J., Li, H., Chang, C., and Fang, S. (2015). Prospects for bioethanol production from macroalgae. *Trends in Renewable Energy*, 1(3), 185–197.

17. US DOE (2014). Algal Biofuels Strategy Proceedings from the March 26–27, 2014, Workshop. US Department of Energy Office of Energy Efficiency and Renewable Energy.

18. Xu, Y., Duan, P., and Wang, B. (2015). Catalytic upgrading of pretreated algal oil with a two-component catalyst mixture in supercritical water. *Algal Research*, 9, 186–193.

19. Sudasinghe, N., Reddy, H., Csakan, N., Deng, S., Lammers, P., and Schaub, T. (2015). Temperature-dependent lipid conversion and nonlipid composition of microalgal hydrothermal liquefaction oils monitored by fourier transform ion cyclotron resonance mass spectrometry. *BioEnergy Research*, 8(4), 1962–1972.

20. Guo, Q., Wu, M., Wang, K., Zhang, L., and Xu, X. (2015). Catalytic hydrodeoxygenation of algae bio-oil over bimetallic Ni–Cu/ZrO$_2$ catalysts. *Industrial & Engineering Chemistry Research*, 54(3), 890–899.

21. Hu, Z., Zheng, Y., Yan, F., Xiao, B., and Liu, S. (2013). Bio-oil production through pyrolysis of blue-green algae blooms (BGAB): product distribution and bio-oil characterization. *Energy*, 52, 119–125.

22. Kim, S. W., Koo, B. S., and Lee, D. H. (2014). A comparative study of bio-oils from pyrolysis of microalgae and oil seed waste in a fluidized bed. *Bioresource Technology*, 162, 96–102.

23. Jena, U. and Das, K. C. (2011). Comparative evaluation of thermochemical liquefaction and pyrolysis for bio-oil production from microalgae. *Energy & Fuels*, 25(11), 5472–5482.

24. Gong, Y. and Jiang, M. (2011). Biodiesel production with microalgae as feedstock: from strains to biodiesel. *Biotechnology Letters*, 33(7), 1269–1284.

25. Ehimen, E. A., Sun, Z. F., and Carrington, C. G. (2010). Variables affecting the in situ transesterification of microalgae lipids. *Fuel*, 89(3), 677–684.

26. Zhou, L. and Lawal, A. (2015). Evaluation of presulfided NiMo/γ-Al$_2$O$_3$ for hydrodeoxygenation of microalgae oil to produce green diesel. *Energy & Fuels*, 29(1), 262–272.

27. Li, Y., Horsman, M., Wang, B., Wu, N., and Lan, C. (2008). Effects of nitrogen sources on cell growth and lipid accumulation of green alga *Neochloris oleoabundans*. *Applied Microbiology and Biotechnology*, 81(4), 629–636.

28. Kleinová, A., Cvengrošová, Z., Rimarčík, J., Buzetzki, E., Mikulec, J., and Cvengroš, J. (2012). Biofuels from algae. *Procedia Engineering*, 42, 231–238.

29. Griffiths, M. J., van Hille, R. P., and Harrison, S. T. L. (2010). Selection of direct transesterification as the preferred method for assay of fatty acid content of microalgae. *Lipids*, 45(11), 1053–1060.

30. Sathish, A. and Sims, R. C. (2012). Biodiesel from mixed culture algae via a wet lipid extraction procedure. *Bioresource Technology*, 118, 643–647.

31. Dejoye Tanzi, C., Abert Vian, M., and Chemat, F. (2013). New procedure for extraction of algal lipids from wet biomass: a green clean and scalable process. *Bioresource Technology*, 134, 271–275.

32. Mendes, R. L., Reis, A. D., and Palavra, A. F. (2006). Supercritical CO_2 extraction of γ-linolenic acid and other lipids from *Arthrospira (Spirulina) maxima*: comparison with organic solvent extraction. *Food Chemistry*, 99(1), 57–63.

33. Reddy, H. K., Muppaneni, T., Patil, P. D., Ponnusamy, S., Cooke, P., Schaub, T., and Deng, S. (2014). Direct conversion of wet algae to crude biodiesel under supercritical ethanol conditions. *Fuel*, 115, 720–726.

34. Halim, R., Gladman, B., Danquah, M. K., and Webley, P. A. (2011). Oil extraction from microalgae for biodiesel production. *Bioresource Technology*, 102(1), 178–185.

35. Thangavel, P. and Sridevi, G. (2014). *Environmental Sustainability: Role of Green Technologies*, Springer, Berlin.

36. Knothe, G. (2010). Biodiesel and renewable diesel: a comparison. *Progress in Energy and Combustion Science*, 36(3), 364–373.

37. Tian, C., Li, B., Liu, Z., Zhang, Y., and Lu, H. (2014). Hydrothermal liquefaction for algal biorefinery: a critical review. *Renewable and Sustainable Energy Reviews*, 38, 933–950.

38. Elliott, D. C., Biller, P., Ross, A. B., Schmidt, A. J., and Jones, S. B. (2015). Hydrothermal liquefaction of biomass: developments from batch to continuous process. *Bioresource Technology*, 178, 147–156.

39. US Department of Energy (2012). Carbon, hydrogen and separation efficiencies in bio-oil conversion pathways (CHASE bio-oil pathways). https://eere-exchange.energy.gov/FileContent.aspx?FileID=6e712603-8cce-4252-92cb-b4c4ae83da71.

40. López Barreiro, D., Prins, W., Ronsse, F., and Brilman, W. (2013). Hydrothermal liquefaction (HTL) of microalgae for biofuel production: state of the art review and future prospects. *Biomass and Bioenergy*, 53, 113–127.

41. Amin, S. (2009). Review on biofuel oil and gas production processes from microalgae. *Energy Conversion and Management*, 50(7), 1834–1840.

42. Guo, Y., Yeh, T., Song, W., Xu, D., and Wang, S. (2015). A review of bio-oil production from hydrothermal liquefaction of algae. *Renewable and Sustainable Energy Reviews*, 48, 776–790.

43. Yeh, T. M., Dickinson, J. G., Franck, A., Linic, S., Thompson, L. T., and Savage, P. E. (2013). Hydrothermal catalytic production of fuels and chemicals from aquatic biomass. *Journal of Chemical Technology & Biotechnology*, 88(1), 13–24.

44. Sudasinghe, N., Dungan, B., Lammers, P., Albrecht, K., Elliott, D., Hallen, R., and Schaub, T. (2014). High resolution FT-ICR mass spectral analysis of bio-oil and residual water soluble organics produced by hydrothermal

liquefaction of the marine microalga *Nannochloropsis salina*. *Fuel*, 119, 47–56.

45. Ross, A. B., Biller, P., Kubacki, M. L., Li, H., Lea-Langton, A., and Jones, J. M. (2010). Hydrothermal processing of microalgae using alkali and organic acids. *Fuel*, 89(9), 2234–2243.

46. Jones, S. B., Zhu, Y., Anderson, D. M., Hallen, R. T., Elliott, D. C., Schmidt, A., Albrecht, K., Hart, T., Butcher, M., and Drennan, C. (2014). *Process design and economics for the conversion of algal biomass to hydrocarbons: whole algae hydrothermal liquefaction and upgrading*, Pacific Northwest National Laboratory.

47. Vardon, D. R., Sharma, B. K., Scott, J., Yu, G., Wang, Z., Schideman, L., Zhang, Y., and Strathmann, T. J. (2011). Chemical properties of biocrude oil from the hydrothermal liquefaction of Spirulina algae, swine manure, and digested anaerobic sludge. *Bioresource Technology*, 102(17), 8295–8303.

48. Duan, P. and Savage, P. E. (2011). Hydrothermal liquefaction of a microalga with heterogeneous catalysts. *Industrial & Engineering Chemistry Research*, 50(1), 52–61.

49. Gelin, F., Gatellier, J. P. L. A., Damsté, J. S. S., Metzger, P., Derenne, S., Largeau, C., and de Leeuw, J. W. (1993). Mechanisms of flash pyrolysis of ether lipids isolated from the green microalga *Botryococcus braunii* race A. *Journal of Analytical and Applied Pyrolysis*, 27(2), 155–168.

50. Miao, X. and Wu, Q. (2004). High yield bio-oil production from fast pyrolysis by metabolic controlling of *Chlorella protothecoides*. *Journal of Biotechnology*, 110(1), 85–93.

51. Grierson, S., Strezov, V., Ellem, G., McGregor, R., and Herbertson, J. (2009). Thermal characterisation of microalgae under slow pyrolysis conditions. *Journal of Analytical and Applied Pyrolysis*, 85(1–2), 118–123.

52. Hu, Z., Ma, X., and Chen, C. (2012). A study on experimental characteristic of microwave-assisted pyrolysis of microalgae. *Bioresource Technology*, 107, 487–493.

53. Sanchez-Silva, L., López-González, D., Garcia-Minguillan, A. M., and Valverde, J. L. (2013). Pyrolysis, combustion and gasification characteristics of *Nannochloropsis gaditana* microalgae. *Bioresource Technology*, 130, 321–331.

54. Pan, P., Hu, C., Yang, W., Li, Y., Dong, L., Zhu, L., Tong, D., Qing, R., and Fan, Y. (2010). The direct pyrolysis and catalytic pyrolysis of *Nannochloropsis* sp. residue for renewable bio-oils. *Bioresource Technology*, 101(12), 4593–4599.

55. Na, J.-G., Park, Y.-K., Kim, D. I., Oh, Y.-K., Jeon, S. G., Kook, J. W., Shin, J. H., and Lee, S. H. (2015). Rapid pyrolysis behavior of oleaginous microalga, *Chlorella* sp. KR-1 with different triglyceride contents. *Renewable Energy*, 81, 779–784.

56. Marcilla, A., Catalá, L., García-Quesada, J. C., Valdés, F. J., and Hernández, M. R. (2013). A review of thermochemical conversion of microalgae. *Renewable and Sustainable Energy Reviews*, 27, 11–19.

57. Brennan, L. and Owende, P. (2010). Biofuels from microalgae — a review of technologies for production, processing, and extractions of biofuels and co-products. *Renewable and Sustainable Energy Reviews*, 14(2), 557–577.

58. Biller, P., Sharma, B. K., Kunwar, B., and Ross, A. B. (2015). Hydroprocessing of bio-crude from continuous hydrothermal liquefaction of microalgae. *Fuel*, 159, 197–205.

59. Jazrawi, C., Biller, P., Ross, A. B., Montoya, A., Maschmeyer, T., and Haynes, B. S. (2013). Pilot plant testing of continuous hydrothermal liquefaction of microalgae. *Algal Research*, 2(3), 268–277.

60. Nagai, M. and Kabe, T. (1983). Selectivity of molybdenum catalyst in hydrodesulfurization, hydrodenitrogenation, and hydrodeoxygenation: effect of additives on dibenzothiophene hydrodesulfurization. *Journal of Catalysis*, 81(2), 440–449.

61. Girgis, M. J. and Gates, B. C. (1991). Reactivities, reaction networks, and kinetics in high-pressure catalytic hydroprocessing. *Industrial & Engineering Chemistry Research*, 30(9), 2021–2058.

62. Duan, P. and Savage, P. E. (2011). Catalytic hydrothermal hydrodenitrogenation of pyridine. *Applied Catalysis B: Environmental*, 108–109, 54–60.

63. Ho, T. C. (1988). Hydrodenitrogenation catalysis. *Catalysis Reviews*, 30(1), 117–160.

64. Sánchez-Delgado, R. A. (2002). *Organometallic Modeling of the Hydrodesulfurization and Hydrodenitrogenation Reactions*, Springer Science & Business Media, Berlin.

65. Zhao, C., Bruck, T., and Lercher, J. A. (2013). Catalytic deoxygenation of microalgae oil to green hydrocarbons. *Green Chemistry*, 15(7), 1720–1739.

66. Arun, N., Sharma, R. V., and Dalai, A. K. (2015). Green diesel synthesis by hydrodeoxygenation of bio-based feedstocks: strategies for catalyst design and development. *Renewable and Sustainable Energy Reviews*, 48, 240–255.

67. Robota, H. J., Alger, J. C., and Shafer, L. (2013). Converting algal triglycerides to diesel and HEFA jet fuel fractions. *Energy & Fuels*, 27(2), 985–996.

68. Murata, K., Liu, Y., Inaba, M., and Takahara, I. (2010). Production of synthetic diesel by hydrotreatment of Jatropha oils using Pt–Re/H-ZSM-5 catalyst. *Energy & Fuels*, 24(4), 2404–2409.

69. Peng, B., Yuan, X., Zhao, C., and Lercher, J. A. (2012). Stabilizing catalytic pathways via redundancy: selective reduction of microalgae oil to alkanes. *Journal of the American Chemical Society*, 134(22), 9400–9405.

70. Peng, B., Yao, Y., Zhao, C., and Lercher, J. A. (2012). Towards quantitative conversion of microalgae oil to diesel-range alkanes with bifunctional catalysts. *Angewandte Chemie International Edition*, 51(9), 2072–2075.

71. Song, W., Zhao, C., and Lercher, J. A. (2013). Importance of size and distribution of Ni nanoparticles for the hydrodeoxygenation of microalgae oil. *Chemistry — A European Journal*, 19(30), 9833–9842.

72. Santillan-Jimenez, E., Morgan, T., Loe, R., and Crocker, M. (2015). Continuous catalytic deoxygenation of model and algal lipids to fuel-like hydrocarbons over Ni–Al layered double hydroxide. *Catalysis Today*, 258, Part 2, 284–293.

73. Morgan, T., Santillan-Jimenez, E., Harman-Ware, A. E., Ji, Y., Grubb, D., and Crocker, M. (2012). Catalytic deoxygenation of triglycerides to hydrocarbons over supported nickel catalysts. *Chemical Engineering Journal*, 189–190, 346–355.

74. Hillen, L. W., Pollard, G., Wake, L. V., and White, N. (1982). Hydrocracking of the oils of *Botryococcus braunii* to transport fuels. *Biotechnology and Bioengineering*, 24(1), 193–205.

75. Kandel, K., Anderegg, J. W., Nelson, N. C., Chaudhary, U., and Slowing, I. I. (2014). Supported iron nanoparticles for the hydrodeoxygenation of microalgal oil to green diesel. *Journal of Catalysis*, 314, 142–148.

76. Kubičková, I., Snåre, M., Eränen, K., Mäki-Arvela, P., and Murzin, D. Y. (2005). Hydrocarbons for diesel fuel via decarboxylation of vegetable oils. *Catalysis Today*, 106(1–4), 197–200.

77. Lee, S.-P. and Ramli, A. (2013). Methyl oleate deoxygenation for production of diesel fuel aliphatic hydrocarbons over Pd/SBA-15 catalysts. *Chemistry Central Journal*, 7(1), 149.

78. Viêgas, C. V., Hachemi, I., Freitas, S. P., Mäki-Arvela, P., Aho, A., Hemming, J., Smeds, A., Heinmaa, I., Fontes, F. B., da Silva Pereira, D. C., Kumar, N., Aranda, D. A. G., and Murzin, D. Y. (2015). A route to produce

renewable diesel from algae: synthesis and characterization of biodiesel via in situ transesterification of *Chlorella* alga and its catalytic deoxygenation to renewable diesel. *Fuel*, 155, 144–154.

79. Elliott, D. C. (2007). Historical developments in hydroprocessing bio-oils. *Energy & Fuels*, 21(3), 1792–1815.

80. Elliott, D. C., Hart, T. R., Neuenschwander, G. G., Rotness, L. J., Olarte, M. V., Zacher, A. H., and Solantausta, Y. (2012). Catalytic hydroprocessing of fast pyrolysis bio-oil from pine sawdust. *Energy & Fuels*, 26(6), 3891–3896.

81. McKetta Jr, J. J. (1992). *Petroleum Processing Handbook*, CRC Press, Boca Raton.

82. Elliott, D. C., Hart, T. R., Schmidt, A. J., Neuenschwander, G. G., Rotness, L. J., Olarte, M. V., Zacher, A. H., Albrecht, K. O., Hallen, R. T., and Holladay, J. E. (2013). Process development for hydrothermal liquefaction of algae feedstocks in a continuous-flow reactor. *Algal Research*, 2(4), 445–454.

83. Albrecht, K. O., Zhu, Y., Schmidt, A. J., Billing, J. M., Hart, T. R., Jones, S. B., Maupin, G., Hallen, R., Ahrens, T., and Anderson, D. (2016). Impact of heterotrophically stressed algae for biofuel production via hydrothermal liquefaction and catalytic hydrotreating in continuous-flow reactors. *Algal Research*, 14, 17–27.

84. Wang, Z., Adhikari, S., Valdez, P., Shakya, R., and Laird, C. (2016). Upgrading of hydrothermal liquefaction biocrude from algae grown in municipal wastewater. *Fuel Processing Technology*, 142, 147–156.

85. Costanzo, W., Hilten, R., Jena, U., Das, K. C., and Kastner, J. R. (2016). Effect of low temperature hydrothermal liquefaction on catalytic hydro-denitrogenation of algae biocrude and model macromolecules. *Algal Research*, 13, 53–68.

86. Patel, B., Arcelus-Arrillaga, P., Izadpanah, A., and Hellgardt, K. (2017). Catalytic hydrotreatment of algal biocrude from fast hydrothermal liquefaction. *Renewable Energy*, 101, 1094–1101.

87. Zhao, B., Wang, Z., Liu, Z., and Yang, X. (2016). Two-stage upgrading of hydrothermal algae biocrude to kerosene-range biofuel. *Green Chemistry*, 18(19), 5254–5265.

88. Duan, P. and Savage, P. E. (2011). Upgrading of crude algal bio-oil in super-critical water. *Bioresource Technology*, 102(2), 1899–1906.

89. Duan, P. and Savage, P. E. (2011). Catalytic hydrotreatment of crude algal bio-oil in supercritical water. *Applied Catalysis B: Environmental*, 104(1–2), 136–143.

90. Duan, P. and Savage, P. E. (2011). Catalytic treatment of crude algal bio-oil in supercritical water: optimization studies. *Energy & Environmental Science*, 4(4), 1447–1456.

91. Duan, P., Bai, X., Xu, Y., Zhang, A., Wang, F., Zhang, L., and Miao, J. (2013). Catalytic upgrading of crude algal oil using platinum/gamma alumina in supercritical water. *Fuel*, 109, 225–233.

92. Bai, X., Duan, P., Xu, Y., Zhang, A., and Savage, P. E. (2014). Hydrothermal catalytic processing of pretreated algal oil: a catalyst screening study. *Fuel*, 120, 141–149.

93. Zhong, W.-c., Guo, Q.-j., Wang, X.-y., and Zhang, L. (2013). Catalytic hydroprocessing of fast pyrolysis bio-oil from *Chlorella*. *Journal of Fuel Chemistry and Technology*, 41(5), 571–578.

94. Nam, H., Kim, C., Capareda, S. C., and Adhikari, S. (2017). Catalytic upgrading of fractionated microalgae bio-oil (*Nannochloropsis oculata*) using a noble metal (Pd/C) catalyst. *Algal Research*, 24, Part A, 188–198.

95. Cinibulk, J., Kooyman, P., Vit, Z., and Zdražil, M. (2003). Magnesia-supported Mo, CoMo and NiMo sulfide catalysts prepared by nonaqueous impregnation: parallel HDS/HDN of thiophene and pyridine and TEM microstructure. *Catalysis Letters*, 89(1–2), 147–152.

96. Vít, Z. (1993). Comparison of carbon- and alumina-supported Mo, CoMo and NiMo catalysts in parallel hydrodenitrogenation and hydrodesulphurization. *Fuel*, 72(1), 105–107.

97. Brunet, S. and Perot, G. (1985). Effect of hydrogen sulfide on the catalytic hydrodenitrogenation of 1,2,3,4-tetrahydroquinoline. Comparison of NiMo/Al_2O_3 and CoMo/Al_2O_3. *Reaction Kinetics and Catalysis Letters*, 29(1), 15–20.

98. Ochoa, R., Lee, W., and Eklund, P. (1995). Catalytic hydrodenitrogenation of quinoline with nanoscale Mo_2N, Mo_2C and MoS_2 synthesized by laser pyrolysis. *Preprints of Papers, American Chemical Society, Division of Fuel Chemistry*, 40(CONF-950402--).

99. Trawczyński, J. (2000). Effect of synthesis conditions on the hydrodesulfurization and hydrodenitrogenation activities of alumina supported Mo and CoMo nitrides. *Applied Catalysis A: General*, 197(2), 289–293.

100. Li, W., Dhandapani, B., and Oyama, S. T. (1998). Molybdenum phosphide: a novel catalyst for hydrodenitrogenation. *Chemistry Letters*, 27(3), 207–208.

101. Sajkowski, D. J. and Oyama, S. T. (1996). Catalytic hydrotreating by molybdenum carbide and nitride: unsupported Mo_2N and $Mo_2CAl_2O_3$. *Applied Catalysis A: General*, 134(2), 339–349.

102. Choi, J.-G., Brenner, J. R., Colling, C. W., Demczyk, B. G., Dunning, J. L., and Thompson, L. T. (1992). Synthesis and characterization of molybdenum nitride hydrodenitrogenation catalysts. *Catalysis Today*, 15(2), 201–222.

103. Colling, C. W. and Thompson, L. T. (1994). The structure and function of supported molybdenum nitride hydrodenitrogenation catalysts. *Journal of Catalysis*, 146(1), 193–203.

104. Choi, J. G., Brenner, J. R;, and Thompson, L. T. (1995). Pyridine hydrodenitrogenation over molybdenum carbide catalysts. *Journal of Catalysis*, 154(1), 33–40.

105. Stinner, C., Prins, R., and Weber, T. (2000). Formation, structure, and HDN activity of unsupported molybdenum phosphide. *Journal of Catalysis*, 191(2), 438–444.

106. Lewandowski, M., Szymańska-Kolasa, A., da Costa, P., and Sayag, C. (2007). Catalytic performances of platinum doped molybdenum carbide for simultaneous hydrodenitrogenation and hydrodesulfurization. *Catalysis Today*, 119(1–4), 31–34.

107. Yuhong, W., Wei, L., Minghui, Z., Naijia, G., and Keyi, T. (2001). Characterization and catalytic properties of supported nickel molybdenum nitrides for hydrodenitrogenation. *Applied Catalysis A: General*, 215(1–2), 39–45.

108. Duan, X., Li, X., Wang, A., Teng, Y., Wang, Y., and Hu, Y. (2010). Effect of TiO_2 on hydrodenitrogenation performances of MCM-41 supported molybdenum phosphides. *Catalysis Today*, 149(1–2), 11–18.

109. Furimsky, E. (2003). Metal carbides and nitrides as potential catalysts for hydroprocessing. *Applied Catalysis A: General*, 240(1–2), 1–28.

110. Vít, Z. and Zdražil, M. (1989). Simultaneous hydrodenitrogenation of pyridine and hydrodesulfurization of thiophene over carbon-supported platinum metal sulfides. *Journal of Catalysis*, 119(1), 1–7.

111. Lewandowski, M., da Costa, P., Benichou, D., and Sayag, C. (2010). Catalytic performance of platinum doped tungsten carbide in simultaneous hydrodenitrogenation and hydrodesulphurization. *Applied Catalysis B: Environmental*, 93(3–4), 241–249.

112. Guerrero-Ruiz, A., Sepulveda-Escribano, A., Rodriguez-Ramos, I., Lopez-Agudo, A., and Fierro, J. L. G. (1995). Catalytic behaviour of carbon-supported FeM (M = Ru, Pt) in pyridine hydrodenitrogenation. *Fuel*, 74(2), 279–283.

113. Hirschon, A. S., Wilson, R. B., and Laine, R. M. (1987). Ruthenium promoted hydrodenitrogenation catalysts. *Applied Catalysis*, 34, 311–316.

114. Hirschon, A. S., Ackerman, L. L., Laine, R. M., and Wilson, R. B. J. (1989). *Use of promoters to enhance hydrodenitrogenation and hydrodeoxygenation catalysis.* US DOE, Washington, DC.

115. Zhang, Y. and Guo, J. (2013). Bio-oil upgrading. In: *Biomass Processing, Conversion and Biorefinery*, B. Zhang and Y. Wang, eds., Nova Science Publishers, Inc., Hauppauge, NY, pp. 201–220.

116. Ozkan, U. S., Ni, S., Zhang, L., and Moctezuma, E. (1994). Simultaneous hydrodesulfurization and hydrodenitrogenation of model compounds over Ni–Mo/γ-Al$_2$O$_3$ catalysts. *Energy & Fuels*, 8(1), 249–257.

117. Lee, Y.-K., Shu, Y., and Oyama, S. T. (2007). Active phase of a nickel phosphide (Ni$_2$P) catalyst supported on KUSY zeolite for the hydrodesulfurization of 4,6-DMDBT. *Applied Catalysis A: General*, 322, 191–204.

118. Ho, T. C., Jacobson, A. J., Chianelli, R. R., and Lund, C. R. F. (1992). Hydrodenitrogenation-selective catalysts I. Fe promoted Mo/W sulfides. *Journal of Catalysis*, 138(1), 351–363.

119. Magarelli, M., Scott, C. E., Moronta, D., and Betancourt, P. (2003). Synthesis, characterisation and HDN activity of iron–vanadium sulphide prepared with the use of a chelating agent. *Catalysis Today*, 78(1–4), 339–344.

120. Szymańska-Kolasa, A., Lewandowski, M., Sayag, C., Brodzki, D., and Djéga-Mariadassou, G. (2007). Comparison between tungsten carbide and molybdenum carbide for the hydrodenitrogenation of carbazole. *Catalysis Today*, 119(1–4), 35–38.

121. Clark, P., Li, W., and Oyama, S. T. (2001). Synthesis and activity of a new catalyst for hydroprocessing: tungsten phosphide. *Journal of Catalysis*, 200(1), 140–147.

122. Oyama, S. T., Clark, P., Wang, X., Shido, T., Iwasawa, Y., Hayashi, S., Ramallo-López, J. M., and Requejo, F. G. (2002). Structural characterization of tungsten phosphide (WP) hydrotreating catalysts by X-ray absorption spectroscopy and nuclear magnetic resonance spectroscopy. *The Journal of Physical Chemistry B*, 106(8), 1913–1920.

123. Zuzaniuk, V., Stinner, C., Prins, R., and Weber, T. (2000). Transition metal phosphides: novel hydrodenitrogenation catalysts. *Studies in Surface Science and Catalysis*, 143, 247–255.

124. Wang, X., Clark, P., and Oyama, S. T. (2002). Synthesis, characterization, and hydrotreating activity of several iron group transition metal phosphides. *Journal of Catalysis*, 208(2), 321–331.

125. Oyama, S. T. (2003). Novel catalysts for advanced hydroprocessing: transition metal phosphides. *Journal of Catalysis*, 216(1–2), 343–352.

126. Liu, C. and Wang, Y. (2013). Biofuel and bio-oil upgrading. In: *Biomass Processing, Conversion and Biorefinery*, B. Zhang and Y. Wang, eds., Nova Science Publishers, Inc., Hauppauge, NY, pp. 221–250.

127. Li, K., Wang, R., and Chen, J. (2011). Hydrodeoxygenation of anisole over silica-supported Ni_2P, MoP, and NiMoP catalysts. *Energy & Fuels*, 25(3), 854–863.

128. Nagai, M. and Miyao, T. (1992). Activity of alumina-supported molybdenum nitride for carbazole hydrodenitrogenation. *Catalysis Letters*, 15(1–2), 105–109.

129. Lewandowski, M. and Sarbak, Z. (2000). The effect of boron addition on hydrodesulfurization and hydrodenitrogenation activity of $NiMo/Al_2O_3$ catalysts. *Fuel*, 79(5), 487–495.

130. Furimsky, E. and Massoth, F. E. (1999). Deactivation of hydroprocessing catalysts. *Catalysis Today*, 52(4), 381–495.

131. Sayag, C., Suppan, S., Trawczyński, J., and Djéga-Mariadassou, G. (2002). Effect of support activation on the kinetics of indole hydrodenitrogenation over mesoporous carbon black composites-supported molybdenum carbide. *Fuel Processing Technology*, 77–78, 261–267.

132. Shu, Y. and Oyama, S. T. (2005). Synthesis, characterization, and hydrotreating activity of carbon-supported transition metal phosphides. *Carbon*, 43(7), 1517–1532.

133. Hillerová, E., Vít, Z., Zdrazˆil, M., Shkuropat, S. A., Bogdanets, E. N., and Startsev, A. N. (1990). Comparison of carbon and alumina supported nickel — molybdenum sulphide catalysts in parallel hydrodenitrogenation and hydrodesulphurisation. *Applied Catalysis*, 67(1), 231–236.

134. Ghampson, I. T., Sepúlveda, C., Garcia, R., García Fierro, J. L., Escalona, N., and DeSisto, W. J. (2012). Comparison of alumina- and SBA-15-supported molybdenum nitride catalysts for hydrodeoxygenation of guaiacol. *Applied Catalysis A: General*, 435–436, 51–60.

135. Clark, P., Wang, X., and Oyama, S. T. (2002). Characterization of silica-supported molybdenum and tungsten phosphide hydroprocessing catalysts by 31P nuclear magnetic resonance spectroscopy. *Journal of Catalysis*, 207(2), 256–265.

136. Harvey, T. G. and Matheson, T. W. (1985). Hydrodenitrogenation catalysed by zeolite-supported ruthenium. *Journal of the Chemical Society, Chemical Communications*, 188–189. DOI: 10.1039/C39850000188.

137. Peeters, E., Cattenot, M., Geantet, C., Breysse, M., and Zotin, J. L. (2008). Hydrodenitrogenation on Pt/silica–alumina catalysts in the presence of H_2S: role of acidity. *Catalysis Today*, 133–135, 299–304.

138. Ramirez, J., Cedeño, L., and Busca, G. (1999). The role of titania support in Mo-based hydrodesulfurization catalysts. *Journal of Catalysis*, 184(1), 59–67.

139. Chen, L., Zhu, Y., Zheng, H., Zhang, C., Zhang, B., and Li, Y. (2011). Aqueous-phase hydrodeoxygenation of carboxylic acids to alcohols or alkanes over supported Ru catalysts. *Journal of Molecular Catalysis A: Chemical*, 351, 217–227.

140. Bykova, M. V., Ermakov, D. Y., Kaichev, V. V., Bulavchenko, O. A., Saraev, A. A., Lebedev, M. Y., and Yakovlev, V. A. (2012). Ni-based sol–gel catalysts as promising systems for crude bio-oil upgrading: guaiacol hydrodeoxygenation study. *Applied Catalysis B: Environmental*, 113–114, 296–307.

141. Damyanova, S., Spojakina, A., and Jiratova, K. (1995). Effect of mixed titania-alumina supports on the phase composition of NiMo/TiO$_2$Al$_2$O$_3$ catalysts. *Applied Catalysis A: General*, 125(2), 257–269.

142. Advanced Motor Fuels (2015). Diesel and gasoline. http://www.iea-amf.org/content/fuel_information/diesel_gasoline#gasoline.

143. Chevron Products Corporation (2007). *Aviation Fuels Technical Review*, Chevron Products Corporation, San Ramon, CA.

144. ASTM D4814 (2015). *Standard Specification for Automotive Spark-Ignition Engine Fuel*, ASTM International, West Conshohocken, PA. DOI: 10.1520/D4814-15A.

145. British Standards Institution (BSI) (2013). *BS EN 228:2012 Automotive Fuels. Unleaded Petrol. Requirements and Test Methods*, BSI, London, UK.

146. ASTM D975 (2015). *Standard Specification for Diesel Fuel Oils*, ASTM International, West Conshohocken, PA. DOI: 10.1520/D0975-15B.

147. British Standards Institution (BSI) (2013). *BS EN 590:2013 Automotive Fuels. Diesel. Requirements and Test Methods*, BSI, London, UK.

148. British Standards Institution (BSI) (2013). *BS EN 14214:2012+A1:2014. Liquid Petroleum Products. Fatty Acid Methyl Esters (FAME) for Use in Diesel Engines and Heating Applications. Requirements and Test Methods*, BSI, London, UK.

149. ASTM D6751 (2015). *Standard Specification for Biodiesel Fuel Blend Stock (B100) for Middle Distillate Fuels*, ASTM International, West Conshohocken, PA. DOI: 10.1520/D6751-15A.

150. ASTM D1655 (2015). *Standard Specification for Aviation Turbine Fuels*, ASTM International, West Conshohocken, PA. DOI: 10.1520/D1655-15C.

151. ASTM D7566 (2015). *Standard Specification for Aviation Turbine Fuel Containing Synthesized Hydrocarbons*, ASTM International, West Conshohocken, PA. DOI: 10.1520/D7566-15A.

152. Popov, S. and Kumar, S. (2013). Renewable fuels via catalytic hydrodeoxygenation of lipid-based feedstocks. *Biofuels*, 4(2), 219–239.

153. Elliott, D. C. and Schiefelbein, G. F. (1989). Liquid hydrocarbon fuels from biomass. *Papers of the American Chemical Society*, 34(4), 1160.

154. Mason, D. M. and Gandhi, K. N. (1983). Formulas for calculating the calorific value of coal and coal chars: development, tests, and uses. *Fuel Processing Technology*, 7(1), 11–22.

155. ASTM D1319 (2014). *Standard Test Method for Hydrocarbon Types in Liquid Petroleum Products by Fluorescent Indicator Adsorption*, ASTM International, West Conshohocken, PA.

156. ASTM D7169 (2011). *Standard Test Method for Boiling Point Distribution of Samples with Residues Such as Crude Oils and Atmospheric and Vacuum Residues by High Temperature Gas Chromatography*, ASTM International, West Conshohocken, PA.

157. ASTM D2887 (2015). *Standard Test Method for Boiling Range Distribution of Petroleum Fractions by Gas Chromatography*, ASTM International, West Conshohocken, PA.

158. Hsieh, P. Y., Widegren, J. A., Fortin, T. J., and Bruno, T. J. (2014). Chemical and thermophysical characterization of an algae-based hydrotreated renewable diesel fuel. *Energy & Fuels*, 28(5), 3192–3205.

159. Fu, J. and Turn, S. Q. (2014). Characteristics and stability of neat and blended hydroprocessed renewable diesel. *Energy & Fuels*, 28(6), 3899–3907.

160. Luning Prak, D. J., Cowart, J. S., Hamilton, L. J., Hoang, D. T., Brown, E. K., and Trulove, P. C. (2013). Development of a surrogate mixture for algal-based hydrotreated renewable diesel. *Energy & Fuels*, 27(2), 954–961.

161. Zhang, B., von Keitz, M., and Valentas, K. (2008). Maximizing the liquid fuel yield in a biorefining process. *Biotechnology and Bioengineering*, 101(5), 903–912.

162. Yang, X., Guo, F., Xue, S., and Wang, X. (2016). Carbon distribution of algae-based alternative aviation fuel obtained by different pathways. *Renewable and Sustainable Energy Reviews*, 54, 1129–1147.

163. NAABB (2014). National Alliance for Advanced Biofuels and Bioproducts Synopsis. http://www.energy.gov/sites/prod/files/2014/07/f18/naabb_synopsis_report_0.pdf.

Chapter 6

Effects of Catalyst Support on Hydroprocessing

Bo Zhang*, Baowei Wang[†], and Changyan Yang*[,‡,§]

*School of Chemical Engineering and Pharmacy,
Wuhan Institute of Technology, China
[†]Key Laboratory for Green Chemical Technology,
School of Chemical Engineering Technology,
Tianjin University, Tianjin, China
[‡]Hubei Key Laboratory for Processing and Application
of Catalytic Materials, Huanggang Normal University,
Hubei, China
[§]Corresponding author: ychy1969@163.com

Abstract

A significant amount of attention has been given to the development of the catalyst supports that might lead to enhanced catalytic hydroprocessing properties over last 30 y. This chapter reviews the support effects of oxides (Al_2O_3, SiO_2, TiO_2, and ZrO_2), carbonaceous materials, and molecular sieves including MCM-41, SBA-15, SAPO, ZSM-5, zeolite Y, and zeolite beta.

Keywords: Support effect, Hydroprocessing, Oxides, Carbon, Molecular sieve

6.1. Introduction

Due to stringent environmental regulations and air pollution problems caused by vehicle exhaust, all countries are required to continue to reduce the sulfur and aromatic content of transport fuels, mainly gasoline and diesel. To fulfill this mission, the oil refining industry must adopt additional hydrogenation (HYD) processes (such as increasing the number of the reactors, reactor volume, reaction pressure and temperature) as well as more hydrotreating catalysts. Over the past decade, biofeedstock (primarily vegetable oil and fats) have been corefined in the oil refineries. Higher fuel standards and low-quality feedstock require the development of new and effective hydrotreating catalysts.

Over 30 y ago, since type I Co(Ni)MoS and type II Co(Ni)MoS active phases were named, a significant amount of attention has been given to the development of the catalyst supports that might lead to enhanced catalytic properties [1]. In the HYD reaction, the strength, specific surface area, pore volume, pore distribution, surface properties of the support, and possible metal-support interactions have important effects on the catalytic activity. Therefore, understanding the role of the catalyst support is closely related to the development of an efficient catalyst for ultra clean fuels [2].

6.2. Oxides

6.2.1. Al_2O_3

Al_2O_3 (especially γ-Al_2O_3) is the most commonly used support for commercial hydrotreating catalysts in the field of petrochemical and environmental catalysis. The γ-Al_2O_3 could provide a surface area of 200–400 m^2/g and a pore volume of 0.5–1 cm^3/g. And it has advantages of good textural properties, thermal stability, mechanical properties, and low cost [3]. The sulfided CoMo/γ-Al_2O_3 and NiMo or NiW/γ-Al_2O_3 are mainly used for hydrodesulfurization (HDS) and hydrodenitrogenation (HDN) reactions, respectively. For most promoted catalysts, the activity behavior is related to the concentration of promoted edge sites (like CoMoS) [4]. For CoMo/γ-Al_2O_3, high-temperature sulfiding studies revealed the existence of a low-temperature (type I) CoMoS structure and a high-temperature (type II) CoMoS structure [5]. The type II CoMoS phase on alumina is present as a multilayer structure, whereas type I exists as a single-slab (monolayer structure) [6]. The differences in the activity of type I and type II CoMoS are explained in terms of the nature of the interactions with the support, which may lead to changes in the electronic properties of the active phase [7].

Al atoms and the metal species supported can have a strong interaction leading to insufficient sulfidation of the oxidic precursors. Moreover, γ-Al$_2$O$_3$ supports have excessive Lewis acid sites and tend to form coke over catalysts. To improve the catalytic activity of the γ-Al$_2$O$_3$ supported catalyst, different promoters including phosphorus (P), fluorine (F), and boron (B) can be added. The use of promoters could modify the acidity of the catalysts, change the dispersion of the active sites, increase the number and species of active sites, improve the ability to adsorb activated hydrogen, and inhibit coking [8–10].

The introduction of the P element can increase the solubility of metal salts during the preparation process, and produce a catalyst with a high metal loading in one step [11]. The introduced P atoms have a strong interaction with Al, and forming AlPO$_4$ on the surface and effectively increasing the dispersion of active metals [12]. Eijsbouts discovered that P improved the HDN activity of NiMo/γ-Al$_2$O$_3$ by inducing the formation of the type II NiMoS structure, especially at high Ni loading, but lowered the HDS activity by inducing a decrease in the dispersion of the NiMoS phases and a segregation of Ni$_3$S$_2$ [13]. Phosphate had very little effect on the Brønsted acidity of alumina and promoted the S- and N-elimination reactions.

The incorporation of F in oxide catalysts is widely used to enhance their activity for acid-catalyzed reactions such as cracking, isomerization, alkylation, polymerization, and disproportionation. The methods for incorporating fluorine include vapor-phase fluorination or impregnation. At low concentrations, fluorine replaces surface hydroxyl and oxide groups, whereas at higher fluorine content other phases, e.g., aluminum fluoride (AlF$_3$) or aluminum hydroxyfluoride (AlF$_2$OH), are formed [14]. The nature of the AlF phase depends on the temperature at which the catalysts have been treated. Fluorinated alumina has a significant number of strong protonic sites and exhibits the high catalytic activity for hydrocarbon reactions. Qu *et al.* studied the roles of F in sulfided NiW/F-γ-Al$_2$O$_3$ and revealed that F improved the reducibility of W^{6+} sulfides, changed the state of Ni species, and improved hydrogen chemisorption [15].

Lewandowskia and Sarbakb studied the catalytic activity of a series of NiMo/γ-Al$_2$O$_3$ modified with borate ions [16]. Modification of the support with borate ions increased the amount of acidity centers with intermediate strength, thus enhancing the acidity of NiMo catalyst. The incorporation of B did not affect the HDS activity, but it considerably increased the HDN activity and the resistance to coking. Yao *et al.* incorporated simultaneously F and B into NiMo/γ-Al$_2$O$_3$ [17]. It's found that the HDS activity decreased in the order NiMo/F, B–Al(5.0) > NiMo/F, B–Al(7.0) > NiMo/F, B–Al(3.5), indicating that there is an optimum amount of F and B for the activity enhancement.

6.2.2. *SiO$_2$*

SiO$_2$-supported CoMo catalysts exhibited a similar or even lower hydrotreating activity compared to the Al$_2$O$_3$-supported catalysts [18]. This can be attributed to the low dispersion of Co and Mo after calcination and consequently to the low concentration of the active phase, caused by the interaction with the supports.

The use of chelating agents including nitrilotriacetic acid, ethylenediaminetetraacetic acid (EDTA), and ethylenediamine during the preparation process without the calcination step can form active hydrotreating catalysts of sulfided NiCo/SiO$_2$ [19] and sulfided NiMo/SiO$_2$ [20]. Their HDS activities were similar to those of the Al$_2$O$_3$-supported catalyst.

Rana *et al.* incorporated ZrO$_2$ into SiO$_2$ to support CoMo (CoMo/ZrO$_2$–SiO$_2$) for hydrotreating of thiophene and cyclohexene [21]. Incorporation of ZrO$_2$ reduced the interaction of active phases with the SiO$_2$ support, overcame poor dispersion on the support surface, and enhanced number of active sites as well as the activity per site. Therefore, ZrO$_2$ counterpart plays a key role to provide better activity through enhanced number of active sites as well as activity per site. The promotional effect increases with decreasing SiO$_2$ content in the support composition. Liu *et al.* studied hydrotreatment of *Jatropha* oil over a

NiMo/SiO_2–Al_2O_3 (8 wt.% Al_2O_3) and showed that SiO_2–Al_2O_3 provided the proper acidity for isomerization and cracking [22].

6.2.3. TiO_2

TiO_2 has attracted attention due to the higher reducibility to a lower valence state of molybdenum and the higher intrinsic HDS activity demonstrated by Mo catalysts supported on this oxide [23]. This high activity has been attributed to the formation of well-dispersed Mo species that arise from the relatively high density of reactive hydroxyl groups on the TiO_2 surface. Additionally, it was found that titanium enhances the reduction and sulfidation of Mo^{6+} oxide species making easier the formation of catalytically active MoS_2.

However, TiO_2 supports have low specific surface areas compared to those of alumina, and the active anatase structure possesses only low thermal stability. These disadvantages make TiO_2 supports alone unsuitable for industrial applications [24]. Therefore, researchers focused on the modification of textural properties of TiO_2 supports. Yoshinaka and Segawa compared molybdenum catalysts supported on γ-Al_2O_3, TiO_2, and TiO_2–Al_2O_3 that was a composite of γ-Al_2O_3 coated by titanium via chemical vapor deposition [25]. X-ray photoelectron spectroscopy (XPS) analyses revealed that a higher TiO_2 loading of the composites support led to higher reducibility for molybdenum species. The HDS of 4,6-dimethyl-dibenzothiophene (4,6-DMDBT) over the Mo/TiO_2–Al_2O_3 catalyst was higher than that of Mo/γ-Al_2O_3. Wei *et al.* prepared the TiO_2–Al_2O_3 mixed-oxide support by impregnating Al_2O_3 with an ethanol solution of $TiCl_4$ or an isoproponol solution of $Ti(C_4H_9O)_4$ [24]. The NiMo/TiO_2–Al_2O_3 catalysts without presulfiding exhibited very high HDS activity.

6.2.4. ZrO_2

Zirconium oxide is of interest because of its special combination of surface properties, which preserve both acidic and basic sites, and reducing and oxidizing properties [26].

Maity *et al.* prepared a series of zirconia-supported molybdenum catalysts [27]. The X-ray powder diffraction (XRD) indicated that Mo was present as a monolayer up to 6 wt.% loading, which resulted in the highest O_2 uptake and catalytic activities. A MoO_3 crystalline growth was observed beyond the 6 wt.% Mo loading. Afanasiev *et al.* synthesized highly dispersed zirconia with tungstate species grafted on the surface (W/ZrO_2), which yielded a specific surface area up to 250 m^2/g due to stabilizing effect of tungstate species [28]. XPS study revealed incomplete sulfidation of tungsten, suggesting some persistent Zr–O–W bonds.

However, ZrO_2 has a low specific surface area and it is more expensive than the traditional oxide materials such as alumina and silica. ZrO_2 was often used to dope other materials such as SBA15 [29], TiO_2 [30, 31], SiO_2 [21], and Al_2O_3 [32, 33].

6.3. Carbon

Because the active metal component of Mo has a various degree of interaction with the supports like Al_2O_3, Mo cannot be completely sulfided and utilized. Research has been conducted to explore the inert support that minimizes the interaction between the active component and the support. Carbon materials have received more research attention, because they often possess a large specific surface area, high porosity, and weak interaction with the active metals. Carbon materials that could be applied as the catalyst supports include activated carbon, carbon black, nanomaterials, fullerenes, graphite, and biochar (Table 6.1). Among them, the activated carbon was predominantly used for supporting hydrotreating catalysts because of the low cost and good catalytic activity.

Carbon-supported sulfided CoMo showed much higher activity for HDS of thiophene than that of the Al_2O_3-supported CoMo [34]. The high HDS activity is usually explained as the electronic structure changes [35]. The essential difference is that Mo–O–Al linkages remain in Al_2O_3-supported CoMo, resulting in type I CoMoS phase. In the case of carbon-supported catalyst, the CoMo component might be more completely sulfided to type II CoMoS phase, which also has a better dispersion on carbon than Al_2O_3. Bouwens *et al.* found that the Mo–S coordination number, the structural ordering, and degree of stacking of the carbon-supported type II CoMoS phase was similar to the type I CoMoS phase

Table **6.1.**   Properties of carbon materials.

Carbon materials	Properties
Activated carbon	Amorphous, noncrystalline form of carbon possessing a large number of micropores and a high surface
Carbon black	Amorphous solid, characterized by degenerate or imperfect graphitic structures
Carbon nanomaterials (nanotubes and nanofibers)	Unusual strength as well as a high electrical and thermal conductivity
Fullerene	Carbon made up of 60 carbons (C_{60}) connected together by hexagons and pentagons as in a soccer ball
Graphite	A naturally occurring form of crystalline carbon, extremely soft, a very low specific gravity, extremely resistant to heat, and nearly inert in contact with almost any other material
Biochar	Made from biomass via pyrolysis, stable solid, rich in carbon

supported on alumina [6]. A Co site, which was exclusively present in the carbon-supported type II CoMoS phase, had a sixfold Co–S coordination with possibly a twofold Co–Mo coordination and showed the highest activity for HDS.

van Veen *et al.* compared activated carbon-supported sulfided CoMo to γ-Al_2O_3 (250 m^2/g) and SiO_2 (265 m^2/g) supports in HDS of thiophene [18]. The support interaction effect decreased in the order $SiO_2 \geq Al_2O_3 > C$. They concluded that even when the active phase (i.e., type II CoMoS) and its dispersion are constant, the support does influence the specific catalytic activity. Either the effect of carbon is positive or that of silica and alumina are negative.

However, the application of activated carbon as the support of hydro-treating catalysts is limited by the significant presence of micropores, which is unfavorable to oil feedstocks containing bulkier molecules. Lee *et al.* prepared nanoporous carbon with a large surface area and mesoporosity, which was used as a support for the hydrotreating catalyst [36]. The activity of CoMoS catalysts for the HDS of dibenzothiophene (DBT) and 4,6-dimethyldibenzothiophene (4,6-DMDBT) showed the order of CoMo/ nanoporous carbon > CoMo/activated carbon > CoMo/A1_2O_3. Dong *et al.* used carbon nanotubes (CNTs) as the support for the CoMoS catalyst [37]. It showed that the application of the CNT support led to a significant increase in the concentration of catalytically active Mo species (Mo^{4+}) at the surface of the functioning catalyst and adsorbed a greater amount of hydrogen generating microenvironments with higher stationary-state concentration of active hydrogen adspecies at the surface of the functioning catalyst.

Prabhu *et al.* studied the hydrotreating activity of NiMo catalysts supported on ordered mesoporous carbon supports (mC), which were functionalized using nitric acid [38]. Functionalization of the carbon support was able to create a large number of surface oxygen groups on the carbon surface, which can act as the anchors for interaction with metallic precursors [39]. Better dispersion of active metals over the functionalized supports was observed through XRD analysis.

Recently, biochar was applied as the support for molybdenum carbide [40, 41] and tungsten carbide [42]. Biochar from fast pyrolysis of lignocellulosic biomass has a lower surface area than activated carbon, but it is a low-cost carbon-rich sustainable material [43]. Furthermore, during the heat treatment process, metal oxides doped into the carbon matrix of biochar can be reduced by reducing gases such as H_2, CO, and CH_4 released from biochar.

However, carbon materials normally are hydrophobic [44] and have a low mechanical strength, which is not suitable for catalyst molding. Their industrial applications are still challenging.

6.4. Molecular sieve

Molecular sieves are aluminosilicate materials with pores of uniform size [45]. These pore diameters are similar in size to small molecules. Thus, smaller molecules can enter or be absorbed, while large molecules are excluded. Molecular sieves can be microporous, mesoporous, or macroporous materials.

6.4.1. *Mesoporous molecular sieve MCM-41*

Mobil Composition of Matter (MCM) is a series of mesoporous materials that were first synthesized by Mobil's researchers in the early 1990s [46]. So far, MCM-41 and MCM-48 are two of the most popular mesoporous molecular sieves that have been applied as supports for catalysts and drug delivery system and as adsorbent in the wastewater treatment. The MCM-41 material possesses a hexagonal array of uniform mesopores, a narrow pore size distribution ranging between 1.5 nm and 20 nm, a high surface area up to more than 1,000 m^2g^{-1}, sorption capacity, and thermal stability, although composed of amorphous silica wall (Figure 6.1). The pore diameter of MCM-41 can be nicely controlled within mesoporous by adjusting the synthesis conditions and/or by employing surfactants with different chain lengths in their preparation. Moreover, MCM-41 has a moderate and tailorable acidity when properly synthesized. These advantages make MCM-41 a distinct catalytic support.

In 1995, Corma *et al.* reported the performance of NiMo (12 wt.% MoO_3 and 3 wt.% NiO) supported on the MCM-41 for hydrotreating of a vacuum gasoil [47], which was compared with that of the same bimetallic catalyst on a USY zeolite. The MCM-41-supported catalyst gave superior HDS, HDN, and hydrocracking activities and showed better selectivity toward diesel-range distillates.

In another study, the MCM-41-supported CoMo catalyst was compared to a commercial Co–Mo/γ-Al_2O_3 catalyst for the desulfurization of a light cycle oil with a sulfur content of 2.19 wt.%. The MCM-41-supported catalyst demonstrated consistently higher activity for the HDS of the refractory dibenzothiophenic sulfur compounds, particularly 4,6-dimethyldibenzothiophene. Meanwhile, the presence of a large concentration of aromatics in the feedstock inhibited the HDS of the substituted dibenzothiophenes [48].

However, when applying MCM-41-supported catalysts containing sulfided mixed oxides of NiMo, NiW, and CoMo in a single-stage hydrotreating process, aromatics' saturation can only be partially accomplished due to thermodynamic limitations [49]. Therefore, the noble-metal-based catalysts are preferred for deep aromatics saturation, and noble metals can

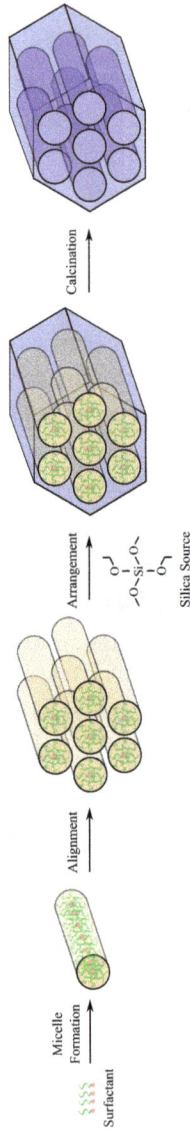

Figure 6.1. Synthesis pathway of MCM-41, illustrated by Hermann Luyken 2014. (available at https://commons.wikimedia.org/wiki/File:MCM-41_Synthesis_English_2014.04.19.svg). (Adapted under a Creative Commons CC0 1.0 Universal Public Domain Dedication.)

work at lower temperatures and avoid the thermodynamic constraints encountered with the sulfided oxides. Because the noble metals are readily poisoned by small amounts of sulfur and nitrogen organic compounds present in the feed, they can be used as the catalyst for the second stage of a hydrotreating process. It's found that the sulfur and nitrogen tolerance of noble metals (e.g., Pt and Pd) can be strongly improved if they are supported on a molecular sieve MCM-41.

A study on the HYD of naphthalene showed that the noble-metal catalysts of Pt/MCM-41 and Pt/MSA (made by ENI) had a superior activity than other conventional Pt-containing supports, such as commercial amorphous silica–alumina, silica, γ-alumina, and zeolite USY. However, the turnover number of the zeolite-based catalyst (i.e., activity per surface exposed Pt) was higher, though it had a lower metal dispersion. This might be due to the contribution of small Pt clusters localized in the supercages of the zeolite, which are subjected to a higher interaction with the strong Brønsted acid sites of the zeolite to form electron-deficient Pt particles with enhanced HYD activity and sulfur tolerance.

Nickel is a cost-effective catalyst for hydrotreating. It's found that Nickel phosphide formation on the pure Si MCM-41 may start with the reduction of NiO to Ni, and the Ni metal assists phosphide formation in the reduction of nickel-rich precursors [50]. Nickel-rich phosphides exhibited much higher HDS activity than catalysts prepared by the *in situ* reduction method.

Because Mo precursor ion and pore wall Si atoms have unfavorable interaction, specifically for pure Si MCM-41, MCM-41 materials were often modified by incorporating aluminum atoms through different methods: wet impregnation with ammonium heptamolybdate (AHM), direct sol-gel method (Pre), and post-synthetic grafting method (Post). The incorporation of Al atoms into the siliceous MCM-41 framework causes a deterioration of the textural characteristics, some loss in the periodicity of the MCM-41 pore structure, and the higher acidity. Klimova *et al.* found that the dispersion of Mo and Ni oxidic species increased with the incorporation of Al in the MCM-41 support due to the strong interaction of Mo and Ni oxidic species with aluminum atoms of the support [51]. However, the strong interaction also caused the formation of $Al_2(MoO_4)_3$, producing an increase in the proportion of Ni and Mo species difficult to reduce.

Méndez *et al.* reported that as Al content increasing, a partial collapse of the MCM-41 structure was observed. If it's used to support NiMo, increasing Al content resulted in the formation of MoO_3, NiO, and $Al_2(MoO_4)_3$. The ^{29}Si-NMR-MAS spectra showed that local arrangement of the Si-O-Si bonds was regular for both pure Si or Al-MCM-41, while the ^{27}Al-MAS-NMR spectra of the aluminosilicates showed structural and nonreticular species. NiMo/MCM-41 showed the excellent hydrodechlorination (HDC) properties, which can be attributed to good dispersion of Ni and Mo active phases and to its bifunctional character, i.e., the participation of both coordinatively unsaturated sites of the NiMoS active phase and Brønsted acid sites of the support [52].

Klimova *et al.* studied HDO of vegetable oil over CoMo/MCM-41 catalysts with various Si/Al ratios [53]. Incorporation of Al into the framework of MCM-41 led to an increase of the conversion of triglycerides and the selectivity to hydrocarbons including *n*-heptadecane and *n*-octadecane. However, the conversion of triglycerides was lower than that achieved over CoMo/γ-alumina.

Park *et al.* synthesized platinum catalysts supported on Al-MCM-41-Post and Al-MCM-41-Pre, and found Pt on Al-MCM-41-Post were more accessible due to the aluminum distribution within the pore wall. The HYD activity, selectivity, and the sulfur tolerance were strongly enhanced by the acidic nature of Al-MCM-41 support [54].

In addition to Al, Ti can also be used to modify MCM-41 supports. Klimova *et al.* incorporated Ti atoms into MCM-41 during the hydrothermal synthetic route or by post-synthetic incorporation via chemical grafting and incipient wetness impregnation [55]. The maximum amount of Ti that can be incorporated during the hydrothermal synthetic process was less than 10 wt.%. Otherwise, it destructed the mesoporous arrangement. While incipient wetness impregnation allowed to prepare samples with higher Ti loading (13.3–38.8 wt.% of TiO_2). Significant decrease in surface area and pore volume was observed when Mo was loaded.

Herrera *et al.* synthesized a series of MCM-41 supports modified with phosphorus (0–5 wt.% P_2O_5) and their respective NiMo/P-MCM-41 catalysts [56]. It's shown that phosphorus insertion caused structure changes of MCM-41. Meanwhile, the acidity of MCM-41 materials increased with the P loading. The dispersion, coordination state, and reduction temperature of oxide and sulfided Mo and Ni species changed when P was incorporated in the MCM-41 support. The small-angle XRD peak of all NiMo catalysts disappeared completely, and this may be due to the obstruction or destruction of the pores after metal deposition.

6.4.2. SBA-15

The Santa Barbara Amorphous-type material (SBA-15) was first produced at the University of California (Santa Barbara) in forms of silica nanoparticles with a hexagonal array of pores [57]. The diameter of pores is in the range of 4.6–30 nm. The original use of this material was molecular sieves, and it has many applications in medicine, biosensors, energy storage, and imaging now [58]. Figure 6.2 shows a scheme for the self-reconstruction process of mesoporous silica SBA-15 materials.

Due to their favorable hydrothermal stability, textural characteristics, and simple surface properties, mesoporous SBA-15 materials are a good candidate for the supports of hydrotreating catalysts such as Co- or Ni-promoted Mo. For example, Ni-promoted sulfided Mo (MoS_2) formed the NiMoS phase, which increased its dispersion on SBA-15 and dramatically reduced the inhibiting effect caused by nitrogenated compounds in the feedstock. It's found that the HDS activity was proportional to the concentration of Mo and Ni on the edges of sulfide particles, and HDN occurred only on accessible Mo cations [60].

The modification of SBA-15 by heteroatom incorporation has been found to yield MoS_2-based catalysts with similar or higher HDS and HDN activity. For example, Al atoms are often introduced into the framework of SBA-15 via the direct synthetic method [61]. This method preserved the pore structure and hydrothermal stability of SBA-15 [62], and majority of Al is in tetrahedral positions [63]. Al-SBA-15 had more Brønsted acid sites and the higher total acid density of the support and the interaction between NiMo and the support increased with the Al content [64]. When a post-synthetic method of chemical grafting (aluminum(III) chloride as alumina source) was used, it led to amorphous silicoaluminate materials with a moderate Brønsted acidity. The dispersion of oxidic and sulfided Mo species increased with the incorporation of Al due to the strong interaction of Mo and Ni species with aluminum atoms, which

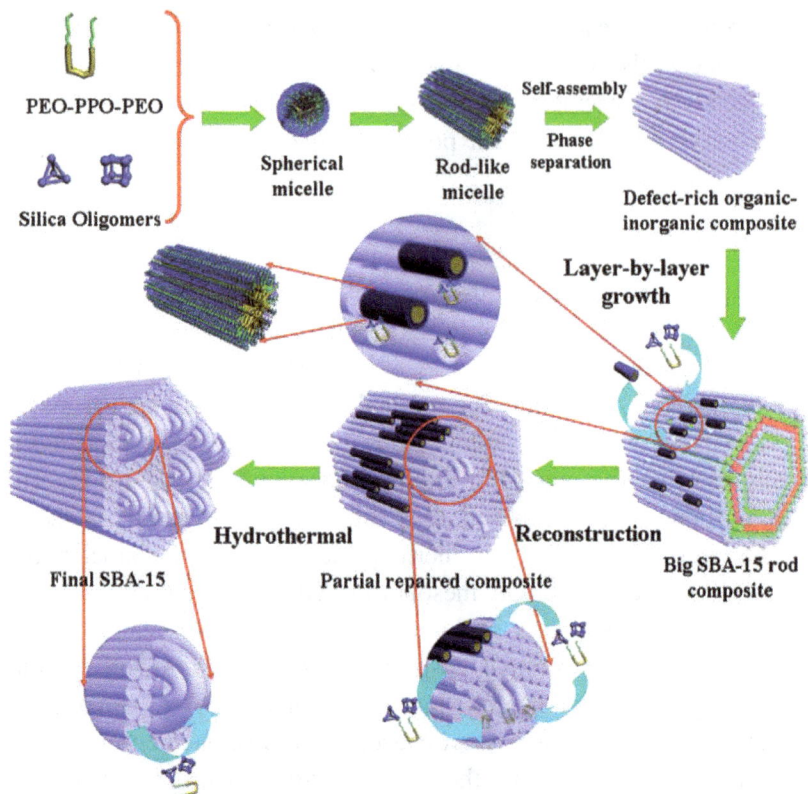

Figure 6.2.   The scheme for the self-reconstruction process of mesoporous silica SBA-15 materials via the self-repairing of small "Pluronic P123/silica" flocs. (Reprinted with permission from Ref. [59]. Copyright © 2011 Royal Society of Chemistry.)

serve as anchoring sites on the support surface. The NiMo/Al-SBA-15 catalysts showed high activity in HDS of DBT and 4,6-DMDBT, which could attribute to the good dispersion of Ni and Mo active phases and the bifunctional character of these catalysts (i.e., possessing both unsaturated sites of NiMoS active phase and Brønsted acid sites) [65].

Sun *et al.* synthesized the Fe-incorporated mesoporous SBA-15 materials with the Fe content up to a Fe/Si molar ratio = 0.05 via prehydrolysis of tetraethyl orthosilicate (TEOS) employing P123 triblock copolymer as the template. The synthesized Fe-SBA-15 showed good hydrothermal stability and enhanced Brønsted and Lewis acidity [66]. Ti [60] and Zr [67] ions were also used to modify SBA-15, and synthesized materials were considered as the suitable support for Mo, CoMo, and NiMo [68].

6.4.3. *Silicoaluminophosphate (SAPO)*

The development of phosphate-based zeolites including aluminophosphates and silicoaluminophosphates (SAPOs) enormously enriched framework versatility and chemical diversity of crystalline microporous materials [69]. Aluminophosphates are composed of tetrahedral Al and P atoms in an exactly alternative manner and their structure are unstable. SAPOs are produced by incorporation of Si atoms into the framework of aluminophosphates, exhibiting high structural stability as well as ion exchange capability that results in Brønsted acidity and redox activity [70].

SAPOs have been widely used as catalysts and adsorbents. Platinum or palladium supported on SAPO-11 or SAPO-41 was identified an excellent hydrocracking, hydrodewaxing, and isomerization catalyst for heavy hydrocarbon oils [71]. Figure 6.3 illustrates the framework structure of SAPO-34, SAPO-46, SAPO-11, and SAPO-41.

Recently, the research of hydrotreating processes is mainly focusing on biofeedstock. Verma *et al.* investigated hydroprocessing of nonedible *Jatropha* oil over sulfided NiMo and NiW catalysts supported on hierarchical mesoporous SAPO-11 [73]. A high yield (about 84 wt.%) of liquid hydrocarbon products was obtained, and the Si content in the support of SAPO-11 did not show significant impact on the hydroprocessing of *Jatropha* oil. Chen *et al.* studied the effects of Si/Al ratio and Pt loading on Pt/SAPO-11 catalysts [74]. The medium acidic sites in SAPO-11 increased with increasing Si/Al ratio, resulting in higher isomerization activity. The deoxygenation, isomerization, and cracking activities strongly depended on the Pt loading.

Kikhtyanin *et al.* studied the hydroprocessing of sunflower oil on the Pd/SAPO-31 catalyst for single-stage production of hydrocarbons in the diesel fuel range [75]. The Pd/SAPO-31 catalyst demonstrated a high initial activity for the hydroconversion of the feed and good isomerization properties, but its deactivation occurred after several hours of operation.

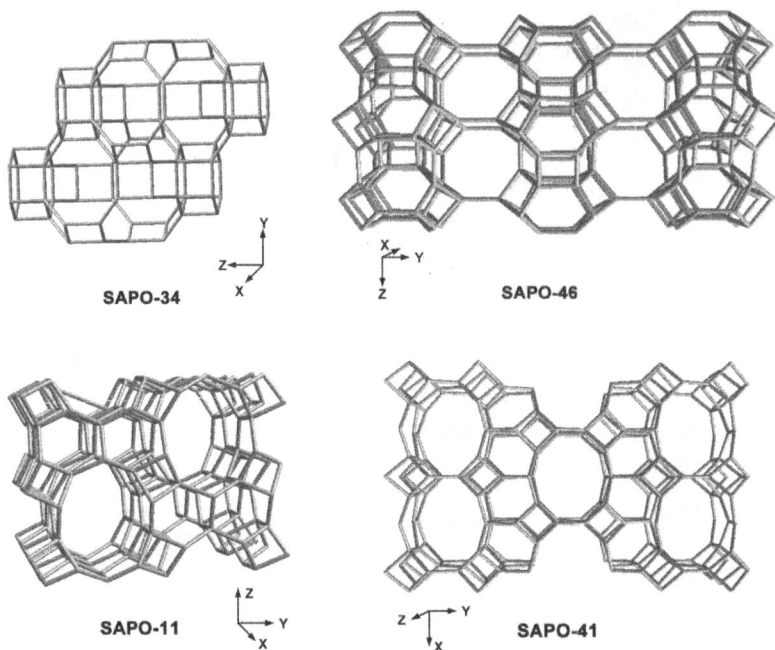

Figure 6.3. Illustration of framework structure of SAPO molecular sieves. (Reprinted with permission from Ref. [72]. Copyright © 2011 American Chemical Society.)

Rabaev *et al.* showed that Pt/SAPO-11–Al$_2$O$_3$ catalysts displayed relatively low hydrothermal stability in hydrotreating of vegetable oils, because water was produced during the process [76]. The hydrothermal deactivation of SAPO-11 was due to partially reversible desilication of its framework, which resulted in the gradual loss of acidity and catalytic activity in isomerization.

6.4.4. ZSM-5

Zeolite Socony Mobil-5 (ZSM-5), is an aluminosilicate zeolite with the chemical formula of Na$_n$Al$_n$Si$_{96-n}$O$_{192}$·16H$_2$O (0 < n < 27). ZSM-5 is composed of several pentasil units, which consist of eight five-membered rings, linked together by oxygen bridges to form pentasil chains. The pentasil chains are interconnected by oxygen bridges to form corrugated

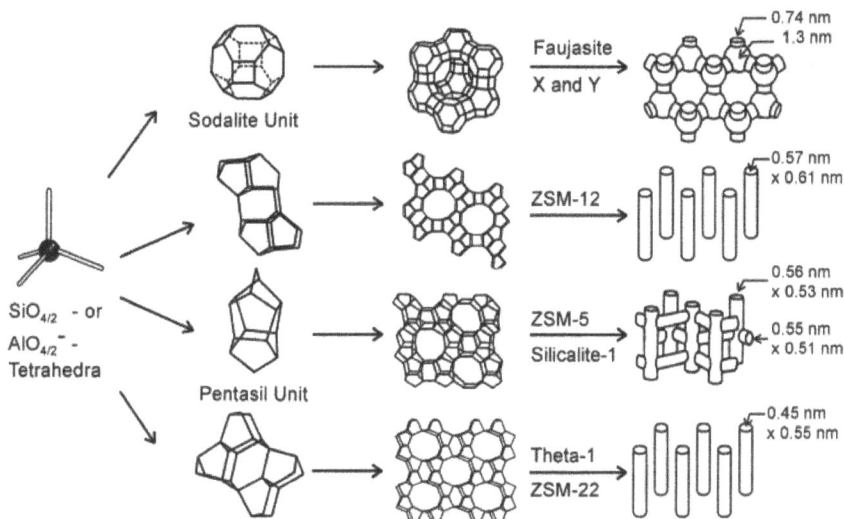

Figure 6.4. Structures of four selected zeolites (from top to bottom: faujasite or zeolite X, Y; zeolite ZSM-12; zeolite ZSM-5 or silicalite-1; zeolite Theta-1 or ZSM-22) and their micropore systems and dimensions. (Reprinted with permission from Ref. [80]. Copyright © 2000 Elsevier.)

sheets, each of which has a structure with straight 10-ring channels running parallel to the corrugations and sinusoidal 10-ring channels perpendicular to the sheets [77]. Adjacent layers of the sheets are related by an inversion point. The estimated pore size of the channel running parallel with the corrugations is 5.4–5.6 Å (i.e., ~0.5 nm) [78]. Figure 6.4 shows the structure of four selected zeolites. The structure of ZSM-5 is orthorhombic at high temperatures, but a phase transition to the monoclinic space group $P2_1/n.1.1$ occurs on cooling below a transition temperature, located between 300 K and 350 K [79]. ZSM-5 has a high silicon to aluminum ratio. Whenever an Al^{3+} cation replaces a Si^{4+} cation, an additional positive charge is required to keep the material charge neutral. With proton (H^+) as the cation, the material becomes very acidic. Thus, the acidity is proportional to the Al content. The very regular three-dimensional structure and the acidity of ZSM-5 can be utilized for acid-catalyzed reactions such as hydrocarbon isomerization and the alkylation of hydrocarbons.

ZSM-5 has been widely used in the petroleum industry as a heterogeneous catalyst and a support material for hydrocarbon isomerization reactions. For example, Pt/HZSM-5 exhibited high and stable catalytic activity for the HDS of thiophen. The Si/Al ratio, Brønsted acid sites on HZSM-5, and spillover hydrogen formed on Pt particles contributed to the HDS activity [81].

However, it's found that HZSM-5-supported CoMo or NiMo catalysts did not improve the HDS of 4,6-DMDBT, because bulky molecules could not diffuse readily into the nanosized channels of the zeolite [82]. During last several years, the mesoporous ZSM-5 zeolite, which possesses high mesoporosity and large mesoporous surface area, was successfully synthesized [83]. The acidic hydroxyl and silanol groups on the surface of mesoporous zeolites weaken the metal-support interaction, resulting in more multistacked MoS_2 active phases and the higher HYD activity [84]. Yu *et al.* mechanically mixed the mesoporous zeolite ZSM-5 (MZSM-5) and γ-Al_2O_3 to form a composite material, which was used to support CoMoS [85]. They concluded that introducing MZSM-5 into γ-Al_2O_3 improved the mesoporous structure and surface properties of the composites as well as mass transfer of the reactant molecules.

Recently, ZSM-5 has been used to modify other materials and form composites that have improved properties as the catalyst support. Wu *et al.* enwrapped nanosized ZSM-5 zeolite crystals with mesoporous KIT-6 silica to synthesize the composite material ZSM-5/KIT-6 (ZK-W), which was used as the catalyst support for NiMo sulfide [86]. The NiMo/ZK-W catalyst showed the high HDS activity to 4,6-DMDBT and a superior isomerization ability.

6.4.5. *Zeolite Y*

The Y zeolite is a type of faujasite, which belongs to the zeolite family of silicate minerals. Faujasite occurs as a rare mineral in several locations worldwide and is also synthesized industrially. Faujasite (sodium form, like NaY) can be synthesized by dissolving alumina sources (like sodium aluminate) and silica sources (like sodium silicate) in a basic environment such as sodium hydroxide solution followed by crystallization at 70–300°C (usually at 100°C). Since the sodium form is not stable, it shall be ion exchanged with ammonium to improve stability. Depending on its silica–alumina ratio, synthetic faujasite zeolites are divided into X and Y zeolites. The X zeolites have a Si/Al ratio of between 1 and 1.5, and the Si/Al ratio of the Y zeolites is higher than 1.5 [87]. Zeolite Y in ammonium form (NH_4Y) can be calcined to convert to proton form (HY) [88].

Zeolite Y shares the faujasite structure. Its framework consists of sodalite cages that are connected through hexagonal prisms. Zeolite Y has a three-dimensional pore structure with pores running perpendicular to each other [89]. The pore is formed by a 12-member oxygen ring with a relatively large diameter of 7.4 Å and leads into a larger cavity of diameter 12 Å. The cavity is surrounded by ten sodalite cages (truncated octahedra) connected on their hexagonal faces. The unit cell is cubic ($a = 24.7$Å) with Fd-3m symmetry [90].

Landau *et al.* reported one of the first attempts to apply zeolites including Y and ZSM-5 as hydrotreating catalysts [82]. The CoMo catalyst supported on HY zeolite was found about three times more active than the commercial CoMo catalyst for the HDS of 4,6-DMDBT. The positive effect of the presence of zeolites on the reactivity of DBT was attributed to a direct action of the acidity on the HDS activity and its effect on the reactivity of 4,6-DMDBT was attributed mainly to demethylation and cracking of the C–C bond of the thiophenic ring connecting the two benzenic rings together. Isoda *et al.* made a series of studies on the subject [91]. The use of CoMo deposited on alumina-containing 5 wt.% of Y zeolite led to a better activity in the HDS of a gas oil feed and in particular to a better conversion of the 4-methyldibenzothiophene (4MDBT) and 4,6-DMDBT contained in the oil than CoMo and NiMo deposited on alumina. Since then, zeolite Y has been extensively studied as a support for hydrotreating catalysts to support metals such as Pt [92], Co [93], Ni–Mo [94], and W–Ni [95].

Different treatments could change the structure of zeolite Y and give different effects on the hydrotreating. Ding *et al.* compared the chemically treated NaY with hydrothermally treated samples [96]. Considerable amounts of Lewis acidic sites, extra-framework aluminum, and mesopores were produced on hydrothermally treated Y zeolites. It's found that the type of zeolites determined the interaction between the metals and the support. Recently, mesoporous zeolites have been applied for the catalytic conversion of large biomolecules into valuable chemicals. Mesoporous zeolite Y showed high alkane selectivity and low aromatic hydrocarbon selectivity [97].

Li *et al.* mixed zeolite Y with the conventional catalyst support of γ-Al$_2$O$_3$ to improve the acidity [98]. Ni was found to be uniformly distributed throughout this mixed catalyst support, while Mo preferentially associated with γ-Al$_2$O$_3$ not the zeolite. Ni or Mo species had higher surface concentrations and higher dispersion and thus could be easily sulfided.

6.4.6. *Zeolite beta*

Beta (BEA) zeolite is an old zeolite discovered before Mobil's ZSM sequence [99]. The structure of zeolite BEA that is very complex was determined in 1988 [100, 101]. Recently, this material became important because of its application for dewaxing operations. Zeolite BEA consists of an intergrowth of two distinct structures termed polymorphs A and B. The polymorphs grow as two-dimensional sheets and the sheets randomly alternate between the two. Both polymorphs have a three-dimensional network of 12-ring pores. The intergrowth of the polymorphs does not significantly affect the pores in two of the dimensions, but in the direction of the faulting, the pore becomes tortuous but not blocked. Figure 6.5 shows the framework structure of BEA zeolites.

Ding *et al.* used zeolite BEA and a mixture of zeolite BEA and Y to support W–Ni [95]. The W–Ni/BEA + Y and W–Ni/BEA catalysts had higher HDS activity than the W–Ni/Y and W–Ni/Al$_2$O$_3$ catalysts. The higher HYD activities of the three zeolite-containing catalysts were associated primarily with enhanced HYD activity and increased acidity. The same research group further compared the hydrotreating activity of Mo–Ni and W–Ni supported on nanosized and microsized zeolite BEA catalysts [103]. They concluded that the nanosized BEA catalyst had higher HDS, HDN, and hydroaromatization (HDA) activities than the microsized catalyst.

Zeolite beta was also used to make composite materials that can be used as the catalyst support. Zhang *et al.* synthesized a beta/KIT-6 composite from zeolite BEA seed solution by a two-step hydrothermal crystallization method [104]. The as-synthesized material-supported NiMo catalysts were prepared and evaluated in the HDS of DBT. This NiMo

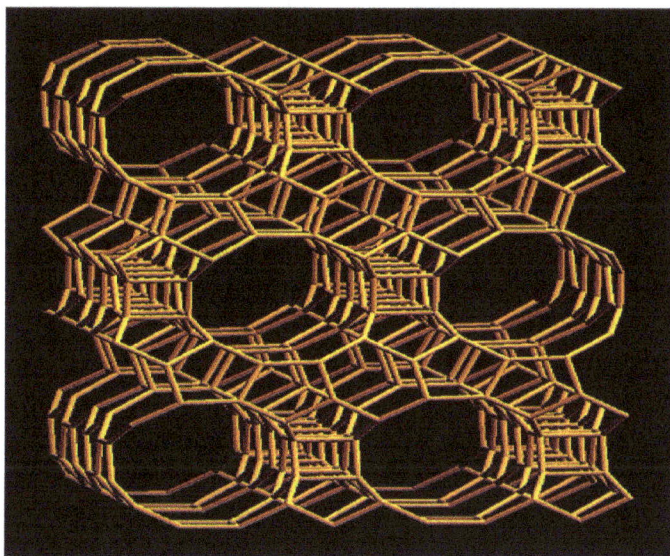

Figure 6.5. Framework structure of zeolite BEA. (Reprinted with permission from Ref. [102]. Copyright © 2014 Royal Society of Chemistry.)

catalyst possessed more acidic sites and the HDS reactions of DBT followed a different network from those over NiMo/γ-Al$_2$O$_3$. Two final products, isodimethyldecalin and alkylcyclohexane, were mainly derived from the isomerization.

Other types of zeolites such as synthetic mordenite and ferrierites are often used as the catalyst for the acid-catalyzed isomerization of alkanes and aromatics in the petrochemical industry.

6.5. Closing remarks

The development of suitable supports for hydrotreating catalysts has been carried out for over 30 y. The support effects include electronic properties, the active phase morphology, the interaction between the active component and the support, the acidity, and the active sites.

As the most used support for commercial hydrotreating catalysts, γ-Al$_2$O$_3$ has its unique properties and economic advantages. The enhanced support effects and catalytic activities were achieved via incorporation of promoters such as P, F, and B. Thus, aluminum-based materials have an irreplaceable position in the hydrotreating catalyst industry. The SiO_2-supported hydrotreating catalysts showed a lower HYD activity, which might be improved by using chelating agents during preparation or incorporation of other oxides. Meanwhile, TiO_2 and ZrO_2 have low specific surface areas, and the interaction exists between Al, Si, Ti, Zr atoms and the supported metal atoms (like Mo and W). The use of the mixed oxides (such as TiO_2–Al_2O_3) increased the specific surface area and lowered the interaction between the support and the active metals.

Carbonaceous materials are of interest because they are inert and show minimized interaction with the supported metals. The most studied material is the activated carbon, and new materials such as mesoporous and nanocarbon materials provided exceptional properties for hydrotreating catalysts. However, activated carbon is not suitable for hydroprocessing large molecules, because it possesses a large number of micropores.

Molecular sieves can provide a large specific surface area and an appropriate pore size. With suitable modifications, molecular sieves may serve as the effective and economical supports for the hydrotreating catalyst.

Abbreviations

AHM	ammonium heptamolybdate
CNT	carbon nanotube
DBT	dibenzothiophene
4MDBT	4-methyldibenzothiophene
4,6-DMDBT	4,6-dimethyldibenzothiophene
HDA	hydroaromatization
HDC	hydrodechlorination
SBA	Santa Barbara Amorphous-type material
TEOS	tetraethyl orthosilicate
XPS	X-ray photoelectron spectroscopy
XRD	X-ray powder diffraction
ZSM	zeolite Socony Mobil

References

1. Breysse, M., Afanasiev, P., Geantet, C., and Vrinat, M. (2003). Overview of support effects in hydrotreating catalysts. *Catalysis Today*, 86(1–4), 5–16.
2. Okamoto, Y., Breysse, M., Murali Dhar, G., and Song, C. (2003). Effect of support in hydrotreating catalysis for ultra clean fuels. *Catalysis Today*, 86(1–4), 1–3.
3. Luck, F. (1991). A review of support effects on the activity and selectivity of hydrotreating catalysts. *Bulletin des Sociétés Chimiques Belges*, 100(11–12), 781–800.
4. Chen, J., Oliviero, L., Portier, X., and Mauge, F. (2015). On the morphology of MoS$_2$ slabs on MoS$_2$/Al$_2$O$_3$ catalysts: the influence of Mo loading. *RSC Advances*, 5(99), 81038–81044.
5. Wivel, C., Clausen, B. S., Candia, R., Mørup, S., and Topsøe, H. (1984). Mössbauer emission studies of calcined Co-Mo/Al$_2$O$_3$ catalysts: catalytic significance of Co precursors. *Journal of Catalysis*, 87(2), 497–513.
6. Bouwens, S. M. A. M., van Zon, F. B. M., van Dijk, M. P., van der Kraan, A. M., de Beer, V. H. J., van Veen, J. A. R., and Koningsberger, D. C. (1994). On the structural differences between alumina-supported comos type I and alumina-, silica-, and carbon-supported comos type II phases studied by XAFS, MES, and XPS. *Journal of Catalysis*, 146(2), 375–393.
7. Topsøe, H. and Clausen, B. S. (1986). Active sites and support effects in hydrodesulfurization catalysts. *Applied Catalysis*, 25(1), 273–293.
8. Lewis, J. M., Kydd, R. A., Boorman, P. M., and van Rhyn, P. H. (1992). Phosphorus promotion in nickel-molybdenum/alumina catalysts: model compound reactions and gas oil hydroprocessing. *Applied Catalysis A: General*, 84(2), 103–121.
9. Sun, M., Nicosia, D., and Prins, R. (2003). The effects of fluorine, phosphate and chelating agents on hydrotreating catalysts and catalysis. *Catalysis Today*, 86(1–4), 173–189.
10. Ferdous, D., Dalai, A. K., and Adjaye, J. (2004). A series of NiMo/Al$_2$O$_3$ catalysts containing boron and phosphorus: Part II. Hydrodenitrogenation and hydrodesulfurization using heavy gas oil derived from Athabasca bitumen. *Applied Catalysis A: General*, 260(2), 153–162.
11. Fitz, C. W. and Rase, H. F. (1983). Effects of phosphorus on nickel-molybdenum hydrodesulfurization/hydrodenitrogenation catalysts of varying metals content. *Industrial & Engineering Chemistry Product Research and Development*, 22(1), 40–44.

12. Gishti, K., Iannibello, A., Marengo, S., MorelliLi, G., and Tittarelli, P. (1984). On the role of phosphate anion in the MoO_3-Al_2O_3 based catalysts. *Applied Catalysis*, 12(4), 381–393.

13. Eijsbouts, S., van Gestel, J. N. M., van Veen, J. A. R., de Beer, V. H. J., and Prins, R. (1991). The effect of phosphate on the hydrodenitrogenation activity and selectivity of alumina-supported sulfided Mo, Ni, and Ni–Mo catalysts. *Journal of Catalysis*, 131(2), 412–432.

14. Ghosh, A. K. and Kydd, R. A. (1985). Fluorine-promoted catalysts. *Catalysis Reviews*, 27(4), 539–589.

15. Qu, L., Jian, M., Shi, Y., and Li, D. (1998). Roles of fluorine in sulfided NiW/Al_2O_3 catalysts. *Chinese Journal of Catalysis*, 19(608–609).

16. Lewandowski, M. and Sarbak, Z. (2000). The effect of boron addition on hydrodesulfurization and hydrodenitrogenation activity of $NiMo/Al_2O_3$ catalysts. *Fuel*, 79(5), 487–495.

17. Yao, S., Zheng, Y., Ding, L., Ng, S., and Yang, H. (2012). Co-promotion of fluorine and boron on $NiMo/Al_2O_3$ for hydrotreating light cycle oil. *Catalysis Science & Technology*, 2(9), 1925–1932.

18. van Veen, J. A. R., Gerkema, E., van der Kraan, A. M., and Knoester, A. (1987). A real support effect on the activity of fully sulphided CoMoS for the hydrodesulphurization of thiophene. *Journal of the Chemical Society, Chemical Communications*, 22, 1684–1686.

19. Thompson, M. S. (1986). Preparation of high activity silica-supported hydrotreating catalysts and catalysts thus prepared. *EP19850201771*.

20. Cattaneo, R., Weber, T., Shido, T., and Prins, R. (2000). A quick EXAFS study of the sulfidation of $NiMo/SiO_2$ hydrotreating catalysts prepared with chelating ligands. *Journal of Catalysis*, 191(1), 225–236.

21. Rana, M. S., Maity, S. K., Ancheyta, J., Murali Dhar, G., and Prasada Rao, T. S. R. (2004). $MoCo(Ni)/ZrO_2$–SiO_2 hydrotreating catalysts: physicochemical characterization and activities studies. *Applied Catalysis A: General*, 268(1–2), 89–97.

22. Liu, Y., Sotelo-Boyás, R., Murata, K., Minowa, T., and Sakanishi, K. (2009). Hydrotreatment of Jatropha oil to produce green diesel over trifunctional Ni–Mo/SiO_2–Al_2O_3 catalyst. *Chemistry Letters*, 38(6), 552–553.

23. Ng, K. Y. S. and Gulari, E. (1985). Molybdena on titania. *Journal of Catalysis*, 95(1), 33–40.

24. Wei, Z. B., Yan, W., Zhang, H., Ren, T., Xin, Q., and Li, Z. (1998). Hydrodesulfurization activity of $NiMo/TiO_2Al_2O_3$ catalysts. *Applied Catalysis A: General*, 167(1), 39–48.

25. Yoshinaka, S. and Segawa, K. (1998). Hydrodesulfurization of dibenzothiophenes over molybdenum catalyst supported on TiO_2–Al_2O_3. *Catalysis Today*, 45(1–4), 293–298.

26. Yamaguchi, T. (1990). Recent progress in solid superacid. *Applied Catalysis*, 61(1), 1–25.

27. Maity, S. K., Rana, M. S., Srinivas, B. N., Bej, S. K., Murali Dhar, G., and Prasada Rao, T. S. R. (2000). Characterization and evaluation of ZrO_2 supported hydrotreating catalysts. *Journal of Molecular Catalysis A: Chemical*, 153(1–2), 121–127.

28. Afanasiev, P., Cattenot, M., Geantet, C., Matsubayashi, N., Sato, K., and Shimada, S. (2002). (Ni)W/ZrO_2 hydrotreating catalysts prepared in molten salts. *Applied Catalysis A: General*, 237(1–2), 227–237.

29. Biswas, P., Narayanasarma, P., Kotikalapudi, C. M., Dalai, A. K., and Adjaye, J. (2011). Characterization and activity of ZrO_2 doped SBA-15 supported NiMo catalysts for HDS and HDN of bitumen derived heavy gas oil. *Industrial & Engineering Chemistry Research*, 50(13), 7882–7895.

30. Maity, S. K., Rana, M. S., Bej, S. K., Ancheyta-Juárez, J., Murali Dhar, G., and Prasada Rao, T. S. R. (2001). TiO_2–ZrO_2 mixed oxide as a support for hydrotreating catalyst. *Catalysis Letters*, 72(1), 115–119.

31. Afanasiev, P. (2008). Mixed TiO_2–ZrO_2 support for hydrotreating, obtained by co-precipitation from Zr basic carbonate and Ti oxosulfate. *Catalysis Communications*, 9(5), 734–739.

32. Damyanova, S., Grange, P., and Delmon, B. (1997). Surface characterization of zirconia-coated alumina and silica carriers. *Journal of Catalysis*, 168(2), 421–430.

33. Wang, J., Yuan, Y., Shuaib, A., Xu, J., and Shen, J. (2015). Effect of ZrO_2 in Ni_2P/ZrO_2-Al_2O_3 catalysts on hydrotreating reactions. *RSC Advances*, 5(91), 74312–74319.

34. Reddy, B. M. and Subrahmanyam, V. S. (1986). Oxygen chemisorption and activity studies on alumina- and carbon-supported hydrodesulphurization catalysts. *Applied Catalysis*, 27(1), 1–8.

35. Harris, S. and Chianelli, R. R. (1986). Catalysis by transition metal sulfides: a theoretical and experimental study of the relation between the synergic systems and the binary transition metal sulfides. *Journal of Catalysis*, 98(1), 17–31.

36. Lee, J. J., Han, S., Kim, H., Koh, J. H., Hyeon, T., and Moon, S. H. (2003). Performance of CoMoS catalysts supported on nanoporous carbon in the hydrodesulfurization of dibenzothiophene and 4,6-dimethyldibenzothiophene. *Catalysis Today*, 86(1–4), 141–149.

37. Dong, K., Ma, X., Zhang, H., and Lin, G. (2006). Novel MWCNT-support for Co-Mo sulfide catalyst in HDS of thiophene and HDN of pyrrole. *Journal of Natural Gas Chemistry*, 15(1), 28–37.

38. Prabhu, N., Dalai, A. K., and Adjaye, J. (2011). Hydrodesulphurization and hydrodenitrogenation of light gas oil using NiMo catalyst supported on functionalized mesoporous carbon. *Applied Catalysis A: General*, 401(1–2), 1–11.

39. Calvo, L., Gilarranz, M. A., Casas, J. A., Mohedano, A. F., and Rodríguez, J. J. (2005). Effects of support surface composition on the activity and selectivity of Pd/C catalysts in aqueous-phase hydrodechlorination reactions. *Industrial & Engineering Chemistry Research*, 44(17), 6661–6667.

40. Li, R., Shahbazi, A., Wang, L., Zhang, B., Hung, A. M., and Dayton, D. C. (2016). Graphite encapsulated molybdenum carbide core/shell nanocomposite for highly selective conversion of guaiacol to phenolic compounds in methanol. *Applied Catalysis A: General*, 528, 123–130.

41. Li, R. and Shahbazi, A. (2015). A review of hydrothermal carbonization of carbohydrates for carbon spheres preparation. *Trends in Renewable Energy*, 1(1), 43–56.

42. Yan, Q., Lu, Y., To, F., Li, Y., and Yu, F. (2015). Synthesis of tungsten carbide nanoparticles in biochar matrix as a catalyst for dry reforming of methane to syngas. *Catalysis Science & Technology*, 5(6), 3270–3280.

43. Xiu, S., Shahbazi, A., and Li, R. (2017). Characterization, modification and application of biochar for energy storage and catalysis: a review. *Trends in Renewable Energy*, 3(1), 86–101.

44. Serp, P. and Machado, B. (2015). *Nanostructured Carbon Materials for Catalysis*, Royal Society of Chemistry, Cambridge.

45. Wikipedia (2017). Molecular sieve. https://en.wikipedia.org/wiki/Molecular_sieve.

46. Kresge, C. T., Leonowicz, M. E., Roth, W. J., Vartuli, J. C., and Beck, J. S. (1992). Ordered mesoporous molecular sieves synthesized by a liquid-crystal template mechanism. *Nature*, 359(6397), 710–712.

47. Corma, A., Martinez, A., Martinezsoria, V., and Monton, J. B. (1995). Hydrocracking of vacuum gasoil on the novel mesoporous MCM-41 aluminosilicate catalyst. *Journal of Catalysis*, 153(1), 25–31.

48. Turaga, U. T. and Song, C. (2003). MCM-41-supported Co-Mo catalysts for deep hydrodesulfurization of light cycle oil. *Catalysis Today*, 86(1–4), 129–140.

49. Corma, A., Martínez, A., and Martínez-Soria, V. (1997). Hydrogenation of aromatics in diesel fuels on Pt/MCM-41 catalysts. *Journal of Catalysis*, 169(2), 480–489.

50. Wang, A., Ruan, L., Teng, Y., Li, X., Lu, M., Ren, J., Wang, Y., and Hu, Y. (2005). Hydrodesulfurization of dibenzothiophene over siliceous MCM-41-supported nickel phosphide catalysts. *Journal of Catalysis*, 229(2), 314–321.

51. Klimova, T., Calderón, M., and Ramírez, J. (2003). Ni and Mo interaction with Al-containing MCM-41 support and its effect on the catalytic behavior in DBT hydrodesulfurization. *Applied Catalysis A: General*, 240(1–2), 29–40.

52. Méndez, F. J., Bastardo-González, E., Betancourt, P., Paiva, L., and Brito, J. L. (2013). NiMo/MCM-41 catalysts for the hydrotreatment of polychlorinated biphenyls. *Catalysis Letters*, 143(1), 93–100.

53. Kubička, D., Bejblová, M., and Vlk, J. (2010). Conversion of vegetable oils into hydrocarbons over CoMo/MCM-41 catalysts. *Topics in Catalysis*, 53(3), 168–178.

54. Park, K.-C., Yim, D.-J., and Ihm, S.-K. (2002). Characteristics of Al-MCM-41 supported Pt catalysts: effect of Al distribution in Al-MCM-41 on its catalytic activity in naphthalene hydrogenation. *Catalysis Today*, 74(3–4), 281–290.

55. Klimova, T., Rodríguez, E., Martínez, M., and Ramírez, J. (2001). Synthesis and characterization of hydrotreating Mo catalysts supported on titania-modified MCM-41. *Microporous and Mesoporous Materials*, 44–45, 357–365.

56. Herrera, J. M., Reyes, J., Roquero, P., and Klimova, T. (2005). New hydrotreating NiMo catalysts supported on MCM-41 modified with phosphorus. *Microporous and Mesoporous Materials*, 83(1–3), 283–291.

57. Zhao, D., Feng, J., Huo, Q., Melosh, N., Fredrickson, G. H., Chmelka, B. F., and Stucky, G. D. (1998). Triblock copolymer syntheses of mesoporous silica with periodic 50 to 300 Angstrom pores. *Science*, 279(5350), 548–552.

58. Mitran, R. A., Berger, D., Munteanu, C., and Matei, C. (2015). Evaluation of different mesoporous silica supports for energy storage in shape-stabilized phase change materials with dual thermal responses. *The Journal of Physical Chemistry C*, 119(27), 15177–15184.

59. Che, R., Gu, D., Shi, L., and Zhao, D. (2011). Direct imaging of the layer-by-layer growth and rod-unit repairing defects of mesoporous silica SBA-15 by cryo-SEM. *Journal of Materials Chemistry*, 21(43), 17371–17381.

60. Gutiérrez, O. Y., Singh, S., Schachtl, E., Kim, J., Kondratieva, E., Hein, J., and Lercher, J. A. (2014). Effects of the support on the performance and promotion of (Ni)MoS$_2$ catalysts for simultaneous hydrodenitrogenation and hydrodesulfurization. *ACS Catalysis*, 4(5), 1487–1499.

61. Suresh, C., Santhanaraj, D., Gurulakshmi, M., Deepa, G., Selvaraj, M., Sasi Rekha, N. R., and Shanthi, K. (2012). Mo–Ni/Al-SBA-15 (Sulfide) catalysts for hydrodenitrogenation: effect of Si/Al ratio on catalytic activity. *ACS Catalysis*, 2(1), 127–134.

62. Betiha, M. A., Hassan, H. M. A., Al-Sabagh, A. M., Khder, A. E. R. S., and Ahmed, E. A. (2012). Direct synthesis and the morphological control of highly ordered mesoporous AlSBA-15 using urea-tetrachloroaluminate as a novel aluminum source. *Journal of Materials Chemistry*, 22(34), 17551–17559.

63. Muthu Kumaran, G., Garg, S., Soni, K., Kumar, M., Sharma, L. D., Murali Dhar, G., and Rama Rao, K. S. (2006). Effect of Al-SBA-15 support on catalytic functionalities of hydrotreating catalysts: I. Effect of variation of Si/Al ratio on catalytic functionalities. *Applied Catalysis A: General*, 305(2), 123–129.

64. Jiang, S., Zhou, Y., Ding, S., Wei, Q., Zhou, W., and Shan, Y. (2016). Effect of direct synthesis Al-SBA-15 supports on the morphology and catalytic activity of the NiMoS phase in HDS of DBT. *RSC Advances*, 6(108), 106680–106689.

65. Klimova, T., Reyes, J., Gutiérrez, O., and Lizama, L. (2008). Novel bifunctional NiMo/Al-SBA-15 catalysts for deep hydrodesulfurization: effect of support Si/Al ratio. *Applied Catalysis A: General*, 335(2), 159–171.

66. Sun, B., Li, L., Fei, Z., Gu, S., Lu, P., and Ji, W. (2014). Prehydrolysis approach to direct synthesis of Fe, Al, Cr-incorporated SBA-15 with good hydrothermal stability and enhanced acidity. *Microporous and Mesoporous Materials*, 186, 14–20.

67. Garg, S., Soni, K., Ajeeth Prabhu, T., Rama Rao, K. S., and Murali Dhar, G. (2016). Effect of ordered mesoporous Zr SBA-15 support on catalytic functionalities of hydrotreating catalysts 2. Variation of molybdenum and promoter loadings. *Catalysis Today*, 261, 128–136.

68. Chandra Mouli, K., Mohanty, S., Hu, Y., Dalai, A., and Adjaye, J. (2013). Effect of hetero atom on dispersion of NiMo phase on M-SBA-15 (M = Zr, Ti, Ti-Zr). *Catalysis Today*, 207, 133–144.

69. Liu, Z., Wakihara, T., Nomura, N., Matsuo, T., Anand, C., Elangovan, S. P., Yanaba, Y., Yoshikawa, T., and Okubo, T. (2016). Ultrafast and continuous flow synthesis of silicoaluminophosphates. *Chemistry of Materials*, 28(13), 4840–4847.

70. Flanigen, E. M. (2009). Molecular sieve zeolite technology — the first twenty-five years. *Pure and Applied Chemistry*, 52(9), 2191–2211.

71. Miller, S. J. (1989). Process for making middle distillates using a silicoalu-minophosphate molecular sieve. *US4859312 A*.

72. Dai, W., Wang, X., Wu, G., Guan, N., Hunger, M., and Li, L. (2011). Methanol-to-olefin conversion on silicoaluminophosphate catalysts: effect of Brønsted acid sites and framework structures. *ACS Catalysis*, 1(4), 292–299.

73. Verma, D., Rana, B. S., Kumar, R., Sibi, M. G., and Sinha, A. K. (2015). Diesel and aviation kerosene with desired aromatics from hydroprocessing of Jatropha oil over hydrogenation catalysts supported on hierarchical mesoporous SAPO-11. *Applied Catalysis A: General*, 490, 108–116.

74. Chen, N., Gong, S., Shirai, H., Watanabe, T., and Qian, E. W. (2013). Effects of Si/Al ratio and Pt loading on Pt/SAPO-11 catalysts in hydrocon-version of Jatropha oil. *Applied Catalysis A: General*, 466, 105–115.

75. Kikhtyanin, O. V., Rubanov, A. E., Ayupov, A. B., and Echevsky, G. V. (2010). Hydroconversion of sunflower oil on Pd/SAPO-31 catalyst. *Fuel*, 89(10), 3085–3092.

76. Rabaev, M., Landau, M. V., Vidruk-Nehemya, R., Goldbourt, A., and Herskowitz, M. (2015). Improvement of hydrothermal stability of Pt/SAPO-11 catalyst in hydrodeoxygenation–isomerization–aromatization of vegetable oil. *Journal of Catalysis*, 332, 164–176.

77. Čejka, J. and van Bekkum, H. (2005). *Zeolites and Ordered Mesoporous Materials: Progress and Prospects: the 1st FEZA School on Zeolites, Prague, Czech Republic, August 20–21, 2005*, Elsevier, Amsterdam.

78. Catlow, C. R. A., Bell, R. G., Gale, J. D., and Lewis, D. W. (1995). Modelling of structure and reactivity in zeolites. In: *Studies in Surface Science and Catalysis*, B. Laurent and K. Serge, eds., Elsevier, Amsterdam, pp. 87–100. DOI: http://dx.doi.org/10.1016/S0167-2991(06)81877-X.

79. Hay, D. G., Jaeger, H., and West, G. W. (1985). Examination of the mono-clinic/orthorhombic transition in silicalite using XRD and silicon NMR. *The Journal of Physical Chemistry*, 89(7), 1070–1072.

80. Weitkamp, J. (2000). Zeolites and catalysis. *Solid State Ionics*, 131(1), 175–188.

81. Sugioka, M., Sado, F., Kurosaka, T., and Wang, X. (1998). Hydrodesulfurization over noble metals supported on ZSM-5 zeolites. *Catalysis Today*, 45(1–4), 327–334.

82. Landau, M. V., Berger, D., and Herskowitz, M. (1996). Hydrodesulfurization of methyl-substituted dibenzothiophenes: fundamental study of routes to deep desulfurization. *Journal of Catalysis*, 159(1), 236–245.

83. Liu, H., Yang, S., Hu, J., Shang, F., Li, Z., Xu, C., Guan, J., and Kan, Q. (2012). A comparison study of mesoporous Mo/H-ZSM-5 and conventional Mo/H-ZSM-5 catalysts in methane non-oxidative aromatization. *Fuel Processing Technology*, 96, 195–202.

84. Fu, W., Zhang, L., Wu, D., Xiang, M., Zhuo, Q., Huang, K., Tao, Z., and Tang, T. (2015). Mesoporous zeolite-supported metal sulfide catalysts with high activities in the deep hydrogenation of phenanthrene. *Journal of Catalysis*, 330, 423–433.

85. Yu, Q., Zhang, L., Guo, R., Sun, J., Fu, W., Tang, T., and Tang, T. (2017). Catalytic performance of CoMo catalysts supported on mesoporous ZSM-5 zeolite-alumina composites in the hydrodesulfurization of 4,6-dimethyldibenzothiophene. *Fuel Processing Technology*, 159, 76–87.

86. Wu, H., Duan, A., Zhao, Z., Li, T., Prins, R., and Zhou, X. (2014). Synthesis of NiMo hydrodesulfurization catalyst supported on a composite of nano-sized ZSM-5 zeolite enwrapped with mesoporous KIT-6 material and its high isomerization selectivity. *Journal of Catalysis*, 317, 303–317.

87. Scherzer, J. (1989). Octane-enhancing, zeolitic FCC catalysts: scientific and technical aspects. *Catalysis Reviews*, 31(3), 215–354.

88. Saceda, J.-J. F., Rintramee, K., Khabuanchalad, S., Prayoonpokarach, S., de Leon, R. L., and Wittayakun, J. (2012). Properties of zeolite Y in various forms and utilization as catalysts or supports for cerium oxide in ethanol oxidation. *Journal of Industrial and Engineering Chemistry*, 18(1), 420–424.

89. Bhatia, S. (1989). *Zeolite Catalysts: Principles and Applications*, CRC Press, Boca Raton.

90. Hriljac, J. A., Eddy, M. M., Cheetham, A. K., Donohue, J. A., and Ray, G. J. (1993). Powder neutron diffraction and 29Si MAS NMR studies of siliceous zeolite-Y. *Journal of Solid State Chemistry*, 106(1), 66–72.

91. Isoda, T., Nagao, S., Ma, X., Korai, Y., and Mochida, I. (1996). Hydrodesulfurization pathway of 4,6-dimethyldibenzothiophene through Isomerization over Y-zeolite containing CoMo/Al$_2$O$_3$ catalyst. *Energy & Fuels*, 10(5), 1078–1082.

92. Sotelo-Boyás, R., Liu, Y., and Minowa, T. (2011). Renewable diesel production from the hydrotreating of rapeseed oil with Pt/zeolite and NiMo/Al$_2$O$_3$ catalysts. *Industrial & Engineering Chemistry Research*, 50(5), 2791–2799.

93. Vissenberg, M. J., de Bont, P. W., Arnouts, J. W. C., van de Ven, L. J. M., de Haan, J. W., van der Kraan, A. M., de Beer, V. H. J., van Veen, J. A. R., and

van Santen, R. A. (1997). Instability of zeolite Y supported cobalt sulfide hydro-treating catalysts in the presence of H$_2$S. *Catalysis Letters*, 47(2), 155–160.

94. Li, D., Nishijima, A., and Morris, D. E. (1999). Zeolite-supported Ni and Mo catalysts for hydrotreatments: I. Catalytic activity and spectroscopy. *Journal of Catalysis*, 182(2), 339–348.

95. Ding, L., Zheng, Y., Zhang, Z., Ring, Z., and Chen, J. (2006). Hydrotreating of light cycled oil using WNi/Al$_2$O$_3$ catalysts containing zeolite beta and/or chemically treated zeolite Y. *Journal of Catalysis*, 241(2), 435–445.

96. Ding, L., Zheng, Y., Zhang, Z., Ring, Z., and Chen, J. (2007). Hydrotreating of light cycle oil using WNi catalysts containing hydrothermally and chemically treated zeolite Y. *Catalysis Today*, 125(3–4), 229–238.

97. Li, T., Cheng, J., Huang, R., Zhou, J., and Cen, K. (2015). Conversion of waste cooking oil to jet biofuel with nickel-based mesoporous zeolite Y catalyst. *Bioresource Technology*, 197, 289–294.

98. Li, D., Nishijima, A., Morris, D. E., and Guthrie, G. D. (1999). Activity and structure of hydrotreating Ni, Mo, and Ni–Mo sulfide catalysts supported on γ-Al$_2$O$_3$–USY zeolite. *Journal of Catalysis*, 188(1), 111–124.

99. PRICE, G. L. (2001). Zeolite Beta. http://www.personal.utulsa.edu/~geoffrey-price/zeolite/beta.htm.

100. Newsam, J. M., Treacy, M. M. J., Koetsier, W. T., and De Gruyter, C. B. (1988). Structural characterization of zeolite beta. *Proceedings of the Royal Society of London. Series A, Mathematical and Physical Sciences*, 420(1859), 375–405.

101. Higgins, J. B., LaPierre, R. B., Schlenker, J. L., Rohrman, A. C., Wood, J. D., Kerr, G. T., and Rohrbaugh, W. J. (1988). The framework topology of zeolite beta. *Zeolites*, 8(6), 446–452.

102. Yu, K., Kumar, N., Roine, J., Pesonen, M., and Ivaska, A. (2014). Synthesis and characterization of polypyrrole/H-beta zeolite nanocomposites. *RSC Advances*, 4(62), 33120–33126.

103. Ding, L., Zheng, Y., Yang, H., and Parviz, R. (2009). LCO hydrotreating with Mo-Ni and W-Ni supported on nano- and micro-sized zeolite beta. *Applied Catalysis A: General*, 353(1), 17–23.

104. Zhang, D., Duan, A., Zhao, Z., and Xu, C. (2010). Synthesis, characterization, and catalytic performance of NiMo catalysts supported on hierarchically porous beta-KIT-6 material in the hydrodesulfurization of dibenzothiophene. *Journal of Catalysis*, 274(2), 273–286.

Chapter 7

Commercial Hydroprocessing Processes for Biofeedstock

Bo Zhang*,‡ and Duncan Seddon†,§

*School of Chemical Engineering and Pharmacy,
Wuhan Institute of Technology, China
†Duncan Seddon & Associates Pty. Ltd.,
116 Koornalla Cres., Mount Eliza, Victoria, 3930, Australia
‡Corresponding author: bzhang_wh@foxmail.com
§seddon@ozemail.com.au

Abstract

Over the past decade, there have been several technologies developed by refinery operators and technology development companies aimed at processing renewable feedstocks into transport fuels. These processes are generally centered on a hydrotreating step in which a renewable feedstock, generally but not necessarily a vegetable oil or fat, is hydrogenated to produce an aliphatic hydrocarbon chain, which after further processing and distillation produces a fuel suitable for use as gasoline, jet fuel or diesel. This process is sometimes complemented by an isomerization step which further improves the properties of the fuel. This paper reviews the development and status of these processes as described in the patent literature and commercial brochures.

Keywords: Neste MY, BP, Ecofining, ENI, Honeywell-UOP, Bio-Synfining, Vegan, HydroFlex

207

7.1. Introduction

The catalytic hydrotreating of vegetable oils as well as suitable fat wastes and residues to produce renewable diesel (also called hydrotreated vegetable oil — HVO — and sometimes green diesel) is already a mature commercial-scale manufacturing process. It is based on oil refining technology and is used to produce biofuels for diesel and jet engines. In this process, the triglyceride is split into three separate carboxylic chains, and hydrogen removes the oxygen from the triglyceride oil molecules creating hydrocarbons and propane. A diagram of the process chemistry of the renewable diesel is shown in Figure 7.1. The hydrocarbons produced are similar to existing diesel fuel components, and this allows blending in any desired ratio with minimal concerns regarding fuel quality.

In general, the hydrogenation (HYD) proceeds to totally reduce the carboxylic group (outcome 1) or decarboxyleate the carboxylic group (outcome 2). In outcome 1, the product hydrocarbon chain has one more carbon atom than in outcome 2. In many cases, the ratio proceeds on a near 50/50 basis so that the resulting chain length distribution is more evenly spread than in the original oil or fat.

Often the catalyst has sufficient activity for the reverse water-gas shift reaction:

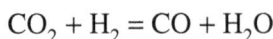

$$CO_2 + H_2 = CO + H_2O$$

This, in effect, produced outcome 3 of Figure 7.1 (decarbonylation) with carbon monoxide found in the product gases.

The glycerol part of the triglyceride, which is often a disposal issue for biodiesel production, is fully reduced to propane and hence bio-LPG.

For reference 500,000 t/y of diesel is approximately 610 million l/y or 161 million gallons/y or 10,500 bbl/d.

7.2. Neste's NEXBTL technology

Over last few years, Neste Oil has become the world's largest producer of renewable diesel that is produced from waste fats, residues, and vegetable oils through the hydrotreating process. Neste's commercial renewable

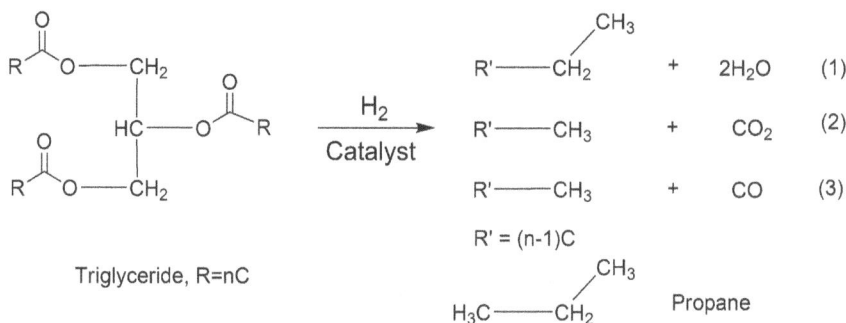

$$100 \text{ Barrel Triglyceride} \rightarrow 99 \text{ Barrel Renewable Diesel} + \text{Propane}$$

Figure 7.1. Process chemistry for the renewable diesel. Reaction (1): hydrodeoxygenation, reaction (2): decarboxylation, and reaction (3): decarbonylation.

diesel has a brand name of Neste MY Renewable Diesel (earlier brand names: NEXBTL and Neste Renewable Diesel) [1, 2]. Neste claims that it is a premium-quality, low-carbon, sulfur-free (<5 mg/kg), oxygen-free, and aromatics-free (<1 wt.%) diesel fuel. Production of Neste's renewable diesel firstly started at the Kilpilahti refinery site in Porvoo, Finland in 2007 [3]. To date, Neste has four facilities in Porvoo (Finland), Rotterdam (Holland), and Singapore with a total capacity of 2 Mt/y. Raw materials include crude palm oil from Southeast Asia (Malaysia and Indonesia), tallow (animal fats) imported from Australia and New Zealand, and corn oil from USA [4]. In 2014, Neste Oil produced nearly 1.3 Mt of renewable diesel [5].

The production process of this renewable diesel consists several steps (Figure 7.2): pretreatment of raw materials, processing of biofuel via the hydrotreatment process, stabilization, and recycling streams [6]. The pretreatment process is to reduce the level of impurities (like P, Ca, and Mg) in the feed to acceptable levels and thus ensure a long catalyst lifetime. The pretreatment is also called bleaching, which begins with the addition of an acid, forming a salt and removal of the salt by precipitation. The resultant feedstock is then fed through silica and/or bleaching earth which act as adsorbents for further reduction of impurities. Spent bleaching earth is disposed off-site. Typically, the impurities will be well below practical

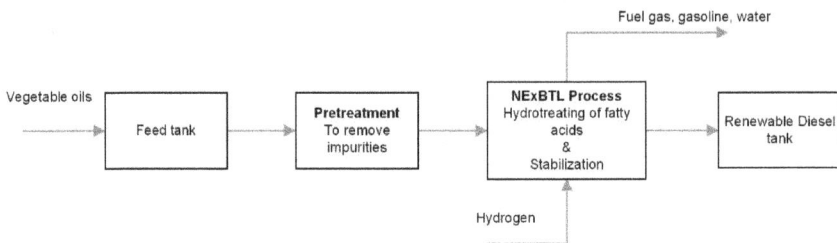

Figure 7.2.   Neste Oil's NExBTL® Renewable Diesel production process.

detection limits of analytical methods (<1 mg/kg) after pretreatment. In the hydrotreating process, the purified feed reacts with hydrogen over catalysts at elevated temperatures and pressures, which is followed by branching or isomerization of the product stream to improve the cold flow properties of the fuel. The hydrotreating of the feed is performed at the temperature between 200°C and 400°C (preferably 280–340°C) [7] and the pressure of 5–10 MPa [8] in the presence of a catalyst of Pd, Pt, Ni, Ni–Mo or Co–Mo supported on carbon, alumina and/or silica [9]. If a sulfided catalyst is used, organic or inorganic sulfur compounds as well as hydrogen may be required. The isomerization step requires a pressure range of 2–15 MPa (preferably 3–10 MPa) and the temperature between 200°C and 500°C (preferably 280–400°C) [10], and the suggested catalysts for the isomerization step are Pt, Pd, or Ni supported on Al_2O_3, SiO_2, SAPO-11, SAPO-41, ZSM-22, ZSM-23, or ferrierite (e.g., Pt/SAPO-11/Al_2O_3, Pt/ZSM-22/Al_2O_3, Pt/ZSM-23/Al_2O_3, and Pt/SAPO-11/SiO_2) [11]. However, for converting fatty acid C_1–C_5 alkyl esters, the reaction conditions for isomerization are an indicated temperature of 250–350°C and a pressure from 0.1 to 5 MPa, with an inert gas/hydrogen mixture in the presence of an acidic catalyst such as faujasite, offeretite, montmorillonite, or mordenite [12]. Following the isomerization process, the product is sent to the stabilization column, in which the light hydro-carbons are separated by stripping with low pressure steam. Finally, a series of gas processing units are used to generate and recycle hydrogen.

The products from Neste's processes include the diesel fuel, small amounts of renewable gasoline components, propane, and isoalkane. Based on a study on the Kilpilahti refinery in Finland, this facility between

Table 7.1. Physical and chemical properties of Neste MY Renewable Diesel [13].

Appearance	Liquid (nonexplosive)
Color	Clear
Odor	Mild
Pour point, cloud point	$<-20°C$, $<-40°C$
Boiling range	180–320°C (EN ISO 3405)
Flash point	>61°C (EN ISO 2719, EC A9)
Vapour pressure	0.087 kPa @ 25°C (EC A4)
Relative density	0.77–0.79 @ 15/4°C (EN ISO 12185, EC A3)
Solubility	Insoluble in water (~0.075 mg/l water @ 25°C) Soluble in the following materials: methanol and hydrocarbons
Partition coefficient	log K_{ow}: >6.5 (EC A8)
Auto-ignition temperature	204°C (EC A15)
Viscosity	Kinematic viscosity 4.0 mm^2/s @ 20°C, 2.6 mm^2/s @ 40°C (OECD 114), dynamic viscosity ≤ 5 mPa s @ 20°C
Cetane number	>70
Energy content	44 MJ/kg

2007 and 2008 produced annually 17,1678 t renewable diesel, 12,268 t propane, and 1,405 t biogasoline [3]. Neste's renewable diesel is a mixture of C_{10}–C_{20} branched and linear alkanes [13]. Its physical properties are summarized in Table 7.1.

Neste MY Renewable Diesel meets following standards: EN 15940:2016 — covering paraffinic diesel fuels and synthetic Fischer–Tropsch products (like GTL, BTL and CTL); EN 590:2013 — European diesel fuel standard though the density of Neste's renewable diesel is below the lower limit; ASTM D975 — the American diesel fuel standard; and Canadian CGSB-3.517. European diesel standards (EN590 and EN 15940) restrict biodiesel (i.e., fatty acid methyl esters or FAME) content to max 7%, and many engine manufacturers do not recommend the use of conventional FAME-type biodiesel due to possible engine damage [14]. Because the chemical nature of the Neste's renewable diesel is hydrocarbons, this limit is not applicable. The EN590 diesel standard sets the density of winter qualities to min. 800 kg/m^3, which limits the blending of Neste's renewable diesel. The ASTM D976 standard in the USA and CAN/CGSB 3.517 standard in Canada allow 100% Neste MY Renewable Diesel to be used. Neste MY Renewable Diesel is currently available as a commercial product in Finland in the form of Neste Pro Diesel, which contains a minimum of 15% Neste MY Renewable Diesel, since 2012 and in Lithuania since 2014. In order to satisfy the market demands in USA, Austria, and Sweden, this product has been delivered from Neste's Porvoo, Rotterdam, and Singapore plants to customers in Europe and North America.

7.3. BP's hydrotreatment processes for renewable diesel

The renewable diesel technology had already been developed in BP's laboratories in the 2000s. They have a series of patents filed and have published data on the conversion of vegetable oils and fats in commercial hydrocracking operations. Several patents filed by BP between 2006 and 2007 described the hydroprocessing process of carboxylic acids or carboxylic acid esters (such as biologically derived fatty acids or fatty acid esters) to produce diesel fuels [15, 16]. In their processes, hydrogen and a hydrocarbon-containing stream were cofed to the first reactor, in which the reaction conditions were maintained at a temperature range of 250–430°C and a pressure of 2–20 MPa in the presence of Ni–Mo or Co–Mo catalysts. The purpose of the first treatment was to produce a hydrocarbon-containing product stream with a reduced concentration of heteroatom-containing organic compounds and/or olefins. Furthermore, this hydrocarbon-containing product stream was hydroprocessed over a sulfided catalyst in the second reactor, in which the reaction conditions were in the temperature range of 200–410°C and the pressure of 2–20 MPa. Suitable catalysts for this hydro-processing process comprised one or more active metals of Pd, Pt, Ni, Ru, Cu, Co, Cr, Mo, and W and an inorganic oxide as the support (such as zirconia, titania or γ-alumina). Preferred catalysts were Ni–Mo or Co–Mo supported on γ-Al$_2$O$_3$. It is claimed that the hydroprocessing of carboxylic acids or its esters produced hydrocarbons. Furthermore, the BP research confirmed that this hydroprocessing process coproduced C$_1$ compounds like carbon monoxide (CO), carbon dioxide (CO$_2$) and methane (CH$_4$) [17]. Because the feedstock hydrocarbons comprised sulfur compounds that might be converted to H_2S in the reactor, H_2S can be separated from the product stream in the vapor fraction of a flash separator. Some noteworthy findings include:

(1) The catalyst of Co–Mo supported on alumina was effective in catalyzing hydrodecarboxylation at lower pressures, e.g., pressures of less than 5 MPa or 3.1 MPa, which generally increases the C1:carboxylate mole ratio.
(2) The presence of carbon monoxide in the vapor fraction can cause loss of catalyst activity through the formation of volatile metal carbonyl species (like nickel carbonyl), some of which can be highly toxic [18].
(3) Low hydrogen partial pressures may also result in the deactivation of the HYD catalysts, so hydrogen is separately and simultaneously injected at two or more different regions of the catalyst bed in the hydroprocessing process [19].

Later, in 2011, a process of cohydrotreating a feedstock of petroleum origin containing sulfur with the biologically derived vegetable oils or animal fats was published. The advantage of this process is to provide the sulfiding of the hydroprocessing catalyst that is used to hydroprocess the sulfur-free biomass-derived feedstock [20]. Another patent filed by the Total Raffinage Marketing provided a similar process with more details [21]. Their process consisted of two parallel hydrotreating units, which were operated at 320–420°C (preferably 340–400°C) and 250–420°C (preferably 280–350°C) under a pressure of 25–150 bar, respectively. Suitable catalysts comprised at least one catalyst based on metal oxides chosen from Group VI-B (Mo, W) and VIII-B (Co, Ni, Pt, Pd, Ru, Rh) supported on a support chosen from alumina, silica/alumina, zeolite, ferrierite, phosphated alumina, and phosphated silica/alumina, preferably NiO, Ni–Mo, Co–Mo, Ni–W, Pt–Pd or a mixture of two or more of these.

In 2007, BP Australia Pty Ltd announced commencement of production of renewable diesel from tallow feedstock at its Bulwer Island refinery in Brisbane [22]. BP claimed that renewable diesel is a product from converting animal or vegetable fats into ultra-low-sulfur diesel in a refinery hydrotreater instead of conventional methanol-esterified biodiesel [23]. Full production of renewable diesel (approximately 130 million l/y) was reached by mid-2008. All diesels produced by the Bulwer Island refinery (about 2.4 billion l/y) contained up to 5% renewable diesel. However, BP has ceased production at the Bulwer Island refinery in May 2015 and converted the site into an import terminal.

7.4. Ecofining™ by ENI/Honeywell-UOP

Ecofining™ is a process developed in the San Donato Milanese laboratories under a collaboration between ENI and Honeywell-UOP. Ecofining™ is a two-stage process: during the first stage of hydrodeoxygenation, vegetable oil, or other biological feedstock, is transformed into a blend of linear C_{16}–C_{18} paraffins; in the second isomerization stage, the paraffin isomers are formed to give the product the necessary cold properties and meet the specifications of diesel fuel, with coproduction of green liquefied petroleum gas (LPG), naphtha, and jet fuel. The process may use a wide range of oily feedstock including rapeseed, soybean, carinata, palm, pennycress, *Jatropha*, camelina, tallow, lard, used cooking oils and algal oil. A simplified scheme of the Eni/UOP Ecofining™ process is shown in Figure 7.3.

The hydroprocessing stage is carried out at a moderate temperature of 270–430°C (~310°C) and a pressure of 3–5 MPa using a hydrotreating catalyst, yielding only linear-paraffins. The hydrotreating catalyst could be one of following: Pd, Pt, Ni, or bimetallic (sulfided Ni–Mo, Co–Mo, Ni–W, or Co–W) supported on alumina, silica, zirconia, titania or mixtures of silica–alumina. The catalysts of Ni–Mo–P on zeolite, Pd/zeolite, and Pt/MSA were also tested [24]. Since linear paraffins have higher melting points, which result in poor cold properties, this requires a

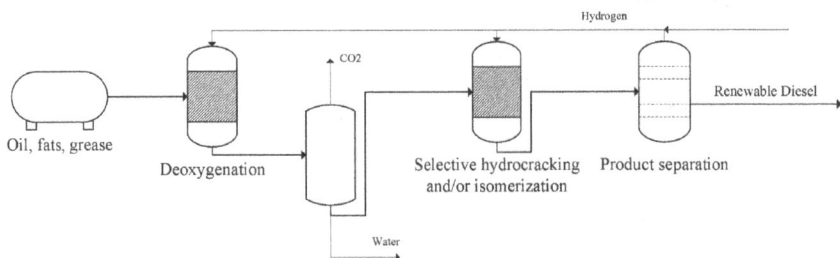

Figure 7.3. Processing oils with the UOP/ENI Ecofining™ process.

hydroisomerization stage that is performed at a temperature ranging from 280°C to 380°C and a pressure ranging from 3 MPa to 5 MPa. The reactions are catalyzed by a bifunctional catalyst (i.e., metal loaded acidic zeolites, sulfonated oxides, SAPOs, mesoporous silica–aluminas) [25]. The active metal could be one of Pt, Pd, Ir, Ru, Rh, and Re. The whole process can be operated at an LHSV ranging from 0.5 h^{-1} to 2 h^{-1}.

In 2015, ENI had converted the oil refinery at Porto Marghera near Venice (Italy) to a biorefinery to produce high quality biofuel from vegetable oil and biomass [26]. The detailed renovation plan can be found in their patent application [27]. Besides its technological core of the Ecofining™ process, the process contains a pretreatment unit to condition the feed. For example, the mixture of fatty acids might be neutralized with ammonia to form ammonium salts prior to hydrodeoxygenation [28]. This bio-refinery is currently fed by palm oil, but the plan is also to use biomass. The maximum biofuels productivity of this site is about 500 kt/y [29]. Currently, ENI is converting the Gela oil refinery to biorefinery.

In the US, Diamond Green Diesel, which was formed via a joint venture between Diamond Alternative Energy LLC (a subsidiary of Valero Energy Corporation) and Darling Ingredients Inc. is the largest producer of renewable diesel [30]. This renewable diesel refinery is using the Ecofining technology and is colocated at the Valero St. Charles Refinery (Louisiana, USA). It has been in operation since June 2013, converting 500 kt/y of animal fats and cooking oil to 400 kt/y of Honeywell Green Diesel™ and 65 kt/y LPG and Naphtha [31].

AltAir Fuels (USA) converted the idled portions of its Paramount petroleum refinery in California to produce renewable jet fuel using UOP's Renewable Jet Fuel Process, which is a variation of the Ecofining technology. Starting in fourth quarter 2015, AltAir Fuels became the first full-scale plant dedicated to producing renewable jet fuel for commercial and military use. Its current productivity capacity is 2,500 bbl/d. United Airlines will purchase 15 million gallons of this jet fuel over a 3-y period, with the option to purchase more. Outside North America, Petrixo Oil & Gas licensed UOP's Renewable Jet Fuel Process and will construct a new facility in the United Arab Emirates to process approximately 500,000 t/y of renewable feedstocks [32].

7.5. Bio-Synfining® process by Renewable Energy Group Inc.

The Bio-Synfining process was originally developed by Syntroleum Co. through a joint venture with Tyson Foods [33]. This process consists of three steps and utilizes a wide variety of feedstocks that are turned into high quality renewable fuels. The first step involves pretreatment of the feedstock, which removes contaminants to create a pure feedstock for the reactor section. Pretreatment section consists of filters to remove physical contaminants, chemical addition followed by high speed centrifugation to remove aqueous phase contaminants, and lastly a solid phase adsorption/filtration step to remove trace phosphorus and metal ion species [34]. The pretreated feedstock is charged to the hydrotreating section, in which the feedstock is processed at a temperature in a range of about 290–379°C under high hydrogen pressure (8–11 MPa) over a catalyst. Their results indicated that the monometallic catalyst often showed lower HYD activity than the bimetallic catalysts such as Ni–Mo, Co–Mo, and Ni–W. So a two-step HDO process using the first molybdenum catalyst and then the second bimetallic catalyst is necessary [33]. The intermediates of this section are pure paraffins that are then isomerized to branched paraffins (i.e., *iso*-paraffins) to yield low cloud point clean fuels. The final separation step of the process is to distil the product stream into three main products — renewable diesel, renewable naphtha, and renewable LPG.

Renewable Energy Group Inc. (REG) acquired the Syntroleum's Bio-Synfining technology and the facility at Geismar, Louisiana, in 2014 and renamed it as Synthetic Fuels Division. This biorefinery has a production capacity of 75 million gallons/y (5,000 bbl/d, 233,000 t/y) and utilizes a wide array of feedstocks including inedible corn oil, used cooking oil, waste vegetable oils, animal fats, greases, and vegetable oils to produce the renewable diesel fuel.

7.6. Vegan® technology by Axens (IFP)

The Vegan® technology is Axens's version of HVO technology that converts liquid lipid sources (like vegetable oil, animal fats, and algal oil) into high quality transportation jet and diesel fuels [35]. The crude lipids may require a pretreatment prior to hydrotreatment to remove species containing heteroelements such as phosphorus, magnesium, calcium, iron or zinc in amounts which may be up to 2,500 ppm, principally in the form of phospholipids, pigments, sterols, and/or soaps [36]. The pretreatment technology patented by Axens is to utilize a fixed bed of adsorbents comprising 100% porous refractory oxide (like activated alumina, or silica–alumina) at a temperature of 130–320°C and a pressure 0.1–7 MPa and with a residence time of 0.1–1 h [37]. Following pretreatment, the feedstock is hydrotreated and isomerized. Hydrotreating of the pretreated effluent is performed in the presence of at least one fixed bed catalyst at a temperature in the range 200–450°C and a pressure of 1–10 MPa with an hourly space velocity in the range 0.1–10 h^{-1} and the hydrogen/feed ratio is in the range 70–1,000 N m^3 of hydrogen/m^3 of feed. The hydroisomerization step is carried out at the temperature between 320°C and 420°C and a pressure of 1–9 MPa with a hydroisomerizing catalyst [38]. As an international provider of catalysts, Axens integrates many of its catalysts into this process. According to Axens's catalyst catalogue, the hydrotreating catalyst is Co–Mo–Ni/Al_2O_3, Co–Mo/Al_2O_3, or Ni–Mo/Al_2O_3 with different shapes of spheres, trilobe extrudates, and quadrilobe extrudates [39]. The Co–Mo–Ni/Al_2O_3 catalyst also shows multifunctions including hydrodemetallization (HDM) and hydrocracking. The hydroisomerizing catalyst comprises 0.05–10% by weight of at least one noble metal of Group VIII (e.g., Pt and Pd) and may also contain tungsten and nickel [38].

In April 2015, France oil company Total S.A. announced an investment of €200m to convert the La Mède oil refinery in southern France to a biorefinery, which will produce 500,000 t of renewable diesel from used cooking oils and other feedstock using Axens's Vegan process [40].

7.7. HydroFlex^{TM} hydrotreating process by Haldor Topsoe

UPM (a Finnish company) is manufacturing the BioVerno renewable diesel, which is produced out of crude tall oil, a natural extract of wood (mainly conifers) as a residue of pulp production. The world production of crude tall oil is approximately 2 Mt/y. Tall oil mainly consists of resin acids and free fatty acids as well as a number of contaminants in smaller concentrations. In order to transform crude tall oil, UPM licensed the HydroFlex™ hydrotreating process developed by Haldor Topsoe (Denmark) [41]. The complete process includes pretreatment of crude tall oil to remove salts, solid particles, and water; application of the Haldor Topsoe HydroFlex™ process for hydrotreating of the pretreated crude tall oil; fractionation of the products [42].

The pretreatment process for purifying crude tall oil comprises evaporating the tall oil at a temperature of 50–250°C and a pressure of 5–10 MPa to separate the feed into two fractions: the first fraction comprising hydrocarbons having a boiling point of up to 250°C (NTP) and water and the second fraction comprising fatty acids, resin acids, neutral substances and residue components. The second fraction is further evaporated to produce a third fraction comprising fatty acids, resin acids and neutral substances having a boiling point under 500°C and a residue fraction [43]. The fatty acids, rosin acids, and sterols can be converted to fuel components via hydrodeoxygenation (HDO) and isomerization reactions. HDO is conducted at a pressure between 5 and 10 MPa and temperature between 280 and 340°C with a Ni–Mo or Co–Mo catalyst. The isomerization reactor is operated at a pressure of 3–10 MPa and temperature between 280°C and 400°C with a Pt/SAPO-11 or Pt/ZSM-23 catalyst. The cracking catalyst for heavier components is more acidic, e.g., Pt/ZSM-5 [44].

Since January 2015, UPM Lappeenranta Biorefinery (Finland), located on the same site as the UPM Kaukas pulp and paper mill, has started commercial production of wood-based renewable diesel. A big portion of the raw material comes from UPM's own pulp mills in Finland. 1 t of dry pulp mill produces up to 50 kg of crude tall oil, which contains up to 70% of free fatty acids. This biorefinery produces approximately 100,000 t/y of BioVerno diesel [45].

Similar to the BioVerno diesel, Preem (a Sweden Company), which licensed the same technology, launched its version of renewable diesel — Preem Evolution Diesel — in the spring of 2011. Its biorefinery utilizes pine oil (i.e., tall oil) as the raw material, but the pretreatment is different from that of UPM. Preem processes crude tall oil through a transesterification process, so the majority of free fatty acids are converted to fatty acid methyl esters (FAMEs), while the resin acids are left almost unconverted. Then the HydroFlex™ hydrotreating process converts so-called crude tall diesel to Preem Evolution Diesel.

7.8. Technology of Cetane Energy LLC

Cetane Energy LLC (USA) owns a pilot plant at Carlsbad, New Mexico, and produces renewable diesel from vegetable oils and wastes [46]. The process developed by Cetane Energy is relatively simple, comprising filtering the biofeedstock, heating the feedstock, introducing hydrogen into the feed treating the feed in a reactor to generate a diesel product. The reactor is packed with a hydrotreating catalyst of Mo/Al_2O_3 as 3/16 inch sized rings or 1/10 inch sized quadrolobe, wherein the treating includes injecting hydrogen into the reactor at multiple locations to cool the diesel product within the reactor. After hydrotreating, the vapors are separated from the liquid, and the liquid is distilled in a distillation column to generate the purified diesel product [47]. In Europe, Renewable Diesel Europe (UK) is the exclusive agent for this patented renewable diesel technology developed by Cetane Energy.

7.9. Conclusions and perspectives

Technically, the process for conversion of vegetable oils and animal fats to HVO comprises the pretreatment units, the hydrotreating units for hydrodeoxygenation and isomerization, and the product separation units. The hydrotreating technologies are mainly covered by patents of Neste Oil, Honeywell-UOP, Haldor Topsoe, Axens, BP, etc. Obviously, the companies who are the suppliers of catalysts hold a more dominant position in this business. Meanwhile, many manufactures developed the pretreatment approaches for their specific feedstock. During the last decade, some companies exited or intended to exit the biofuels business. However, the intellectual properties owned by them are still valuable and may turn into new businesses in the future. The intellectual property owned by ConocoPhillips (US) can be found in Ref. [48] and Brazilian Petrobras' process can be found in Refs. [49, 50].

References

1. Neste (2017). Neste MY Renewable Diesel. https://www.neste.com/na/en/customers/products/renewable-products/neste-my-renewable-diesel.
2. Neste (2017). Neste rebrands its renewable diesel to Neste MY Renewable Diesel in North America. https://www.neste.com/fi/en/neste-rebrands-its-renewable-diesel-neste-my-renewable-diesel-north-america.
3. Nikander, S. (2008). Greenhouse gas and energy intensity of product chain: case transport biofuel. Ph.D., Helsinki University of Technology.
4. Neste Oil (2015). NExBTL® Renewable Diesel Singapore Plant: NORTH AMERICAN TECHNICAL CORN OIL PATHWAYS DESCRIPTION. https://www.arb.ca.gov/fuels/lcfs/2a2b/apps/nes-co-rd-rpt-072915.pdf.
5. Neste Oil (2015). Neste Oil claims world leadership in biofuels from waste, residue. http://www.biodieselmagazine.com/articles/324835/neste-oil-claims-world-leadership-in-biofuels-from-waste-residue.
6. Neste Oil (2013). NExBTL® Renewable Diesel Singapore Plant: TALLOW PATHWAY DESCRIPTION. https://www.arb.ca.gov/fuels/lcfs/2a2b/apps/neste-aus-rpt-031513.pdf.
7. Myllyoja, J., Aalto, P., Savolainen, P., Purola, V. M., Alopaeus, V., and Grönqvist, J. (2007). Process for the manufacture of diesel range hydrocarbons. *WO2007003709 A1*.
8. Jakkula, J., Niemi, V., Nikkonen, J., Purola, V. M., Myllyoja, J., Aalto, P., Lehtonen, J., and Alopaeus, V. (2007). Process for producing a hydrocarbon component of biological origin. *EP1396531 B1*.
9. Myllyoja, J., Aalto, P., and Harlin, E. (2012). Process for the manufacture of diesel range hydrocarbons. *US8278492 B2*.
10. Myllyoja, J., Aalto, P., Savolainen, P., Purola, V. M., Alopaeus, V., and Grönqvist, J. (2007). Process for the manufacture of diesel range hydrocarbons. *US20070010682 A1*.
11. Myllyoja, J., Aalto, P., Savolainen, P., Purola, V. M., Alopaeus, V., and Grönqvist, J. (2014). Process for the manufacture of diesel range hydrocarbons. *US8859832 B2*.
12. Koivusalmi, E. and Jakkula, J. (2008). Process for the manufacture of hydrocarbons. *US7459597 B2*.
13. Neste (2016). NEXBTL Renewable Diesel safety data sheet. https://www.neste.com/sites/default/files/attachments/nexbtl_renewable_diesel_version3.0_safetysheet.pdf.
14. Neste (2016). *Neste Renewable Diesel Handbook*. https://www.neste.com/sites/default/files/attachments/neste_renewable_diesel_handbook.pdf.

15. Gudde, N. J. and Townsend, J. A. (2007). Hydrogenation process. *WO2007138254 A1*.
16. BP Oil International Limited (2008). Process for hydrogenation of carboxylic acids and derivatives to hydrocarbons. *EP1911734 A1*.
17. Gudde, N. J. (2008). Hydrogenation process. *WO2008040980 A1*.
18. Gudde, N. J. and Townsend, J. A. (2014). Process for hydrogenation of carboxylic acids and derivatives to hydrocarbons. *US8742184 B2*.
19. Gudde, N. J. (2013). Hydrogenation process. *US8552234 B2*.
20. Shabaker, J. W. (2014). Renewable diesel refinery strategy. *US8884086 B2*.
21. Mayeur, V., Vergel, C., Morvan, G., Mariette, L., and Hecquet, M. (2011). Process for hydrotreating a diesel fuel feedstock, hydrotreating unit for the implementation of the said process, and corresponding hydrorefining unit. *US20110047862 A1*.
22. Australian Institute of Petroleum (2007). Biofuels factsheet. http://www.aip.com.au/sites/default/files/download-files/2017-10/BioFuelFactSheet.pdf.
23. Murphy, M. (2007). BP turns animal for renewable diesel production. http://www.theage.com.au/news/business/bp-turns-animal-for-renewable-diesel-production/2007/04/15/1176575680869.html.
24. Perego, C. (2015). From biomass to advanced biofuel: the greendiesel case. http://www.sinchem.eu/wp-content/uploads/2015/01/15-Perego-ENI.pdf.
25. Perego, C., Sabatino, L. M. F., Baldiraghi, F., and Faraci, G. (2013). Process for producing hydrocarbon fractions from mixtures of a biological origin. *US8608812 B2*.
26. ENI (2016). Green Refinery. https://www.eni.com/en_IT/innovation/technological-platforms/green-refinery.page.
27. Rispoli, G. F., Bellussi, G., Calemma, V., and de Angelis, A. R. (2015). Biorefinery and method for revamping a conventional refinery of mineral oils into said biorefinery. *WO2015181279 A1*.
28. Bellussi, G., Molinari, D., Assanelli, G., and de Angelis, A. R. (2015). Process for the production of hydrocarbon fractions from mixtures of a biological origin. *WO2015107487 A1*.
29. ENI (2014). Green Refinery: reinventing petroleum refineries. https://www.eni.com/docs/en_IT/enicom/publications-archive/company/operations-strategies/refining-marketing/eni_Green-Refinery_esecutivo.pdf.
30. Honeywell UOP (2017). Honeywell Green Diesel™. https://www.uop.com/processing-solutions/renewables/green-diesel/.
31. Diamond Green Diesel (2017). What is Green Diesel? https://www.diamondgreendiesel.com/what-is-green-diesel.

32. Honeywell UOP (2017). Honeywell Green Jet Fuel™. https://www.uop.com/processing-solutions/renewables/green-jet-fuel/.
33. Abhari, R. and Havlik, P. (2011). Hydrodeoxygenation process. *US8026401 B2*.
34. Renewable Energy Group (2017). Bio-Synfining. http://www.regi.com/technologies/bio-synfining.
35. Axens (2016). Vegan. http://www.axens.net/product/technology-licensing/11008/vegan.html.
36. Karleskind, A. and Wolff, J. P. (1996). *Oils and Fats Manual: A Comprehensive Treatise: Properties, Production, Applications*, Intercept, Andover, UK.
37. Dandeu, A., Coupard, V., and Chapus, T. (2016). Process for converting feeds derived from renewable sources with streatment of feeds by hot dephosphatation. *US9447334 B2*.
38. Chapus, T. and Dupassieux, N. (2011). Method of converting feedstocks coming from renewable sources into high-quality gas-oil fuel bases. *US7880043 B2*.
39. Axens (2017). Catalysts. http://www.axens.net/our-offer/by-products/catalysts-and-adsorbents/catalysts.html.
40. Kotrba, R. (2015). Total to convert oil refinery to renewable diesel production. http://www.biodieselmagazine.com/articles/355201/total-to-convert-oil-refinery-to-renewable-diesel-production.
41. Haldor Topsoea (2015). Wood-based renewable diesel bio-refinery goes on-stream in Finland. http://www.topsoe.com/news/2015/03/wood-based-renewable-diesel-bio-refinery-goes-stream-finland.
42. UMP (2017). Advanced biofuel production. http://www.upmbiofuels.com/biofuel-production/advanced-biofuel-production/Pages/Default.aspx.
43. Nousiainen, J., Laumola, H., Rissanen, A., Kotoneva, J., and Ristolainen, M. (2015). Process and apparatus for purifying material of biological origin. *EP2643442 B1*.
44. Knuuttila, P., Kukkonen, P., and Hotanen, U. (2010). Method and apparatus for preparing fuel components from crude tall oil. *WO2010097519 A2*.
45. UMP (2015). UPM Lappeenranta Biorefinery is in commercial production. http://www.upm.com/About-us/Newsroom/Releases/Pages/UPM-Lappeenranta-Biorefinery-is-in-commercial-production-001-Mon-12-Jan-2015-11-30.aspx.
46. Cetane Energy LLC (2014). About Us. http://www.cetaneenergy.com/aboutus.html.
47. Richard, A. and Smith, J. (2013). Systems and methods of generating renewable diesel. *US8563792 B2*.

48. Ghonasgi, D. R., Edward, L. S. I. I., Yao, J., and Xu, X. (2011). Hydrotreating and catalytic dewaxing process for making diesel from oils and/or fats. *US7955401 B2.*

49. Gomes, J. (2006). Vegetable oil hydroconversion process. *US20060186020 A1.*

50. Silva, M., de Rezende Pinho, A., Huziwara, W. K., da Silva Neto, A. P., Khalil, C. N., Cabral, J. A. R., Leite, L. C. F., Casavechia, L. C., and de Carvalho Silva, R. (2012). Catalytic cracking process for production of diesel from seeds of oleaginous plants. *US8231777 B2.*

Chapter 8

Renewable Fuels and Fuel Regulations and Standards

Duncan Seddon

Duncan Seddon & Associates Pty. Ltd.,
116 Koornalla Cres., Mount Eliza, Victoria, 3930, Australia
seddon@ozemail.com.au

Abstract

In order to continually reduce vehicle pollution and greenhouse gas emissions, fuel standards and regulations are in a state of constant change and tightening. These changes and future trends have an impact on the take-up of renewable fuels for both gasoline and diesel fuels. This chapter discusses current and future fuel standards and illustrates how the hurdles experienced by ethanol and bio-diesel blends have been overcome. Nevertheless, ethanol and bio-diesel still remain and are likely to remain a minor part of the world fuel market.

Ideally new renewable fuels should be drop-in fuels with minimum compatibility issues with conventional fuels. This is illustrated by bio-refinery operations to produce green diesel which is compatible with conventional diesel.

There may be a role for a new renewable additive to improve gasoline octane as octane requirements rise. The market for such an additive could justify large-scale manufacture.

Keywords: Fuel regulation, Fuel standards, Fuel quality, Gasoline, Diesel, Ethanol, Biofuels, Biodiesel, FAME, Green diesel, Compliance

8.1. The objective of fuel standards

The performance of any motor vehicle is dependent upon the quality of fuel being used. Over the years, fuel has been produced not only to ensure optimum performance but also to ensure the fuel does not cause engine damage, both in the short and the long term. Many of these fuel standards were agreed on a country by country basis between the local fuel suppliers, usually the major refiners, and the original engine (vehicle) manufacturers (OEMs).

In recent times, these industry standards have been incorporated into regulated government standards. These regulated standards have the further objective of minimizing air pollution produced from fugitive vehicle emissions and, more importantly, from the vehicle exhaust. The development of these standards requires tripartite discussions and agreement between the government regulators, OEMs, and the refining industry. Thus to achieve a particular emission objective, OEMs designed vehicles with added components, such as catalytic converters, and used a fuel of agreed quality provided by the oil refineries.

This regulated system facilitates the continuing improvement in vehicle emissions by continued improvement in engine design coupled with improved fuel quality.

Furthermore, international agreement between various jurisdictions facilitates the harmonization of outcomes across countries. This benefits the major OEMs since it facilitates the production of vehicles in large export-oriented manufacturing plant, which can produce vehicles that perform to a given level in terms of engine integrity, mileage, and emissions, in countries with the same or similar standards of fuels.

The requirement to continue to improve vehicle emissions and fuel standards has been promoted to date by the European Union (EU), the US, and Japan. These standards have similar objectives but differ in detail to satisfy local issues. Other countries tend to follow and adopt, often with local minor changes, one of these widely reported authorities. Following the EU standards (Euro standards) is common.

In addition to these standards, the world auto industry have produced their own requirements for fuel standards in the "World-Wide Fuel Charter" (WWFC). This has been published since 1998 and is regularly updated.[1] Usefully, this charter gives technical reasoning for the adoption of a particular fuel parameter with reference to more detailed technical articles. The WWFC defines categories of fuel quality which reflects the regulations of a particular jurisdiction. The current WWFC envisages five categories as follows:

1. Markets with no or first level requirements for emission control,
2. Markets with requirements for emission control or other market demands,
3. Markets with more stringent requirements for emission control or other market demands,
4. Markets with advanced requirements for emission control and enable sophisticated NO_x (nitrogen oxides) and particulate matter after-treatment technologies,
5. Markets with highly advanced requirements for emission control and fuel efficiency.

These WWFC categories are, with some minor differences, required by vehicles performing to various jurisdictional emission standards as detailed in Table 8.1.

8.2. What the standards require

For ease of discussion, the WWFC category fuels are discussed.

8.2.1. *Gasoline (petrol)*[2]

The characteristics of the WWFC category fuels for gasoline are listed in Table 8.2.

---

[1] World Wide Fuel Charter, September 2013 available from European Automobile Association (ACEA), Alliance of Automobile Manufacturers (Auto Alliance), Truck and Engine Manufacturers Association (EMA), Japan Automobile Manufacturers Association (JAMA), or their websites.

[2] Gasoline is the preferred industry term for petrol.

Table 8.1.   WWFC categories and emission standards for vehicles.

WWFC	EU	USA	California	Japan
Category 1	Euro-1	Tier-0	—	—
Category 2	Euro-2/II and 3/III	Tier-1	—	—
Category 3	Euro-4/IV	LEV	LEV or ULEV	JP 2005
Category 4	Euro-4/IV, 5/V and 6/VI	Tier-2/3	LEV II	JP 2009
Category 5	—	US 2017 fuel economy	LEV III	—

Note: Corresponding country fuel standards in Europe follow the naming of the vehicle emission standards.

Note that as the level of the category increases, more requirements for the fuels become specified. Most developed countries and many of the world's developing countries require category 3 fuels or higher, so it is the specification of categories 3, 4, and 5 fuels that is most relevant to a new fuel source such as a renewable fuel.

Three octane grades are envisaged, characterized by the research octane number (RON) as 91 RON, 95 RON, and 98 RON. Note the motor octane number (MON) is 10 or less than the RON value. This discriminates against excessive use of some components where RON minus MON >10. At the time of writing, the use of MON in characterizing fuel is disputed, and greater flexibility allows more components with poor MON to be used. Note also that for category 5 fuel, which is for better fuel economy, the lowest grade is 95 RON and implies the use of higher compression engines to achieve the required economy.

Renewable fuels containing high levels of linear alkanes and *cyclo*-alkanes will have difficulty achieving these RON requirements without the use of significant levels of octane boosting components which will result in other compliance issues.

Oxygen stability is an issue with highly unsaturated components especially conjugated olefins. Mono-olefins are capped at 10% in the higher grades, so renewable fuels containing high levels of unsaturated (olefinic) compounds will require hydrogenation to achieve the standards.

Aromatics and benzene are formed in the pyrolysis of biomass and the like and from the conversion of renewable fuels such as ethanol using zeolite catalysts. Components derived from these sources would have to

Table 8.2. WWFC categories for gasoline.

Category	Units	ASTM or test method	1	2	3	4	5
91 RON–RON (min)	—	D2699	91.0	91.0	91.0	91.0	—
91 RON–MON (min)	—	D2700	82.0	82.5	82.5	82.5	—
95 RON–RON (min)	—	D2699	95.0	95.0	95.0	95.0	95.0
95 RON–MON (min)	—	D2700	85.0	85.0	85.0	85.0	85.0
98 RON–RON (min)	—	D2699	98.0	98.0	98.0	98.0	98.0
98 RON–MON (min)	—	D2700	88.0	88.0	88.0	88.0	88.0
Oxidation stability (min)	min	D525	360	480	480	480	480
Sulfur (max)	mg/kg	D2622, D5453	1,000	1,000	30	10	10
Trace metal (max)	mg/kg		1 or ND[a]	1 or ND	1 or ND	1 or ND	1 or ND
Oxygen (max)	% m/m	D4815	2.7	2.7	2.7	2.7	2.7
Olefins (max)	% v/v	D1319	—	18.0	10.0	10.0	10.0
Aromatics (max)	% v/v	D1319	50.0	40.0	35.0	35.0	35.0
Benzene (max)	% v/v	D5580, D3606	5.0	2.5	1.0	1.0	1.0
Sediment (max)	mg/l	D5452	—	1	1	1	1
Unwashed gums (max)	mg/100ml	D381	70	70	30	30	30
Washed gums (max)	mg/100ml	D381	5	5	5	5	5
Density (min)	kg/m^3	D4052	715	715	715	715	720
Density (max)	kg/m^3	D4052	780	770	770	770	775
Copper corrosion (max)	Rating	D130	Class 1	Class 1	Class 1	Class 1	Class 1

(Continued)

Table 8.2.   (*Continued*)

Category	Units	ASTM or test method	1	2	3	4	5
Silver corrosion	Rating	D7671	—	—	—	Class 1	Class 1
Appearance	—	D4176	Clear	Clear	Clear	Clear	Clear
Carburettor cleanliness (min)	Merit	CEC F-03-T	8.0	—	—	—	—
Fuel injector cleanliness (max): method 1	% Flow loss	D5598	10	5	5	5	5
Fuel injector cleanliness (max): method 2	% Flow loss	D6421	10	10	10	10	10
Particulate size distribution		ISO 4406	—	—	18/16/13	18/16/13	18/16/13
Intake valve sticking		CEC F-16-T	—	Pass	Pass	Pass	Pass
Intake valve cleanliness (min)	Merit	CEC F-04-A	9.0	—	—	—	—
Intake valve cleanliness (max)	mg/valve	CEC F-05-A	—	50	30	30	30
Intake valve cleanliness (max)	mg/valve	ASTM D5500	—	100	50	50	50
Intake valve cleanliness (max)	mg/valve	ASTM D6201	—	90	50	50	50
Combustion chamber deposits (max)	% of base fuel	ASTM D6201	—	140	140	140	140
Combustion chamber deposits (max)	mg/engine	CEC-F-20-A-98	—	3,500	2,500	2,500	2,500
Combustion chamber deposits (max)	% Mass @ 450°C	TGA-FLTM BZ154-01	—	20	20	20	20

Note: [a]ND: not detected.

take cognizance of the increasing demand for fuels to contain lower levels of aromatics. One aspect of this is that aromatics have a high octane number, so this specification limits the ability to boost octane using aromatic components such as toluene.

Many conventional (petroleum derived) fuels can be treated with chemical additives such as detergents and antioxidants to achieve the requirements to maintain clean burning characteristics of a finished gasoline. For fuels derived from renewable sources and containing components other than those in conventional gasoline, it will be important to verify that good combustion character can be maintained.

For the higher grades, the sulfur level is very low and down to 10 ppm for categories 4 and 5. It is important to note that the testing point for the fuels is at the forecourt when the fuel is dispensed to the customer. Since sulfur is ubiquitous in the environment, this effectively means that no sulfur should be present in the produced gasoline, that is below 5 ppm. Because sulfur is ubiquitous and is used in many agricultural operations, renewable fuels have to be produced sulfur free and steps have to be taken to achieve this.

8.2.1.1. *Volatility*

Gasoline is a volatile product. The volatility is an important characteristic of a finished gasoline, and the character will depend upon the ambient temperature which is illustrated in Table 8.3.

Volatility of gasoline outside the appropriate class specification results in poor starting and rough running resulting in customer complaints. This can be a major issue for renewable fuels, and components produced with high vapor pressure will have limited use.

8.2.1.2. *Solvation power*

Many chemical components containing functional groups have high solvation power. Such materials:

• Attack seals in the fuel distribution system causing fuels leaks and possible fires,
• Mobilize sediment in the fuel tank and lines causing blockages in filters and injectors or erosion in pumps and injectors,

Table 8.3. Gasoline volatility requirements (from WWFC).

Class	A	B	C	D	E
Ambient temperature (°C)	>15	5–15	−5 to 5	−5 to −15	<−15
Vapor pressure (kPa)	45–60	55–70	65–80	75–90	85–105
T_{10} (°C) (max)[a]	65	60	55	50	45
T_{50} (°C)	77–100	77–100	77–100	77–100	65–100
T_{90} (°C)	130–175	130–175	130–175	130–175	130–175
End point (°C) (max)	205	205	205	205	205
E_{70} (%)[b]	20–45	20–45	20–47	25–50	25–50
E_{100} (%)	50–65	50–65	50–65	55–70	55–70
E_{180} (%) (min)	90	90	90	90	90
Driveability index (DI) (max)[c]	570	565	560	555	550

Note: [a]T_n is the temperature in °C at which n% of the gasoline is distilled.
[b]E_m is the volume percent evaporated at m °C.
[c]Driveability index = $1.5 \times T_{10} + 3 \times T_{50} + T_{90}$.

- Strips or stains paintwork if spilled during filling operations,
- Can result in chemical attack on advanced engineering alloys, especially titanium alloys.

Because of these issues, components of high solvation power are opposed by the OEMs. This could affect furan and furan derivatives made from renewable sources.

8.2.2. *Diesel*

The corresponding WWFC diesel fuel categories are given in Table 8.4.

Cetane number is the equivalent of octane number for diesel engines. However, unlike octane, it is costly to measure and cetane index which is determined by the physical properties of the fuel is commonly used instead. Unfortunately, cetane index is only applicable to hydrocarbon (conventional) fuels and cannot usefully be used if nonhydrocarbons are present which can be a particular problem for renewable fuels.

In broad terms, the larger the diesel engine, the poorer the diesel fuel that can be used in terms of cetane. The dieselization of the small vehicle

Table 8.4. WWFC categories for diesel fuel.

Category	Units	ASTM or test method	1	2	3	4	5
Cetane number (min)	—	D613	48.0	51.0	53.0	55.0	55.0
Cetane index (min)	—	D4737	48.0 (45.0)	51.0 (48.0)	53.0 (50.0)	55.0 (52.0)	55.0 (52.0)
Density @ 15°C (min)	kg/m³	D4052	820	820	820	820	820
Density @15°C (max)	kg/m³	D4052	860	850	840	840	840
Viscosity @ 40°C (min)	mm²/s	D445	2.0	2.0	2.0	2.0	2.0
Viscosity @ 40°C (max)	mm²/s	D445	4.5	4.0	4.0	4.0	4.0
Sulfur	mg/kg	D5453, D2622	2,000	300	50	10	10
Trace metal (max)	mg/kg	D7111	—	1 or ND	1 or ND	1 or ND	1 or ND
Total aromatics (max)	% m/m	D5186	—	25	20	15	15
PAH (di+, tri+)	% m/m	D5186	—	5	3.0	2.0	2.0
T_{90} (max)	°C	D86	—	340	320	320	320
T_{95} (max)	°C	D86	370	355	340	340	340
Final boiling point (max)	°C	D86	—	365	350	350	350
Flash point (min)	°C	D93	55	55	55	55	55
Carbon residue (max)	% m/m	D4530	0.30	0.30	0.20	0.20	0.20
Cold filter plugging point (CFPP), low-temperature flow test (LTFT), or cloud point (CP) (max)	°C	D6371, D4539, D2500	<Ambient	<Ambient	<Ambient	<Ambient	<Ambient
Water (max)	mg/kg	D6304	500	200	200	200	200
Oxidation stability (method 1) (max)	g/m³	D2274	25	25	25	25	25

(*Continued*)

Table 8.4. (*Continued*)

Category	Units	ASTM or test method	1	2	3	4	5
Oxidation stability (method 2a) (min)	h	EN15751	30	35	35	35	—
Oxidation stability (method 2b) (max)	mg KOH/g	D644, D22274	0.12	0.12	0.12	0.12	—
Oxidation stability (method 2c) (min)	min	EN16091	60	65	65	65	—
Foam volume (max)	ml	NF M 07-075	—	—	100	100	100
Foam vanishing time (max)	s	NF-M 07-075	—	—	15	15	15
Biological growth	—	NF M 07-070	—	None	None	None	None
FAME (max)	% v/v	D7371	5	5	5	5	None
Ethanol/methanol	% v/v	D4815	None	None	None	None	None
Total acid number (max)	mg KOH/g	D664	—	0.08	0.08	0.08	0.08
Ferrous corrosion	—	D665	—	Light	Light	Light	Light
Copper corrosion	Rating	D130	Class 1	Class 1	Class 1	Class 1	Class 1
Ash (max)	% m/m	D482	0.01	0.01	0.01	0.001	0.001
Particulate contamination (max)	—	D6217, D7321	10	10	10	10	10
Particulate size distribution	—	D7619	—	18/16/13	18/16/13	18/16/13	18/16/13
Appearance	—	D4176	Clear	Clear	Clear	Clear	Clear
Injector cleanliness, method 1 (max)	% Air flow loss	CEC (PF-023) TBA	—	85	85	85	85
Injector cleanliness, method 2 (max)	% Power loss	CEC-F-098	—	—	—	2	2
Lubricity (high-frequency reciprocating rig (HFRR) wear scar dia. @ 60°C) (max)	Micron	D6079	460	460	460	400	400

fleet (engine capacities 2 l or less) has resulted in higher cetane requirements. This is reflected in the specifications where a progressive increase in cetane is required. This is probably not required for large prime mover (20 l capacity) engines.

Like advanced gasoline engines, advanced diesel engines require no sulfur in the fuel and the comments made above regarding the ubiquitous nature of sulfur in the environment apply to diesel and to diesel from renewable sources. Removal of sulfur results in poor lubricity, which is important for diesel engine pumps. However, this can be ameliorated with additives.

The components required for high cetane stand opposite to those for high octane. Linear paraffins are preferred components, and aromatics and olefins should be avoided. However, an overriding consideration is the flash point which is closely monitored and specified for safety reasons. Blending of components of different flash points is nonlinear with a resultant mixture tending to have the flash point near to the flash point of the component with the lowest value. Renewable components with low flash point are unacceptable as a diesel blendstock. This is possibly the major reason for the exclusion of ethanol and methanol from diesel fuels in the WWFC categories.

It is important in high category fuels to minimise solid material. This is typically not an issue with gasoline since the components are generally distilled. But a renewable material of higher boiling point and acceptable flash point for diesel blending often decomposes on distillation. Biodiesel components are often purified by water washing and decanting and filtering. Inefficiencies in these processes can result in particulate and trace metal components contaminating the blendstock.

Note that tests for foaming are added in the higher categories. How renewable fuels would fare under this requirement is unknown.

8.3. How do renewable fuels fit into the standards

8.3.1. *Gasoline*

The use of ethanol as a gasoline-blending component illustrates the key points.

Ethanol has been used as a vehicle fuel since the invention of the Otto (gasoline) engine. However, since the introduction of cheap and widely available petroleum fuels, its use as a motor fuel has greatly reduced. In some jurisdictions, high levels of ethanol are used, for example 85% ethanol mixed with 15% gasoline, commonly called E85. In general, these require special engines for the motor vehicle. For the most part, we are concerned with the introduction of ethanol into the fuel as a blend component at a much lower level which can be used by the majority of gasoline vehicles without adaption of the engine.

Ethanol has a high RON and so is a useful component in 91 RON and 95 RON grades. However, it has a poor MON, and this could limit its applicability to the 98 RON grade. It contains a high level of oxygen which reduces its calorific value and hence reduces the fuel economy. Ethanol fuels containing significant quantities of ethanol (10% ethanol, E10) are generally sold cheaper than fuels that do not contain this additive.

There are several other issues which detract ethanol as a "drop-in" fuel, that is, compromise its fungibility with hydrocarbon gasoline:

1. It has a high vapor pressure which results in volatility issues.[3] This is exacerbated by anomalous vapor-blending curves — a small amount of ethanol results in a large increase in vapour pressure and hence an increasing tendency to cause vapour lock and poor running. This can be overcome by blending less volatile hydrocarbon components but is costly for the refiner as he/she has to find alternative uses for them.
2. Ethanol is miscible with water and is very hygroscopic. Contact with water results in separation of components into a hydrocarbon phase and an ethanol/water phase. To get round this issue, ethanol blends are handled in dry distribution systems which, since most refineries have water present throughout the refining process, requires ethanol to be blended at the last stage in the distribution system and stored in water-free tanks on the retail forecourt.
3. Ethanol is easily oxidized to acetic acid. This occurs when ethanol comes into contact with air (wine into vinegar) and is accelerated by

---

[3] WWFC contains extensive discussion on the volatility of ethanol blends.

certain bacteria. Oxidation increases corrosion rates, especially in storage tanks which can lead to fuel leakage into the environment.
4. Ethanol is an excellent solvent and as well as attacking seals can attack paintwork.

A common approach to use ethanol fuels is to produce a separate standard for ethanol blends. Generally an ethanol blend is produced by mixing fuel ethanol produced to a separate standard (e.g., ASTM D4806) and a gasoline that meets the gasoline standard. However, following this method, the resulting gasoline could have too high a vapor pressure and result in high fugitive emissions. On the retail forecourt, the resulting ethanol blend is stored separately from other fuels and explicitly labeled.

One development to avoid the compatibility (fungibility) issues is the development of renewable routes to butanol which has a lower vapor pressure than ethanol and is more fungible with conventional gasoline.

8.3.2. *Diesel*

The introduction of renewable fuels into diesel fuel is illustrated by the use of fatty acid methyl esters, commonly referred to as FAMEs (or biodiesels), into the fuel.

Vegetable oils and animal fats comprise fatty acid esters of glycerol. The fatty acids are mixed with generally 15–18 carbon atoms and can be unsaturated or saturated. Although there exists a body of enthusiastic supporters for using these as diesel fuel, oils and fats are not recommended by OEMs because the high viscosity causes pump wear and poor injector performance. The preferred blendstock is FAME which is produced by transesterification of the glycerol esters with methanol to produce the methyl ester of the fatty acids and glycerol as a by-product.

The mixed FAMEs cover a wide range of composition which is dependent on the source vegetable oil or fat. Also geographic, climatic, and seasonal issues can significantly modify the final composition from any given source, e.g., palm oil. In general, saturated or slightly unsaturated fatty acids are preferred because poly unsaturated oils are easily oxidized. Furthermore, because the production of FAME does not involve a distillation step, the product can be contaminated with fatty acids themselves, glycerol, and particulates passing through the transesterification process. Because of this variability, FAME for diesel blending is produced to a standard (e.g., ASTM D6751) which quantifies the quality of the FAME and sets limits to the quantum of impurities.

FAME (as biodiesel) can be used in diesel engines, but as in the case with ethanol, the engine requires modification, and so it is not a "drop-in" fuel. FAME will not meet the WWFC diesel category standards, especially with regard to density and viscosity. However, it is generally agreed that small amounts of FAME (e.g., 5% or 7%) can be blended into an excess of conventional diesel without issue, and the blend is referred to as B5 or B7. FAME has a lower calorific value than pure hydrocarbon diesel, but at this level of blend, there is little difference in energy content of the finished product and price differentiation is not warranted.

In some jurisdictions, higher concentrations of FAME exist (e.g., 20% FAME or B20). These higher concentrations require separate fuel standards, and the use of these higher concentrations is not universally supported by the OEMs.

8.4. Bio-refineries — green diesel

Neither ethanol nor FAME is drop-in fuel, and special standards are required for ethanol–gasoline and FAME–diesel blends. FAME can only be used is small amounts in conventional fuels, and again higher concentrations (e.g., B20) require special standards.

Because of this, there is an increasing interest in producing a renewable drop-in fuel or blendstock that does not have the drawbacks of ethanol or FAME. This is the basis of the biorefinery which converts vegetable oil and/or animal fats (and certain other renewable feedstock) into a drop-in fuel, particularly diesel fuel often referred to as "green diesel" (see Chapter 7).

A typical biorefinery uses vegetable oil as feedstock, which is available in sufficient production volume such as palm oil. Production is typically 500 kt/y or about 10,000 bbl/d. The biorefinery uses conventional oil-refinery processes (or adapted processes) to convert the triglyceride esters into naphtha (for gasoline blending), jet fuel, and diesel. The carboxylic groups of the ester are fully reduced or eliminated as carbon dioxide with the remaining C_3 unit of the glycerol being fully reduced to propane.

This process is generally carried out in two steps. In the first step, the vegetable oil is hydrogenated to produce a mixed hydrocarbon stream, carbon dioxide, and water; the hydrocarbon stream is separated. In the second step, the hydrocarbon stream is again mixed with hydrogen and the hydrocarbon stream is passed through an isomerization (ISOM) unit. This has the effect of turning the dominantly linear paraffins produced from the vegetable oils into branched isomers. This considerably increases the octane value of the naphtha (gasoline) stream and produces a jet fuel and diesel with good cold-flow properties. The typical properties of the products are compared in Table 8.5.

Note the green diesel is a good comparison with conventional ULSD. The major issue with green diesel, and other synthetic diesel fuels with high aliphatic content such as those produced by the Fischer–Tropsch process, is the low density when compared to the WWFC requirements of a minimum of 0.82 kg/m^3. In some jurisdictions, there is pressure to accept these low-density fuels as special clean burning fuels. The alternative is to either use them as cold weather (alpine) diesel fuels where low density is often accepted or use them as high-quality blendstock to bring inferior fuel into specification. Inferior fuel generally has a high level of aromatics which have high density but are poor in cetane. Note the very high cetane number of green diesel helps in this regard.

Table 8.5. Properties of mineral diesel, biodiesel and green diesel.[4]

	ULSD[a]	Biodiesel[b]	Green diesel
Renewable content	0%	100%	100%
Oxygen content	0%	11%	0%
Density	0.84	0.88	0.78
Calorific value (MJ/kg)	43	38	44
Cloud point (°C)	−5	−5 to +15	−20
Distillation range (°C)	200–350	340–355	200–320
Polyaromatics (wt.%)	11	0	<2
Cetane number	51	50–65	70–90
Oxidation stability	Baseline	Poor	Excellent

Note: [a]ULSD: ultra-low-sulfur diesel.
[b]FAME.

[4] Adapted from ENI promotional literature for Venice Biorefinery.

8.4.1. *Existing refineries*

It is not necessary to build a dedicated biorefinery to achieve this result. Conventional refinery operations can be adapted to achieve a similar result. The unit operations in a modern refinery are very large when compared to a typical biorefinery, so the renewable feedstock could be mixed with conventional stocks to produce a product containing a known quantum of renewable component. The way this is best achieved is in refineries which contain a hydrocracking unit because this operation generally contains a catalyst system which will convert the triglyceride esters into hydrocarbons, carbon dioxide, and water. Hydrocrackers generally operate at high temperatures and high hydrogen pressures with acidic catalysts, and ISOM of the hydrocarbon chain also occurs.

8.5. Future developments

There is also a move in many jurisdictions to legislate for tightening vehicle emission standards for NO_x and particulates. In effect, this means tightening fuel specifications toward a maximum of 10 ppm sulfur for both gasoline and diesel which assists in NO_x and particulate control. This effectively means zero sulfur in the fuel. Lowering the quantum of aromatic components also helps in controlling particulate emissions.

Furthermore, there is an increasing tendency to legislate maximum greenhouse emissions (particularly carbon dioxide and methane). In effect, this means improving fuel economy. In some ways, this objective goes counter to the NO_x and particulate measures in that fuel economy is improved in the case of gasoline by higher octane, which is enhanced by aromatics and olefins and by increasing the density of diesel fuel, and further improved by increasing the aromatics' content.

How this will turn out in future years is unknown, but whatever be the final approach, there will be an increase in demand for branched aliphatic molecules for both gasoline and diesel to obtain sufficient octane for the gasoline pool and a diesel with acceptable cold-flow properties.

For gasoline, reducing aromatics and olefins results in an octane shortfall for many refineries. In most jurisdictions, this is assisted by the use of ethers, particularly methyl *tert*-butyl ether (MTBE) and the ethanol derivative, ethyl *tert*-butyl ether (ETBE). In jurisdictions where ethanol is the preferred octane booster, there may be issues with obtaining higher octane grades.

This octane shortfall is now sparking renewed interest in octane enhancers other than ethers or alcohols. In part, this is being helped by attempts by oil majors to formulate aviation gasoline without having to resort to lead additives. This means that chemicals are being tried which might (more probably will) be unacceptable to the general vehicle gasoline fleet. Two such additives are *n*-methyl aniline and *secondary*-butyl acetate which have been banned in some jurisdictions.

8.5.1. *Issues for renewable fuels*

As witnessed by the use of ethanol and biodiesel (FAME) blends, nonfungibility causes problems for the OEM and the logistics of fuel distribution — there is limited space on retail fuel forecourts for selling multiple grades and additives which have fungibility issues.

Also, marketing of a new fuel or additive can be problematic. In some jurisdictions, despite years of use and price discounts, many general motorists (customers) take exception to ethanol blends which are avoided if at all possible. Renewable fuels should avoid this trap.

The preferred renewable fuel or additive would be a drop-in type with complete interchangeability (fungibility) with conventional fuels that does not compromise on the pertinent fuel standard.

8.5.2. *Oxygenates*

As noted above, as fuel standards tighten, there is likely to arise an octane shortfall especially for higher (95 RON) gasoline. This gives an opportunity to develop octane boosters which offer the potential of large tonnage manufacture from renewable sources for widespread use in fuels, provided fungibility and vehicle incompatibility issues can be avoided.

Chapter 9

Spent Hydroprocessing Catalysts Management

Changyan Yang*,† and Bo Zhang*,‡

*School of Chemical Engineering and Pharmacy,
Wuhan Institute of Technology, Hubei, China
†Hubei Key Laboratory for Processing and Application of Catalytic
Materials, Huanggang Normal University, Hubei, China
‡Corresponding author: bzhang_wh@foxmail.com

Abstract

As the energy demand continues to rise and the environmental regulations are more stringent, both the use of refinery catalysts and the quantity of spent hydroprocessing catalysts are increasing. When the performance of the catalysts is unable to meet the desired level, they will be unloaded as the spent catalysts, which might be subject to a series of treatments. If the spent hydroprocessing catalysts meet the requirements of the regenerability, they are regenerated via oxidative regeneration and resulfided or reactivated via the rejuvenation process that combines the thermal and chemical treatments. When the catalytic activity cannot be restored, metals are recovered from the spent catalysts, and remaining materials can be used to prepare other useful products. Finally, the residue of spent hydroprocessing catalysts is stabilized and disposed by using an environmentally sound method. The goals of the treatment of spent hydroprocessing catalysts are always to maximize the life of the catalysts and minimize the generation of hazardous wastes.

Keywords: Spent hydroprocessing catalyst, Deactivation, Regulations, Transportation, Regeneration, Rejuvenation, Metal reclamation, Land disposal

9.1. Introduction

Due to the development of unconventional oil and gas reservoirs, shale oil is becoming a major feed for many refineries. Meanwhile, biofeedstock such as vegetable oil and fats have been corefined in some oil refineries. The markets for hydrotreating catalysts and total refinery catalysts in the oil refining sector are expected to reach projected market values of $2 billion and $6 billion by 2019, respectively [1]. The major driving force for these markets includes the surging energy demand and stringent environmental regulations [2]. The increased use of refinery catalysts can help refiners meet fuel standards, manage operational efficiency, and enhance conversion and selectivity. Accordingly, the total quantity of spent hydrotreating catalysts generated worldwide was estimated to be in the range of 150,000–170,000 t/y in 2007 [3]. With an anticipated 5% annual increase in catalyst consumption, the generation of spent hydro-processing catalysts was predicted to be 200,000 t annually [4].

9.2. Deactivation of hydroprocessing catalysts

Hydroprocessing catalysts generally include hydrotreating and hydroc-racking catalysts. Hydrotreating catalysts and most commonly used cata-lysts are molybdate supported on alumina (Al_2O_3) and promoted by cobalt (Co), nickel (Ni), or tungsten (W), while hydrocracking catalysts are bifunctional, consisting of active metals (like Mo, Pt, Ru) supported on a zeolite (e.g., ZSM-5). These catalysts are normally used in the fixed bed applications. The catalyst life varies for different applications: 1–2 y for hydroprocessing of atmospheric gas oils or vacuum gas oils, 0.5–1 y for hydrotreating resid, 5–10 y for a naphtha hydrotreater using straight run feed [3], and ~2 y for hydrotreating biofeedstock [5].

Table 9.1. Characteristics of the spent commercial catalysts. (Reproduced with permission from Ref. [8]. Copyright © 1996 Elsevier.)

Type	CoMo	NiMo	NiMo	CoMo
Metals amount	Medium	Medium	High	Medium
Unit feed	AGO	AGO	VGO	AGO
C (wt.%)	5.7	7.5	15.8	10.9
S (wt.%)	6.7	7.5	9.3	10.4
LOI (wt.%)	12.4	16.3	19.8	16.6

Note: AGO: atmospheric gas oil, VGO: vacuum gas oil, LOI: loss on ignition.

Typically, deactivation of hydroprocessing catalysts is due to coking, sintering, and contaminations (metal deposition and poisoning). Coke formation is the most common problem of a fixed bed of catalysts. If a hydrotreating catalyst with the high activity was placed at the entrance of the feed, coke deposition can occur within a matter of seconds. The carbon content on spent hydrodesulfurization catalysts largely varies between 5 wt.% and 25 wt.%, with an average of 10 wt.% [6]. Sintering may be due to the damage of active phase structure and dispersion. Contamination is caused by the adsorption of various chemicals (e.g., vanadium (V) and Ni) on the active sites [7]. Hydroprocessing catalysts can tolerate different amounts of metal deposition and poisoning: about 10 wt.% C, 2–3 wt.% Ni + V (100 wt.% Ni + V for vacuum gas oil and residue hydroprocessing catalysts), up to 15 wt.% Si, and 0.5 wt.% Pb. Table 9.1 gives the characteristics of the spent commercial catalysts unloaded from different types of hydroprocessing units [8]. Figure 9.1 shows schematically some possible structures that may be present in deactivated catalysts.

Deactivation caused by coke formation and sintering may be reversed via the regeneration process. Refining and petrochemistry catalysts reusable after regeneration can be found in the literature [3]. When the activity of a catalyst is too low to meet product specifications, the reactor is shut down and the catalyst is unloaded as the spent catalyst. Spent hydrotreating catalysts generated in the petroleum refining industry are routinely recycled by regenerating the catalyst so that it may be used again as a catalyst. When regeneration is no longer possible, these spent catalysts are

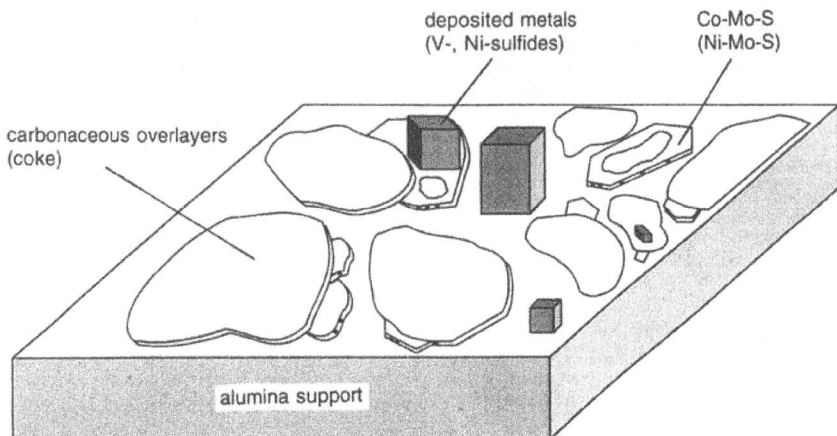

Figure 9.1.   Schematic model for the surface of a deactivated catalyst. (Reproduced with permission from Ref. [9]. Copyright © 1996 Springer.)

either treated and disposed of as listed hazardous wastes or sent to Resource Conservation and Recovery Act (RCRA)-permitted reclamation facilities, where metals, such as vanadium, molybdenum, cobalt, and nickel are reclaimed from the spent catalysts.

9.3. Regulations

In 1998, the US Environmental Protection Agency (EPA) added spent hydrotreating and hydrorefining catalysts (waste codes: K171 — spent hydrotreating catalyst from petroleum-refining operations, including guard beds used to desulfurize feeds to other catalytic reactors and K172 — spent hydrorefining catalyst from petroleum refining operations, including guard beds used to desulfurize feeds to other catalytic reactors) to the list of RCRA hazardous wastes [10], due to the facts that (a) these spent catalysts can ignite spontaneously in contact with air, exhibiting pyrophoric properties and (b) these materials were shown to pose unacceptable risk (i.e., toxicity) to human health and the environment when mismanaged [11]. This indicates that they need to be stored, handled, and disposed according to the set rules. On-site bulk pads approved by RC RA and the US Department of Transportation (DOT) approved containers have to be used for storage and transportation purposes. Certain catalyst shipments are also governed by national and state laws [12].

In EPA's final rule 2015 for definition of solid wastes, EPA has added a regulatory definition of the "contained" standard as it applied to the generator-controlled exclusion (40 CFR 261.4(a)(23)) and to the verified-recycler exclusion (40 CFR 261.4(a)(24)) [11]. This new definition includes a requirement to address the risk of fires and explosions. This provision addresses the pyrophoric properties of the spent petroleum catalysts as well as other types of ignitability or reactivity for the purposes of the generator-controlled exclusion and the verified-recycler exclusion. Therefore, EPA has conditionally excluded spent hydrotreating and hydrorefining catalysts generated in the petroleum refining industry from the definition of solid waste when these hazardous secondary materials are reclaimed [13].

9.4. Transportation of the spent hydroprocessing catalysts

Although EPA believes that the spent hydroprocessing catalysts have pyrophoric behaviors, the spent hydroprocessing catalysts more qualify for self-heating properties that are defined by the United Nations (UN) self-heating test [14]. Pyrophoric materials are liable to ignite spontaneously on exposure to air. Spent catalysts are not pyrophoric, except the special cases in which the top of a reactor is heavily contaminated by iron in the form of iron sulfide, which can result in a spontaneous combustion under air. Most spent hydroprocessing catalysts fall under that category of self-heating substances and are classified under the Class 4.2, UN no. 3190 category, as "inorganic, self-heating substances". The ability of a substance to undergo oxidative self-heating is determined by exposure of it to air at temperatures of 100°C, 120°C, or 140°C in a 25 mm or 100 mm wire mesh cube. The material is said to be of self-heating category if the temperature exceeds 200°C at one moment during a 24-h test. Consequently, the spent hydroprocessing catalysts need to be stored and transported in solid and tight packaging, such as drums and metallic bins [15]. Further, to ensure the safe handling of all products and wastes, the workers involved in the production, packaging, and transportation should use respiratory protection. There is a tendency to use larger packaging (e.g., big bags and containers), enabling faster reactor loading and minimizing packaging costs.

9.5. Options for the spent hydroprocessing catalysts

Figure 9.2 shows a scheme of the catalyst life cycle [3]. When the performance of the catalysts cannot meet the desired level, they will be unloaded. The special precautions (e.g., under a vacuum) must be taken during the catalyst withdrawal. Otherwise, it may result in more hazardous characteristics of spent hydroprocessing catalysts, such as corrosiveness and flammability. The unloaded catalyst will be examined for its regenerability. The regenerated spent hydroprocessing catalysts will be re-evaluated. Catalysts of a good quality are pooled and resulfided for reusing, while catalysts of a lower quality than the one specified for the pool could be reused in less critical applications. At the end of the catalyst life, metals in the spent catalyst will be recovered and the residues are disposed by using an environmentally sound method.

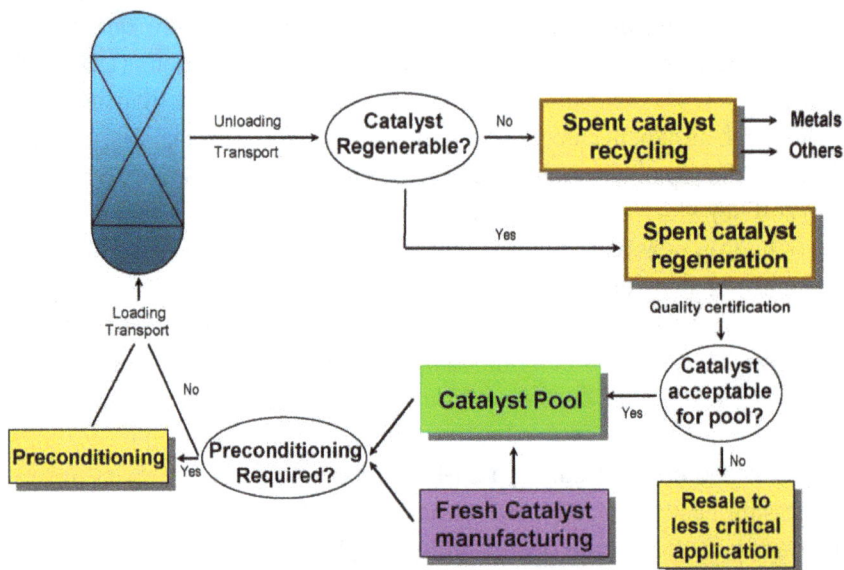

Figure 9.2. Hydroprocessing catalyst life cycle. (Reproduced with permission from Ref. [3]. Copyright © 2007 Elsevier.)

9.5.1. *Regeneration*

Oxidative regeneration is mainly used to burn off the coke deposited on the spent hydrotreating catalysts and transform metal sulfides back into their oxides. Yoshimura and Furimsky found the removal of sulfur and carbon reached 80 and 20%, respectively, when burning the sulfided spent CoMo catalyst at 250°C [16]. A more complete removal requires burning at a high temperature of ~450°C. Oxidation of sulfidic sulfur and organic sulfur resulted in the release of SO_2 at 250°C and 450°C. Oxidation of carbon happens around 450°C too, and MoO_3 species participate in oxidation reactions [17]. Their further study on the spent NiW catalysts showed the removal of sulfidic sulfur as SO_2 around 227–327°C and the removal of carbon as CO_2 or CO around 377–577°C [18]. More severe oxidation conditions were needed for the NiW catalyst because of the lower oxidation activity of NiO and WO_3 compared to Co_3O_4 and MoO_3. Oxidative regeneration could recover the metal species on the surface, but the structural properties cannot be recovered to the same level as the fresh catalysts. For example, Mo can only be partially redispersed through solid–solid wetting of MoO_3 on γ-Al_2O_3. The reaction mechanism of the oxidative regeneration of CoMo and NiW catalysts [17] is summarized in Table 9.2.

This process is affected by multiple factors including O_2 supply, the composition of the coke, and mass transfer. Air generally provides a good

Table 9.2. Reactions during the oxidative regeneration.

Organic carbon	$C + \frac{1}{2}O_2 \rightarrow CO$
Organic carbon	$C + O_2 \rightarrow CO_2$
Organic hydrogen	$H + \frac{1}{4}O_2 \rightarrow \frac{1}{2}H_2O$
Sulfided Mo	$MoS_2 + O_2 \rightarrow MoO_3$
Sulfided W	$WS_2 + O_2 \rightarrow WO_3$
Sulfided Co	$Co_9S_8 + O_2 \rightarrow MoO$
Sulfided Ni	$Ni_3S_2 + O_2 \rightarrow NiO$
Organic sulfur	$S + O_2 \rightarrow SO_2$
Organic nitrogen	$N + O_2 \rightarrow NO_x$

O_2 source. However, uncontrolled temperature increase may lead to sintering of the catalyst. Under certain circumstances, diluted air may be more suitable. Massoth and Menon discovered that the deposits with a high H/C ratio was removed first and concluded that the removal of hydrogen from deposits was more rapid than that of carbon [19]. In addition, the form of the catalysts (pellets or extrudates) established the mass transfer limitations, which may be minimized by grinding the deactivated catalyst to a fine particle size [20, 21].

Deactivation due to sintering, segregation, and metal deposition can only be restored by oxidative regeneration to a limited extent, which is largely dependent on the degree of sintering and the amount of contaminants present. Typically, the spent catalyst should not contain more than 2–3 wt.% of metal contaminants (mostly Ni and V) to be suitable for oxidative regeneration [22].

The oxidative regeneration can be carried out *in situ* or *ex situ*. The conventional *in situ* technique burns off the coke and resulfides the catalyst in the hydrotreating reactors. Regeneration is performed by injecting a stream of diluted air with nitrogen or steam and removes coke and reversible poisons like sulfur and nitrogen by oxidizing them at temperatures between 450°C and 550°C into gaseous CO, CO_2, SO_x, and NO_x. The process parameters such as air concentration and temperature are carefully controlled to prevent runaway combustion [23]. Sulfiding converts the metal oxides impregnated onto the catalyst support into the corresponding metal sulfides and forms H_2S that can enter the process water to be removed. According to environmental regulations, acidic gases must be neutralized, and the neutralization process is time consuming and requires injection systems, trained operators, and time. In addition, *in situ* regeneration requires long unit downtime and gives poor activity recovery due to uneven gas flow [24].

The *ex situ* (i.e., off-site) oxidative regeneration has been widely accepted by the petroleum refining industry in 1990s, because it provides benefits on safety, time savings, and less environmental problems caused by generating SO_x and CO_x. Better activity recovery can be achieved by the *ex situ* regeneration, because it allows to perform more than one cycle with the same catalyst batch [8]. The reactor corrosion due to the formation of acidic gases is eliminated. The chance for accidents, hot spots, and reactor malfunction is lower. Dedicated catalyst specific regeneration procedures can be applied and the fines can be removed by screening. The costs of the oxidative *ex situ* regeneration are around 20% of the fresh catalyst price [7].

9.5.2. *Industrial regeneration process*

Companies offering the *ex situ* regeneration service include Eurecat and Porocel, which are using rotating kiln and belt oven technologies, respectively. Figure 9.3 shows a simplified process scheme of the Eurecat regeneration process, which uses a Roto-Louvre oven enabling an excellent contact between gas and solids [25]. The air fed to the oven is first preheated through a heat exchanger, and a gas burner serves as the secondary heater. The exhaust gases from the oven are filtered to remove fines and finally scrubbed by soda to remove sulfur dioxide before releasing. Various grading or physical separation equipment is required to handle different spent catalyst mixtures before or after the regeneration. Separation of particles of various diameters is also often necessary.

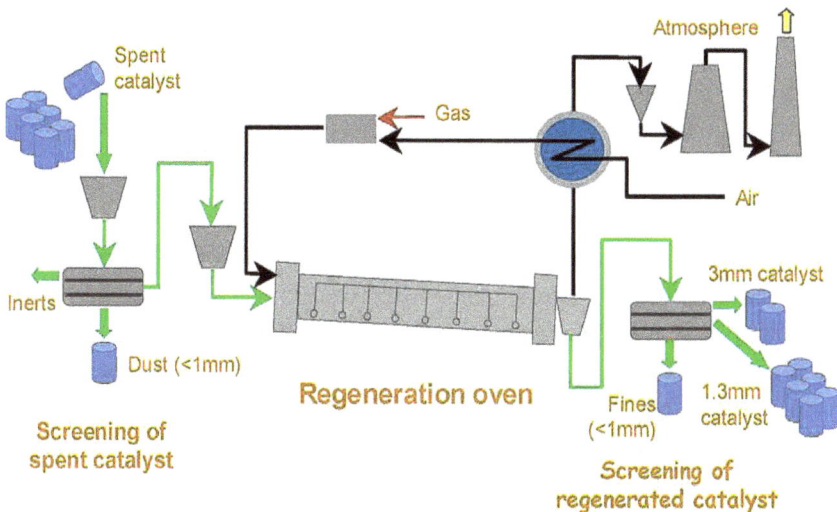

Figure 9.3. Simplified process scheme of the Eurecat regeneration process. (Reproduced with permission from Ref. [3]. Copyright © 2007 Elsevier.)

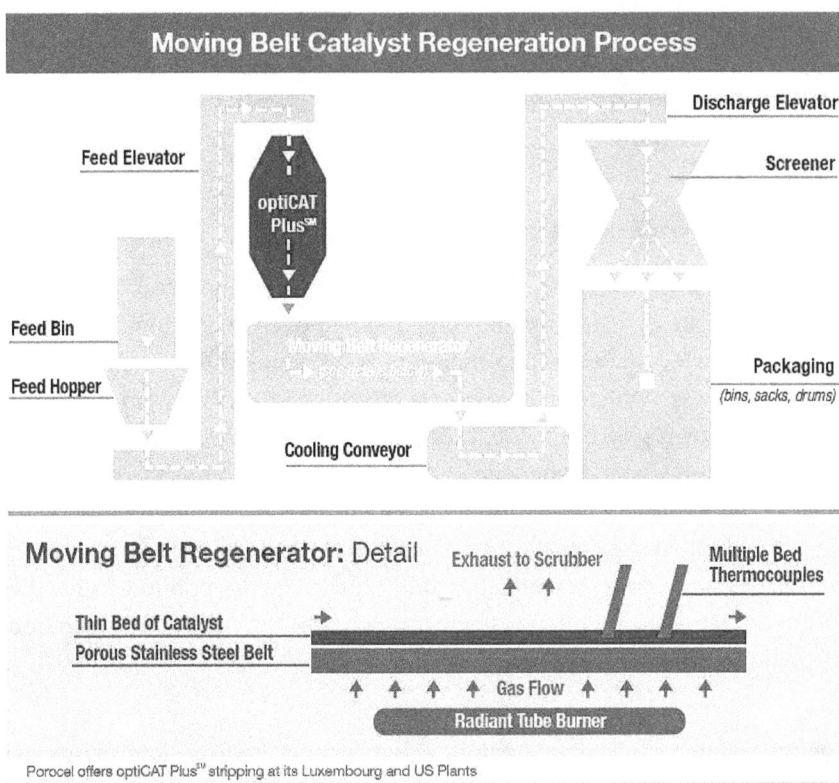

Figure 9.4.   Simplified process scheme of the Porocel regeneration process. (Available at http://www.porocel.com/userfiles/image/Catalyst%20Regen%20Image.JPG, Public domain.)

According to Porocel's statement, it has a 30,000 t/y catalyst regeneration capacity [26]. A pretreatment of hydrocarbon stripping (called opti-CAT Plus^{SM}) is applied prior to the regeneration. The pretreatment process significantly reduces the amount of coke deposited and removes reactive sulfur. Porocel's moving belt regeneration technology transports a thin bed of material through multiple regeneration zones on a porous, stainless steel belt (Figure 9.4). Radiant tube burners heat the air and combustion gasses that flow through the catalyst bed, removing the remaining carbon and sulfur embedded deep within the catalyst or adsorbent pores. Several plows along the regeneration belts gently turn over the catalyst to ensure it regenerates evenly. Multiple thermocouples in each zone monitor the bed temperatures. The company claims that the advantages offered by this combined process include excellent heat transfer; temperature control; minimized residence time and attrition; minimized problems of channeling, hot spots and sintering; and improved yields [27].

9.5.3. *Reusing of the spent catalysts*

Reuse of a spent catalyst after regeneration is preferred and will maximize the life of a catalyst. Thus, the *ex situ* regeneration is a better solution. The companies doing catalyst regeneration business might have multiusers of the same type of catalysts. For example, the regenerated catalysts may be suitable for less demanding refinery operations. Typically, the regenerated gas oil hydrotreating catalyst might be used for hydrotreating of kerosene, and the regenerated kerosene hydrotreating catalyst can be applied for naphtha hydrotreating [28].

It's also possible to use at least a small portion of spent catalysts for preparation of useful materials like fused alumina, Anorthite glass-ceramics, and abrasive material [29]. In addition, depending on the remaining porosity and surface area, spent catalysts may still have potential, especially in some gas–solid applications (e.g., they are used as a H_2S clean-up sorbent) [30]. This kind of catalyst management services can optimize the global catalyst inventory through a pool where each site or unit will take the required catalyst quantity corresponding to their need [3, 7].

9.5.4. *Rejuvenation*

Additional processes for the rejuvenation of spent catalysts, also referred to as revitalization or reactivation, may be required. Oxidative regeneration is typically performed on older generation catalysts known as type I, with recovered activity in the range of 70–85% of fresh activity, depending on the degree of metals contamination and surface area. For more recent generation catalysts containing the highly active type II sites, the oxidative regeneration of type II catalysts only results in a mediocre activity recovery. One possible reason for this low-activity recovery is that these catalysts have not been exposed to the temperatures that are required to restore their activities. The rejuvenation process developed by Porocel consists of two steps: an initial thermal regeneration to remove carbon and sulfur, followed by a proprietary chemical treatment to remove the inactive crystalline compound such as β-$CoMoO_4$ or $NiMoO_4$, redisperse the metals, and restore the type II active sites for maximum activity recovery [31]. These treatments typically use some oxygen-containing compounds that play the role of chelating agents and thus may help redisperse the metals [32]. Rejuvenation of the spent catalyst may restore greater than 90% of fresh catalyst activity, providing the spent catalyst meets certain physical and chemical criteria.

In some cases, the primary objective of the rejuvenation is to selectively leach out metals of interest from the support while leaving most of the support intact. Marafi *et al.* developed selective extraction methods using an oxalic acid–ferric nitrate reagent to extract deposited V from the spent catalyst prior to decoking [33]. Menoufy and Ahmed reported that a mixture of 4% oxalic acid and 5% H_2O_2 was the most efficient leaching solvent to facilitate the total metal removal and keep the dissolution of supports to a minimum. Characterization of the rejuvenated catalyst revealed the unchanged crystalline phase of the catalyst [34].

9.5.5. *Metal reclamation*

The spent catalyst reaches the end of cycle if the desired level of activity cannot be restored or the mechanical properties strongly deteriorate during regeneration. As the prices of metals fluctuate strongly, the exact way of disposal depends on the economics and the local environmental regulation [35]. Hydrotreating catalysts usually contain enough metals to achieve recovery, which removes toxic components and makes further disposal of residues possible [36].

Spent hydrotreating catalysts mainly contain metal sulfides and oxides supported on alumina, which are heavily coated in carbonaceous materials. There are two types of reclamation processes: hydrometallurgy and pyrometallurgy.

The hydrometallurgical reclamation involves the solubilization of metals via roasting, followed by a selective leaching of metals of interest [37]. Literature regarding the hydrometallurgical reclamation of metals is

Table 9.3.   Reactions during the metal reclamation.

Roasting	$V_2O_5 + Na_2CO_3 \rightarrow 2NaVO_3 + CO_2$
Roasting	$MoO_3 + Na_2CO_3 \rightarrow NaMoO_4 + CO_2$
Precipitation	$NaVO_3 + NH_4Cl \rightarrow NH_4VO_3 \downarrow + NaCl$
Precipitation	$NaMoO_4 + CaCl_2 \rightarrow CaMoO_4 + 2NaCl$

extensive and has been reviewed by Furimsky [37] and Marafi and Stanislaus [29, 38].

During a common metal reclamation, the spent catalysts are first ground and roasted at 600°C to remove coke and produce metal oxides. The products are then roasted with sodium carbonate (Na_2CO_3) at 750°C to form soluble salts. Salts leach into water at 100°C, which is followed by adding ammonium chloride or calcium chloride to form the precipitates. Reclamation reactions are summarized in Table 9.3.

In a similar nickel-reclaiming process, the spent catalyst containing large amounts of nickel was first roasted with caustic soda at 500°C to convert its alumina to water-soluble aluminate, which can be digested by water to remove the aluminate and form the nickel residues [39].

Alternatively, a leaching with sulfuric acid, which is followed by a series of extraction processes, can also achieve an excellent separation of Mo, V, Co, Ni, Al, and Fe.

The selective leaching of metals depends on the chemistry. For example, a two-stage leaching process without roasting uses the alkali leach for MoO_3 and V_2O_5 removal and a subsequent acid leach for Ni and Co oxides removal [40]. A better separation of vanadium from molybdenum species was achieved by the use of a quaternary ammonium salts dissolved in toluene [41]. Villarreal *et al.* recovered sodium vanadate and molybdate by leaching with aqueous NH_3 or NaOH solutions after preliminary treatment of catalyst with CS_2 or organic solvents [42]. Angelidis *et al.* applied a slight alkali solution of $NaHCO_3$ for the selective dissolution of Re from a catalyst containing Pt and Re [43].

Al-Sheeha *et al.* compared three different metal recovery methods including base leaching, acid leaching, and roasting using industrial spent atmospheric resid desulfurization (ARDS) catalysts that contained high levels of metals (Figure 9.5) [44]. All three methods were optimized and

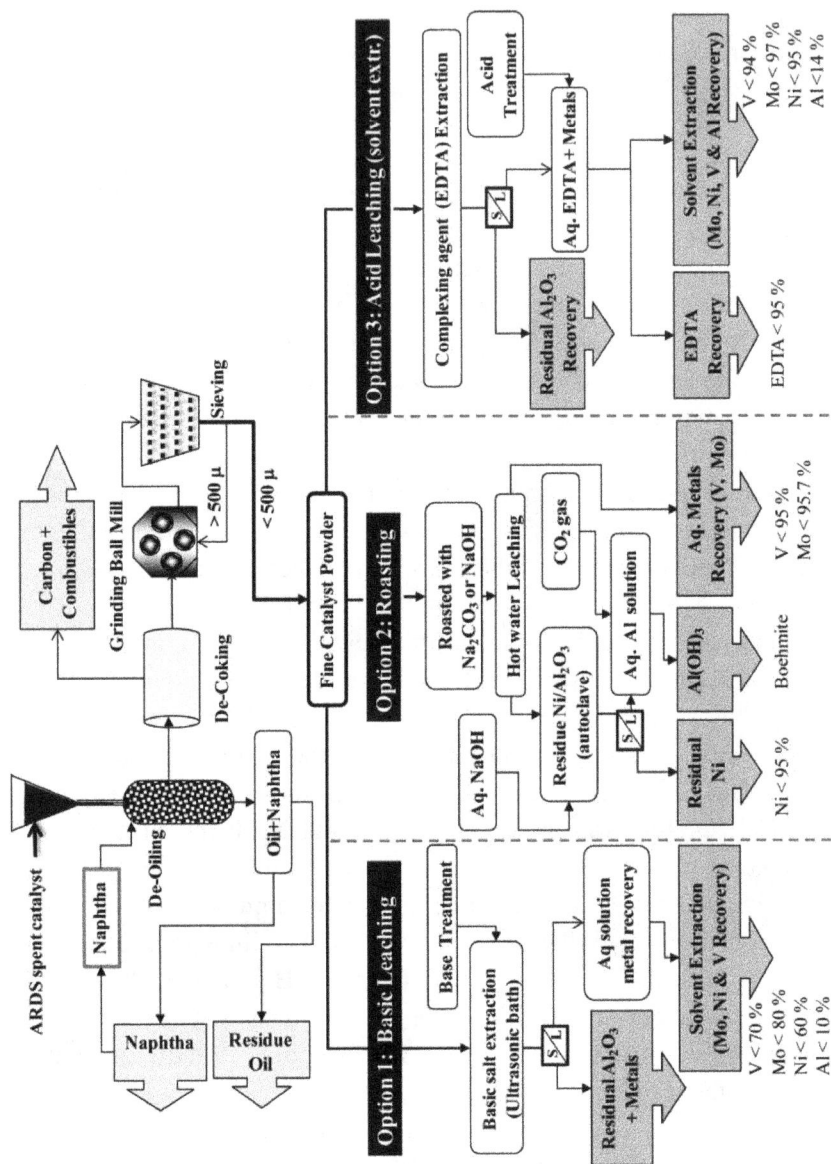

Figure 9.5. Flow diagram of spent hydroprocessing catalyst recovery processes by using three different options. (Reproduced with permission from Ref. [44]. Copyright © 2013 American Chemical Society.)

resulted in high efficiencies of metal recovery. Their research demonstrated the possibility of the total recycling process, which could reduce the disposal of hazardous wastes and provide economic benefits [45].

The pyrometallurgical process starts with melting dry catalysts in a furnace at temperatures around 1,200–1,500°C [46]. Heavy metals sink to the bottom as alloys containing the alumina or silica support, which are further separated from the slag. Pyrometallurgical processes for recovering metals (like Pt and Pd) in hydrocracking catalysts involve chlorination at high temperatures (900–950°C) for recovery of platinum and other metals as volatile chlorides, which is followed by heating at high temperatures (800°C) in a gas flow containing water [47].

The possibilities and the economic variations for metal recovery apparently depend on individual systems. Fox example, the recovery of small amounts of precious metal is probably more advantageous than large amounts of aluminum [37]. Thus, individual assessment is required. Residues after leaching operations are considered as hazardous wastes and must be handled with care. Encapsulation or fixation may be required.

9.5.6. *Land disposal*

Disposal of a spent catalyst or absorbent by means like landfill is always not an environmentally preferred option and is subject to specific and stringent regulations. Landfill does not remove or destroy any hazardous material, which may pose long-term health and environmental hazards, and could be affected by future regulatory changes [48].

Due to the flammability and leachability of the spent hydroprocessing catalyst, they must be converted into an inert nonleachable solid before being disposed as the solid waste [4]. The roasted catalyst can be encapsulated in an impervious layer of sealant such as bitumen, polyethylene, or concrete. The encapsulants must be stable over long periods of time and must withstand mechanical shear and weather changes [35]. However, some encapsulants are flammable and may undergo a long-term deterioration [22]. Additionally, the use of encapsulants adds considerable amounts of materials to the spent catalysts, thereby increasing the volume and cost of disposal.

The catalysts can also be stabilized by via chemical reactions with alumina, cement, or silicate glass at high temperatures, forming stable/nonreactive inorganic compounds (i.e., silicates and aluminates). In this way, the catalysts are encapsulated simultaneously without significantly increasing the volume and mass of the deposits, and the leachability of the hydroprocessing catalysts is also reduced [22]. The disadvantage of this stabilization method is that the particles may slowly deviate and break.

The stabilized spent hydroprocessing catalysts can be disposed in landfills. Landfills for nonliquid hazardous wastes are selected and designed to minimize the chance of release of hazardous waste into the environment. Design standards for hazardous waste landfills require double liner; double leachate collection and removal systems; leak detection system; run on, runoff, and wind dispersal controls; and construction quality assurance program [49].

Historically, about 25% of the spent hydrotreating and hydrorefining catalysts generated by refiners has been buried in landfills. But due to limitations of stabilization treatment methods and more stringent environmental regulations, there is a rapid reduction of approved landfills and the amount of catalysts disposed in this way. The potential future liability of landfills is estimated at about US$ 200/t [50].

9.6. Conclusions

As the energy demand continues to rise and the environmental regulations are more stringent, the increased use of refinery catalysts results in an increasing quantity of spent hydrotreating catalysts. When the performance of the catalysts cannot meet the desired level, they will be unloaded as the spent catalysts. The treatment processes for the spent hydroprocessing catalysts have following options:

(1) The spent catalysts can be regenerated via oxidative regeneration and resulfided if it meets the requirements of the regenerability.

(2) The spent catalysts can be reactivated via the rejuvenation process that combines thermal and chemical treatments.

(3) Metals in the spent catalyst can be recovered if the catalytic activity cannot be restored.

(4) The spent catalysts can be utilized to make other useful materials.
(5) The stabilized residues of spent hydroprocessing catalysts can be disposed in landfills.

The goals of the treatment of spent hydroprocessing catalysts are always to maximize the life of the catalysts and minimize the generation of hazardous wastes.

Recommended reading

Eijsbouts, S. (1999). Life cycle of hydroprocessing catalysts and total catalyst management. *Studies in Surface Science and Catalysis*, 127, 21–36. DOI: 10.1016/S0167-2991(99)80391-7.

Furimsky, E. (1996). Spent refinery catalysts: environment, safety and utilization. *Catalysis Today*, 30(4), 223–286. DOI: 10.1016/0920-5861(96)00094-6.

Marafi, M. and Stanislaus, A. (2008). Spent hydroprocessing catalyst management: a review: Part II. Advances in metal recovery and safe disposal methods. *Resources, Conservation and Recycling*, 53(1), 1–26. DOI: 10.1016/j.resconrec.2008.08.005.

Marafi, M., Stanislaus, A., and Furimsky, E. (2010). *Handbook of Spent Hydroprocessing Catalysts*, Elsevier, Amsterdam.

Trimm, D. L. (2001). The regeneration or disposal of deactivated heterogeneous catalysts. *Applied Catalysis A: General*, 212(1–2), 153–160. DOI: 10.1016/S0926-860X(00)00852-8.

References

1. Markets and Markets (2014). Refinery Catalysts Market worth $6,707.92 Million by 2019. http://www.marketsandmarkets.com/PressReleases/global-refinery-catalyst-market.asp.
2. Processing Magazine (2014). Refining catalyst market worth over $6 Billion. http://www.processingmagazine.com/refinery-catalysts-market-worth-over-6-billion/.
3. Dufresne, P. (2007). Hydroprocessing catalysts regeneration and recycling. *Applied Catalysis A: General*, 322, 67–75.
4. Chiranjeevi, T., Pragya, R., Gupta, S., Gokak, D. T., and Bhargava, S. (2016). Minimization of waste spent catalyst in refineries. *Procedia Environmental Sciences*, 35, 610–617.

5. Jones, S. B., Zhu, Y., Snowden-Swan, L. J., Anderson, D., Hallen, R. T., Schmidt, A. J., Albrecht, K., and Elliott, D. C. (2014). Whole Algae Hydrothermal Liquefaction: 2014 State of Technology. Pacific Northwest National Laboratory (PNNL), Richland, WA (US).

6. Dufresne, P. and Girardier, F. (1997). Off-site treatment of refining catalysts. *International Journal of Hydrocarbon Engineering*, 2, 31.

7. Eijsbouts, S., Battiston, A. A., and van Leerdam, G. C. (2008). Life cycle of hydroprocessing catalysts and total catalyst management. *Catalysis Today*, 130(2–4), 361–373.

8. Dufresne, P., Valeri, F., and Abotteen, D. S. (1996). Continuous developments of catalyst off-site regenerationand presulfiding. *Studies in Surface Science and Catalysis*, 100, 253–262.

9. Topsøe, H., Clausen, B. S., and Massoth, F. E. (1996). Hydrotreating catalysis. In: *Catalysis: Science and Technology*, J. R. Anderson and M. Boudart, eds., Springer, Berlin, pp. 1–269.

10. Treviño, C. (1998). New EPA rule will affect spent catalyst management. http://www.ogj.com/articles/print/volume-96/issue-41/in-this-issue/general-interest/new-epa-rule-will-affect-spent-catalyst-management.html.

11. US EPA (2015). Final Rule: 2015 Definition of Solid Waste (DSW). https://www.gpo.gov/fdsys/pkg/FR-2015-01-13/pdf/2014-30382.pdf.

12. Markets and Markets (2014). Global Catalyst Regeneration Market — Forecast to 2019. https://www.marketresearch.com/product/sample-8437448.pdf.

13. ASTSWMO (2015). 2015 Definition of Solid Waste Final Rule Summary Analysis. http://www.astswmo.org/Files/Policies_and_Publications/Hazardous_Waste/DSW%20Final%20Rule%20Summary%20July%202015.pdf.

14. Kallada, S. (2015). Class 4.2: Pyrophoric and Self-Heating Substances. http://www.shashikallada.com/class-4-2-pyrophoric-self-heating-substances/.

15. Flo-Bin Rentals (2014). Catalyst Bins — Welcome to Flo-Bin Rentals. http://www.cckxleasing.com/.

16. Yoshimura, Y. and Furimsky, E. (1986). Oxidative regeneration of hydrotreating catalysts. *Applied Catalysis*, 23(1), 157–171.

17. Furimsky, E. and Yoshimura, Y. (1987). Mechanism of oxidative regeneration of molybdate catalyst. *Industrial & Engineering Chemistry Research*, 26(4), 657–662.

18. Yoshimura, Y., Sato, T., Shimada, H., Matsubayashi, N., Imamura, M., Nishijima, A., Yoshitomi, S., Kameoka, T., and Yanase, H. (1994). Oxidative regeneration of spent molybdate and tungstate hydrotreating catalysts. *Energy & Fuels*, 8(2), 435–445.

19. Ozawa, Y. (1969). Regeneration of coked catalyst in adiabatic fixed beds at lower temperatures. *Industrial & Engineering Chemistry Process Design and Development*, 8(3), 378–383.
20. Massoth, F. E. (1967). Oxidation of coked silica-alumina catalyst. *Industrial & Engineering Chemistry Process Design and Development*, 6(2), 200–207.
21. Mickley, H. S., Nestor, J. W., and Gould, L. A. (1965). A kinetic study of the regeneration of a dehydrogenation catalyst. *The Canadian Journal of Chemical Engineering*, 43(2), 61–68.
22. Eijsbouts, S. (1999). Life cycle of hydroprocessing catalysts and total catalyst management. *Studies in Surface Science and Catalysis*, 127, 21–36.
23. Leprince, P. (2001). *Petroleum Refining. Vol. 3 Conversion Processes*, Editions Technip, Paris.
24. Vukovic, J. and Neuman, D. (2004). Ex-situ Catalyst Treatments Reduce Refinery Emissions and Unit Downtime: Update on the Economic and Environmental Benefits of TRICAT's Ex-situ Regeneration and Pre-activation Technologies. *5th EMEA Catalyst Technology Conference*, Cannes, France.
25. Abotteen, S. and Dufresne, P. (2005). Effective use of catalysts through catalyst regeneration. https://www.eurecat.com/pdf/effective-use-of-catalyst-regeneration.pdf.
26. Porocel (2017). Catalyst Regeneration. http://www.porocel.com/13-regeneration_catalyst_services/.
27. Porocel (2017). The Significance of optiCAT PlusSM. http://www.porocel.com/index.php?page_id=23&page_safename=opticat_plus.
28. Marafi, M. (2008). Spent Catalyst Waste Minimization and Utilization. http://thefutureenergy.org/wp-content/uploads/2018/02/2008Spent-Catalyst-Waste-Minimization-and-Utilization.pdf.
29. Marafi, M. and Stanislaus, A. (2008). Spent catalyst waste management: a review: Part I—Developments in hydroprocessing catalyst waste reduction and use. *Resources, Conservation and Recycling*, 52(6), 859–873.
30. Furimsky, E. (1997). Activity of spent hydroprocessing catalysts and carbon supported catalysts for conversion of hydrogen sulphide. *Applied Catalysis A: General*, 156(2), 207–218.
31. Porocel (2017). Catalyst Rejuvenation. http://www.porocel.com/14-rejuvenation_catalyst_services/.
32. Stanislaus, A., Marafi, M., and Absi-Halabi, M. (1993). Studies on the rejuvenation of spent catalysts: effectiveness and selectivity in the removal of foulant metals from spent hydroprocessing catalysts in coked and decoked forms. *Applied Catalysis A: General*, 105(2), 195–203.

33. Marafi, M., Stanislaus, A., and Absi-Halabi, M. (1994). Heavy oil hydro-treating catalyst rejuvenation by leaching of foulant metals with ferric nitrate-organic acid mixed reagents. *Applied Catalysis B: Environmental*, 4(1), 19–27.

34. Menoufy, M. F. and Ahmed, H. S. (2008). Treatment and reuse of spent hydrotreating catalyst. *Energy Sources, Part A: Recovery, Utilization, and Environmental Effects*, 30(13), 1213–1222.

35. Trimm, D. L. (2001). The regeneration or disposal of deactivated heterogeneous catalysts. *Applied Catalysis A: General*, 212(1–2), 153–160.

36. Trimm, D. L. (1989). Deactivation, regeneration and disposal of hydroprocessing catalysts. *Studies in Surface Science and Catalysis*, 53, 41–60.

37. Furimsky, E. (1996). Spent refinery catalysts: environment, safety and utilization. *Catalysis Today*, 30(4), 223–286.

38. Marafi, M. and Stanislaus, A. (2008). Spent hydroprocessing catalyst management: a review: Part II. Advances in metal recovery and safe disposal methods. *Resources, Conservation and Recycling*, 53(1), 1–26.

39. Tsuen-Ni, L., Jing-Chie, L., and Teh-Chung, H. (1983). Chemical reclaiming of nickel sulfate from nickel-bearing wastes. *Conservation & Recycling*, 6(1), 55–62.

40. Cotton, F. A., Wilkinson, G., Murillo, C. A., Bochmann, M., and Grimes, R. (1988). *Advanced Inorganic Chemistry*, Wiley, New York.

41. Olazabal, M. A., Orive, M. M., Fernández, L. A., and Madariaga, J. M. (1992). Selective extraction of vanadium (V) from solutions containing molybdenum (VI) by ammonium salts dissolved in toluene. *Solvent Extraction and Ion Exchange*, 10(4), 623–635.

42. Villarreal, M. S., Kharisov, B. I., Torres-Martínez, L. M., and Elizondo, V. N. (1999). Recovery of vanadium and molybdenum from spent petroleum catalyst of PEMEX. *Industrial & Engineering Chemistry Research*, 38(12), 4624–4628.

43. Angelidis, T. N., Rosopoulou, D., and Tzitzios, V. (1999). Selective rhenium recovery from spent reforming catalysts. *Industrial & Engineering Chemistry Research*, 38(5), 1830–1836.

44. Al-Sheeha, H., Marafi, M., Raghavan, V., and Rana, M. S. (2013). Recycling and recovery routes for spent hydroprocessing catalyst waste. *Industrial & Engineering Chemistry Research*, 52(36), 12794–12801.

45. Marafi, M. and Rana, M. (2016). Refinery waste: the spent hydroprocessing catalyst and its recycling options. *WIT Transactions on Ecology and the Environment*, 202, 219–230.

46. Marafi, M., Stanislaus, A., and Furimsky, E. (2010). Metal reclamation from spent hydroprocessing catalysts. In: *Handbook of Spent Hydroprocessing Catalysts*, Elsevier, Amsterdam, pp. 269–315.
47. Marafi, M. and Furimsky, E. (2017). Hydroprocessing catalysts containing noble metals: deactivation, regeneration, metals reclamation, and environment and safety. *Energy & Fuels*, 31(6), 5711–5750.
48. European Catalysts Manufacturers Association (2001). Guidelines for the Management of Spent Catalysts. https://hnlkg4f5wdw34kx1a1e9ygem-wpengine.netdna-ssl.com/wp-content/uploads/2017/07/5_Guidelines_for_the_management_of_spent_catalysts.pdf.
49. US EPA (2017). Hazardous Waste Management Facilities and Units. https://www.epa.gov/hwpermitting/hazardous-waste-management-facilities-and-units.
50. Marafi, M. and Stanislaus, A. (2003). Options and processes for spent catalyst handling and utilization. *Journal of Hazardous Materials*, 101(2), 123–132.

Chapter 10

Hydrogen Production

Qi Qiu[*,‡] and Bo Zhang[†,§]

*College of Chemistry and Environmental Engineering,
Shenzhen University, Guangdong, China
†School of Chemical Engineering and Pharmacy,
Wuhan Institute of Technology, Hubei, China
‡qqiu@szu.edu.cn
§Corresponding author: bzhang_wh@foxmail.com

Abstract

A significant amount of hydrogen is required for hydroprocessing processes to satisfy the needs of petroleum refinery, natural gas cleaning, and biofuel upgrading. The first section of this chapter introduces the background of hydrogen production technologies. The second section reviews resources that can be used to manufacture hydrogen. The third section provides an overview of the current development of methane reforming, gasification, electrolysis, and other technologies. The last section concludes the chapter and presents the future trends.

Keywords: Hydrogen production, Methane, Coal, Biomass, Steam reforming, Dry reforming, Partial oxidation, Autothermal reforming, Trireforming, gasification, electrolysis

10.1. Introduction

The flow sheet for many hydrotreating or hydrodesulfurization (HDS) processes is similar to the process shown in Figure 10.1. The feed is mixed with compressed recycle gas and make-up hydrogen and then sent on to the reactor containing the catalyst. After passing through the reactor, the effluent is heat exchanged with reactor feed, cooled, and sent to a high-pressure separator. The recycle gas from the top of the separator might be cleaned for the removal of H_2S and amine. The use of a recycle gas minimizes the loss of the valuable hydrogen. For simple processes, by-product hydrogen from catalytic reforming with a purity of about 87% can be utilized. For severe hydrotreating and hydrocracking, a dedicated hydrogen production unit is required to produce a large volume of hydrogen with >99.5% purity [1].

The consumption of hydrogen is especially high when treating heavier feeds. Additionally, the recent developments in biofuel technologies indicate that hydrotreating and/or hydrocracking processes must be

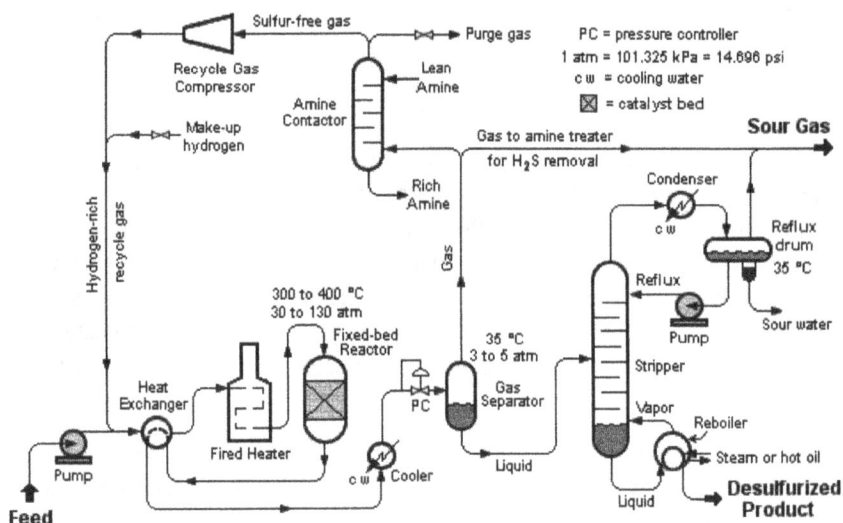

Figure 10.1. A typical HDS unit in a petroleum refinery, by M. Beychok, 2006, https://commons.wikimedia.org/wiki/File:HDS_Flow.png. (Adapted under a GNU Free Documentation License.)

applied to upgrade the low-quality biofuels to either drop-in fuels or fine chemicals. Therefore, large quantities of hydrogen are needed for hydro-processing processes to satisfy the needs of petroleum refinery, natural gas cleaning, and biofuel upgrading. Today, 80–85% of the global hydrogen production is accomplished via steam reforming (SR) of natural gas [2], while 95% of the hydrogen produced in the United States is made via steam–methane reforming [3]. Meanwhile, a significant number of technologies have been explored to make hydrogen a less costly chemical [4].

Hydrogen production pathways of US Department of Energy (US DOE) include a wide portfolio of processes over a range of time frames and production scales (Figure 10.2); this focuses on a target of producing hydrogen at less than $4/kg [5]. In general, hydrogen can be produced via thermochemical processes, electrolysis, and biological processes [6].

Thermochemical processes adopt heat and chemical reactions to release hydrogen from organic materials such as fossil fuels and biomass. Typical thermochemical processes for hydrogen production include

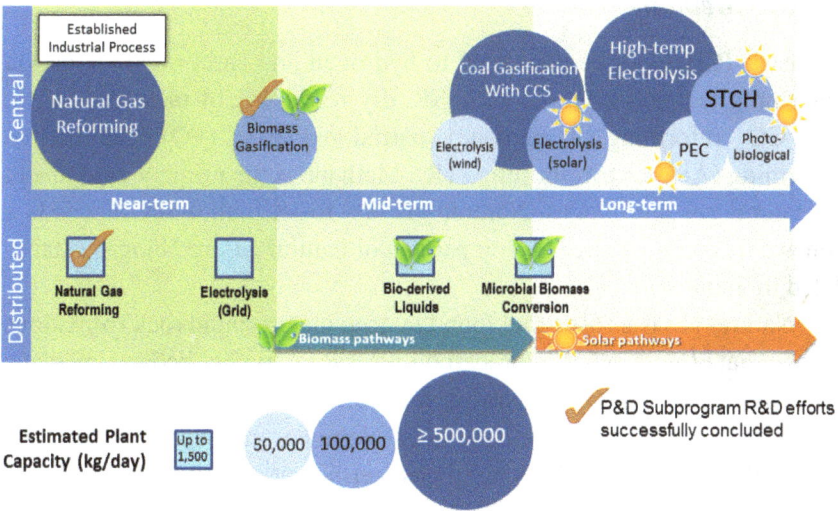

Figure 10.2. Hydrogen production pathways illustrated by US DOE, https://energy.gov/sites/prod/files/styles/borealis_default_hero_respondxl/public/h2_production_pathways.png. Public domain.

natural gas reforming, coal gasification, biomass gasification, biomass-derived liquid reforming, and solar thermochemical hydrogen.

Electrolysis uses electricity to split water (H_2O) into hydrogen (H_2) and oxygen (O_2). Electricity generated via clean sources such as wind and solar is preferred. New solar technologies such as photolytic processes hold a potential for sustainable hydrogen production with low environmental impact. Similarly, microorganisms such as bacteria and algae can produce hydrogen through biological metabolism from renewable resources, enabling a long-term potential for sustainable, low-carbon hydrogen production.

10.2. Hydrogen resources

Hydrogen can be made from a diverse range of resources. Currently, hydrogen is mostly produced from fossil fuels, specifically natural gas. Renewable sources such as wind, solar, geothermal, and biomass are also used to generate hydrogen [7].

10.2.1. *Methane*

Methane (CH_4) can be converted to hydrogen and carbon monoxide via reforming technologies including SR, dry reforming of methane (DRM, i.e., CO_2 reforming of methane), partial oxidation (PO), autothermal reforming (AR), or trireforming (TR). Methane is the primary component of natural gas and biogas. Natural gas is a fossil fuel, while biogas is a renewable energy source. A comparison of natural gas and biogas is tabulated in Table 10.1.

Natural gas is a fossil fuel found in deep underground rock formations or associated with other hydrocarbon reservoirs in coal beds as methane clathrates. It is formed when layers of decomposing plant and animal matter are exposed to intense heat and pressure under the surface of the Earth over millions of years [8]. Natural gas consists primarily of methane, but commonly includes varying amounts of other higher alkanes, and sometimes carbon dioxide (CO_2), nitrogen (N_2), hydrogen sulfide (H_2S), or helium. Some natural gases contain large amounts of CO_2 (widespread), N_2 (US Rockies), and H_2S (Canada) — well over 30%.

Table 10.1. A comparison of the composition of typical natural gas and biogas.

Character	Unit	Natural gas	Biogas
Methane (CH_4)	vol.%	81–89	60–70
Other hydrocarbons	vol.%	3.5–9.4	0
Hydrogen (H_2)	vol.%	—	0
Carbon dioxide (CO_2)	vol.%	0.67–1	30–40
Nitrogen (N_2)	vol.%	0.28–14	2–6
Oxygen (O_2)	vol.%	0	0.5–1.6
Hydrogen sulfide (H_2S)	ppm	0–2.9	0–2,000
Ammonia (NH_3)	ppm	0	100

Biogas, also known as swamp gas, bog gas, and marsh gas, is emitted naturally by wetlands as well as human activities. The methane in the biogas could be as high as 60–70% by volume [9] with carbon dioxide (CO_2) in a range of 30–40%. The remaining constituents are nitrogen (N_2), oxygen (O_2), hydrogen sulfide (H_2S), ammonia (NH_3), siloxane, etc. The US Environmental Protection Agency (EPA) reported that natural sources deliver about 36% of methane emissions, and human-related sources create the majority of methane emissions, accounting for 64% of the total [10]. Methane emissions related to human sources come from the raising of livestock, landfills, as well as the production, transportation, and use of fossil fuels. Methane is one of the greenhouse gases (GHGs) with other GHGs including carbon dioxide (CO_2), water vapor (H_2O), and nitrous oxides. GHGs cause the "greenhouse effect" and global warming [11]. Methane has an average lifetime of 28 y in the atmosphere with its 100-y global warming potential ~28 times higher than that of carbon dioxide [12].

Methane makes the biogas a combustible fuel gas and can be used as an energy source to replace the fossil natural gas for heating or cogenerating heat and electricity [13]. In developing countries, biogas is an attractive energy alternative for regions relying heavily on traditional biomass for their energy needs. Biogas produced on a very small scale for households is used mainly for cooking and water heating [14]. Biogas produced on the large industrial scale [16] can either be burnt in the biogas engine for cogeneration or upgraded to meet natural gas standards for injection into the natural gas network as biomethane (i.e., bionatural gas) or directly as gaseous biofuel in gas engine-based captive fleets such as heavy duty trucks [17].

Biogas is typically produced through a process called anaerobic digestion (AD) [18]. AD is a series of biological processes, in which bacteria break down organic wastes in the absence of oxygen. The AD process is often divided into four stages (namely hydrolysis, acidogenesis, acetogenesis, and methanogenesis) for illustrating the sequence of microbial events that occur during the digestion process and the production of methane (Figure 10.3). The hydrolysis step degrades both insoluble organic materials and high molecular weight compounds such as carbohydrates, fats, proteins, and nucleic acids, into soluble organic substances (e.g., sugars, fatty acids, and amino acids). The soluble compounds

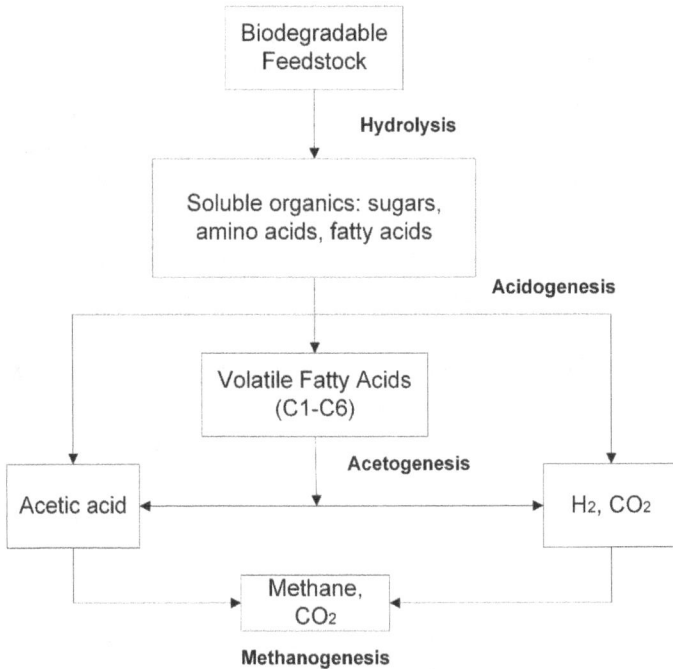

Figure 10.3. Sequential stages in the AD process.

formed during hydrolysis are further digested by a large diversity of facultative anaerobes and anaerobes through many fermentative processes during the second stage of acidogenesis. During acidogenesis, volatile fatty acids (VFA) are produced along with CO_2, hydrogen gas (H_2), alcohols, ammonia, hydrogen sulfide, and other by-products. In the third stage of acetogenesis, the higher organic acids and alcohols produced by acidogenesis are further split by acetogens to produce mainly acetic acid as well as CO_2 and H_2. This conversion is controlled largely by the partial pressure of H_2 in the mixture. The last step of AD, methanogenesis, produces methane by two groups of methanogenic microorganisms: the first group splits acetate into methane and carbon dioxide, and the second group uses hydrogen as electron donor and carbon dioxide as acceptor to produce methane. The most common reason for the upsets of AD is due to the inhibition of the methane-forming bacteria in the last stage.

10.2.2. *Coal and biomass*

Coal is a combustible sedimentary rock usually occurring in rock strata in layers or veins called coal beds or coal seams [19]. Coal is composed primarily of carbon, along with variable quantities of other elements such as hydrogen, sulfur, oxygen, and nitrogen. Coal can be converted into a variety of products including power, liquid fuels, chemicals, and hydrogen. Hydrogen is produced via coal gasification by first reacting coal with oxygen and steam under high pressures and temperatures. US DOE anticipates that the coal gasification for hydrogen production together with carbon capture, utilization, and storage could be deployed in the midterm time frame [20].

Biomass often refers to plants or plant-based materials that convert solar energy into chemical energy through photosynthesis. As an industry term, biomass is employed to obtain energy by burning. Biomass gasification is a relatively mature technology that adopts a controlled process involving heat, steam, and oxygen to convert biomass to hydrogen and other products [21]. Like coal, biomass gasification can form hydrogen, carbon monoxide, and carbon dioxide at high temperatures (>700°C) under a controlled amount of oxygen and/or steam without combustion. Other possible ways for hydrogen production from biomass may involve first conversion of biomass to ethanol or bio-oil, which could be reformed to hydrogen. In addition, some microorganisms have the ability of digesting biomass and releasing hydrogen.

10.2.3. *Solar and wind*

Solar and wind energy are renewable resources that could generate electricity, which may be used to split water to produce hydrogen via electrolysis. The current grid electricity is not the ideal source of electricity for electrolysis because most of the electricity is generated by technologies that result in greenhouse gas emissions and are energy intensive [22]. Electricity generation using renewable or nuclear energy technologies is a feasible option to overcome these limitations.

Meanwhile, sunlight can induce the photoelectrochemical water splitting reaction or the photolytic biological processes that turn water or organic matter into hydrogen.

10.3. Hydrogen production processes

10.3.1. *SR*

SR is an important technology pathway for near-term hydrogen production. Currently, most of the hydrogen produced in the United States is made by natural gas reforming in large central plants and on-site hydrogen production units in refineries. SR of hydrocarbon feedstocks was introduced to industry over 80 y ago to produce H_2-enriched gases. Because of the higher H/C ratio and less carbon deposition than naphtha, methane became a more favorable feedstock about 50 y ago. A historical review can be found in the literature [23] on the SR technology.

The reaction (1) shows the major reaction of SR of hydrocarbon (like naphtha), which transforms a liquid hydrocarbon stream into a gaseous mixture constituted by CO and H_2 [24]. SR of methane is a mature technology, and the major chemical reactions include reaction (2) and water-gas shift reaction (3). Firstly, methane reacts with a high-temperature steam (700–1,000°C) under 0.3–2.5 MPa pressure in the presence of a catalyst to produce hydrogen, carbon monoxide, and a relatively small amount of carbon dioxide. Subsequently, the carbon monoxide reacts with steam using a catalyst to form carbon dioxide and more hydrogen in the water-gas shift reaction. To obtain pure hydrogen product, a pressure-swing adsorption process is normally adopted to remove carbon dioxide and other impurities from the gas stream. Although water-gas shift reaction generates a small amount of heat, SR is still endothermic, i.e., it needs heat provided by external sources.

SR reactions:

$$C_nH_{2n+2} + nH_2O \rightleftharpoons nCO + (2n + 1)H_2 \tag{1}$$

$$CH_4 + H_2O \rightleftharpoons CO + 3H_2 \ (\Delta H^\theta = +206 \text{ kJ/mol } CH_4) \tag{2}$$

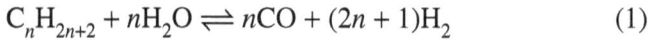

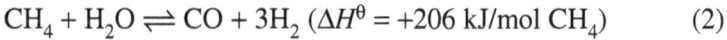

Water-gas shift reaction:

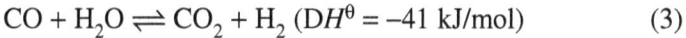

$$CO + H_2O \rightleftharpoons CO_2 + H_2 \ (DH^\theta = -41 \text{ kJ/mol}) \tag{3}$$

SR can also be utilized to produce hydrogen from other biomass-derived fuels. For example, biomass can easily be converted into a number of liquid fuels, including methanol, ethanol, biodiesel, and pyrolysis oil, which could be transported and employed to generate hydrogen [25] as shown in reactions (4)–(7).

Methanol reforming:

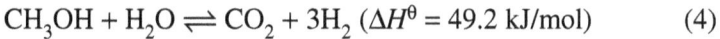

$$CH_3OH + H_2O \rightleftharpoons CO_2 + 3H_2 \ (\Delta H^\theta = 49.2 \text{ kJ/mol}) \tag{4}$$

Ethanol reforming [26]:

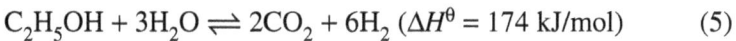

$$C_2H_5OH + 3H_2O \rightleftharpoons 2CO_2 + 6H_2 \ (\Delta H^\theta = 174 \text{ kJ/mol}) \tag{5}$$

SR of methyl oleate as a model substance for biodiesel [27]:

$$C_{19}H_{36}O_2 + 17H_2O \rightleftharpoons 19CO + 35H_2 \ (\Delta H^\theta = 2645 \text{ kJ/mol}) \tag{6}$$

The overall steam-reforming reaction of bio-oil [28]:

$$C_nH_mO_k + (2n - k) H_2O \rightleftharpoons nCO_2 + (2n + m/2 - k)H_2 \tag{7}$$

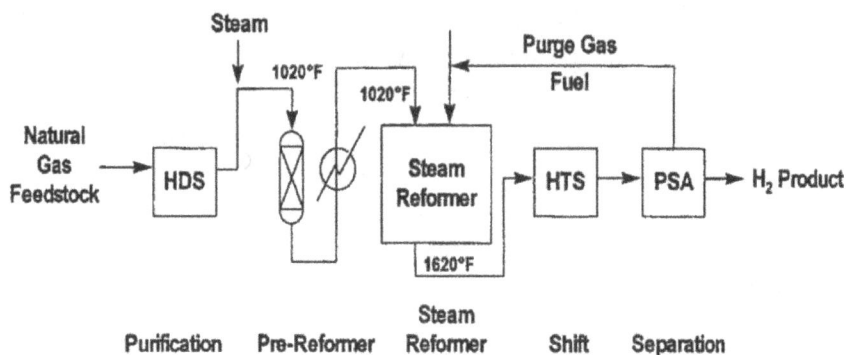

Figure 10.4. Simplified process flow diagram for SR of methane. (Reprinted with permission from Ref. [29]. Copyright © 2001 American Chemical Society.)

The industrial steam-reforming process typically consists of a heated furnace (reformer) and downstream cleaning units. A simplified process flow diagram for SR of methane is shown in Figure 10.4. The feed is pre-treated for HDS, and then mixed with superheated process steam. The catalyst (normally containing nickel) is placed in a number of high-alloy reforming tubes placed in a furnace. The outer diameters of the tubes range typically 10 cm with a length of 10–13 m. Commercial processes operate with a significant temperature gradient along the catalyst bed with the typical inlet temperatures in the catalyst bed of 450–650°C and the outlet temperature of the reformers approaching 800–950°C [29]. The steam reformers now have capacities up to 300,000 N m^3 of H$_2$ (or syngas)/h and are operated at average heat fluxes exceeding 100,000 kcal/m^2/h (0.12 MW/m^2) [30]. The furnace consists of a box-type radiant section including burners and a convection section to recover the waste heat of the flue gases leaving the radiant section.

In some processes, an adiabatic prereformer is added to solve the problem of carbon deposition, a critical issue shortening the lifetime of commercial catalysts. The advantages of a prereformer include (1) converting all higher hydrocarbons and bringing the methane reforming and shift reactions into equilibrium [31]; (2) preheating the feed, and thus reducing the size of the tubular reformer; and (3) allowing for feedstock flexibility ranging from natural gas and refinery off-gas to liquid fuels [32].

Following the steam reformer, the process may use a shift reactor, in which a portion of CO in the reformed gas is used for additional hydrogen generation via the water-gas shift reaction (3). The process is exothermic and limited by the chemical equilibrium. The CO content at the reactor outlet can be brought down to 2.5%, 0.5%, and 0.2% for high-temperature shift (HTS) at 300–450°C, medium-temperature shift at 220–270°C, and low-temperature shift at 180–250°C, respectively [33]. The low/medium temperature shift prefers Cu/Fe/Cr and Cu/Zn based catalysts, while HTS prefers Fe/Cr catalysts for the good activity from Fe and reduced sintering from Cr [34].

A pressure-swing adsorption (PSA) process is also applied after SR to remove carbon dioxide and other impurities from the gas stream, leaving essentially pure hydrogen. SR yields gases with the H_2:CO ratio of ~3. A lower H_2:CO ratio of 1 is yielded if CO_2 replaces the steam. The gas products can be employed directly for synthesis of methanol, ammonia, and synthetic hydrocarbon fuels [35].

For SR, nickel catalysts are usually selected. Although biogas/methane may cause carbon formation on the catalysts, this carbon formation could be easily inhibited by adding excess steam or other oxidants (like CO_2 or air) [36, 37]. Accordingly, the reformer is often operated with a higher steam/carbon relationship than theoretical necessity to prevent the formation of carbon deposition.

To date, researchers are still actively seeking new technologies to further improve the SR process. For example, Italian scientists conducted a project to develop the steam reformer powered by concentrating solar power (CSP) plants using molten salts as the heat transfer fluid, and this process is focused on performing SR at a temperature range of 400–550°C [38]. Meanwhile, new catalysts are under development suitable for low-temperature SR [39].

10.3.2. *Dry reforming*

DRM, which converts two GHGs CO_2 and CH_4 into synthesis gas (CO and H_2) at 700–900°C, is very promising for both industrial and environmental implication. The DRM reaction (reaction (8) in Table 10.2) was first studied by Fischer and Tropsch in 1928 [40], which is

Table 10.2. Overall reactions in DRM system.

Reaction	Name	Chemical reaction	ΔH^{θ} (kJ/mol)	$\Delta G^{\theta} = 0$, T (K)	Reaction temperature (°C)
(8)	DRM	$CH_4 + CO_2 \rightleftharpoons 2CO + 2H_2$	247	$61{,}770 - 67.32T$	640[a]
(9)	RWGS	$CO_2 + H_2 \rightleftharpoons CO + H_2O$	41	$-8{,}545 + 7.84T$	820[b]
(10)	Boudouard reaction	$2CO \rightleftharpoons CO_2 + C(s)$	−172	$-39{,}810 + 40.87T$	700[b]
(11)	Methane cracking	$CH_4 \rightleftharpoons 2H_2 + C(s)$	75	$21{,}960 - 26.45T$	557[a]
(12)	CO reduction	$CO + H_2 \rightleftharpoons C + H_2O$	−131	—	—

Note: [a]Lower limit.
[b]Upper limit.

thermodynamically favored at low pressures with most studies performed at 1 atmosphere pressure or lower. Biogas is a suitable substitution for this reaction as it contains both CH_4 and CO_2 at a ratio of 1:1–1.5, which is close to the stoichiometry of this chemical reaction.

The DRM process, being highly endothermic, requires high energy input. DRM is always accompanied by another four side reactions including reverse water-gas shift (RWGS) reaction, Boudouard reaction, methane cracking, and CO reduction (Table 10.2). Because the chemical potential of carbon deposition for the stoichiometric dry reforming reaction is significantly higher than that in the equivalent SR reaction, the carbon deposition occurring in Boudouard reaction, methane cracking, and CO reduction often causes catalyst deactivation, a major problem inhibiting the industrial application of the DRM technology [23]. Nevertheless, despite its notorious deactivation issues, DRM still has a 20% lower operating cost compared with the other reforming processes [40].

Because of the inevitable catalyst deactivation due to carbon deposition, a high coking-resistant catalyst is needed to further commercialize the DRM process [41]. An ideal catalyst should demonstrate high activity towards preferred products and high stability over a long period. Generally, noble- and nonnoble-metal catalysts were widely used for this application. The most studied metal catalyst was nickel (Ni). Nickel-based catalysts are attractive because of their high activity and low cost, but they are likely to be deactivated by coke formation [42]. Noble metals such as Pt, Ru, Rh, Pd, and Ir are very reactive catalysts for DRM. Additionally, noble metals are less sensitive to carbon deposition than nickel, probably due to their lower carbon solubility [43]. Although the cost of noble metal is a barrier for the process development, the use of these metals for doping might improve the methane reforming activity of bimetallic catalysts. A comprehensive review of noble-metal catalysts for DRM can be found in the literature [44].

Table 10.3 summarizes representative studies of the DRM to syngas during 2007–2017 [34]. Studies on CO_2 reforming of methane were normally conducted between 700 and 950°C at varying CH_4/CO_2 ratio (within a range of 0.5–4) and types of catalysts. From the thermodynamic point of view, the RWGS equilibrium enables higher CO_2 conversion than that of CH_4. DRM is useful in making synthesis gas with H_2/CO ratios of

Table 10.3.   Recent studies on DRM. (Adapted with permission from Ref. [34]. Copyright © 2016 Elsevier.)

Input CH_4/CO_2	Metals	Temp. (°C)	Output H_2/CO	CH_4 conversion (%)	CO_2 conversion (%)	Carbon formation (mg/g catalyst)	Ref.
0.5	Ce-Gd-O	800	1.07	50	88	NR	[54]
0.8	Rh-Al	700	1	42	NR	NR	[55]
1	NiCo/CeZrO$_2$	800	0.84	79	84	0.24–8.2% of catalyst	[53]
1	WC	900	0.96	95	95	None	[52]
1	Ni–La$_2$O$_2$CO$_3$	700	0.86	70	82	1.2–10.3% of catalyst	[56]
1	Ni–Al	700	0.67	19	31	NR	[57]
1	Ni–Pb–Al	700	0.88	60	78	NR	[57]
1	Ni–Pb–1P–Al	700	0.77	55	71	NR	[57]
1	Ni	700	1	54	66	41	[58]
1	Co	700	1	75	67–80	20–268	[58]
1	Ni–Co	700	1	56–71	83	290	[58]
1	Pt–Ru	700	<0.5	90	48	NR	[59]
1	La–NiMgAlO	700	0.8	80	85	NR	[60]
1	Ni	750	NR	32	36	3.6% of inlet C	[61]
1	Ni/Si	750	1	73	89	Negligible	[62]
1	Ce-Gd-O	800	0.96	68	72	NR	[54]

1	Ru/ZrO$_2$–SiO$_2$	800	1.07	95.8	89.8	None	[63]
1	Ni/porous γ-Al$_2$O$_3$	850	0.82	99	NR	NR	[64]
1	Pt–Al	900	NR	NR	NR	22% of inlet C	[65]
1	Ni–La–Al	950	NR	99	90	NR	[66]
1	NiO–MgO	700	0.87	67	77	NR	[67]
1.5	Ni–Al	750	0.9	49	81	NR	[68]
1.5	Ni–Mg–Al	750	0.86	59	70	NR	[68]
1.5	Ni–La–Mg–Al	750	0.95	61	70	NR	[68]
1.5	Rh–Ni–Mg–Al	750	1	58	85	NR	[68]
1.5	Rh–Ni–La–Mg–Al	750	1.06	50	94	NR	[68]
1.5	Rh–Ni	800	1	65	100	NR	[68]
1.5	Ni–Al	850	0.55	72	96	180	[69]
1.5	Ni–Ce–Al	850	0.65	73	97	170	[69]
2	Ce–Gd–O	800	0.84	66	46	NR	[54]
2.1	Ni	750	NR	21	29	3.6% of inlet C	[61]

Note: NR: not reported.

1 or less. This is useful for the OXO synthesis and promoting long-chain products in the Fischer–Tropsch process [45]. Most studies fabricated new catalysts to enhance the dry reforming and minimize the carbon formation. Some catalysts comprise active metal(s) and supports such as C, SiO_2, Al_2O_3, MgO, ZrO_2, ZrO_2-SiO_2, CeZrO, and CeO_2 [46]. Generally, CH_4 is activated on metal, and CO_2 is activated on acidic or basic supports. If the supports were inert materials, such as C and SiO_2, both reactants are activated by metal [44]. The catalysts include crystalline oxide catalysts such as pyrochlores (formula $A_2B_2O_7$) [47] or perovskite (ABO_3) [48], mesoporous catalysts [49], and hydrotalcite catalysts [50]. Moreover, the activity of catalysts towards DRM could be increased by promoters [51].

Carbon deposition is a major drawback of DRM, and is observed in almost all studies. Although several studies announced that the carbon formation was either not detected [52] or negligible [53], those examined were only done in bench-scale reactors for limited time (400–500 h). Until the results can be further confirmed at the pilot scale and commercial scale, there are still many technique barriers to be overcome.

Recently, nonthermal plasma catalysis was introduced into the DRM process research. Plasma catalysis possesses nonequilibrium properties and requires low power input. Tu *et al.* reported that the DRM reaction was induced by a plasma below 300°C with a Ni/Al_2O_3 catalyst [70]. Because the reactions happened at a lower temperature, this technology may have the prospect of minimizing the carbon deposition.

10.3.3. *Partial oxidation, autothermal reforming, and trireforming*

In partial oxidation, methane reacts with a limited amount of oxygen. This limited oxygen cannot oxidize the reactants completely to carbon dioxide and water [3]. As shown in reaction (13), methane is oxidized to hydrogen and carbon monoxide with less than the stoichiometric amount of oxygen available. With the air as the oxygen source, the reaction products contain additional nitrogen. A small amount of carbon dioxide and other compounds also exist in the final products. Subsequently, the carbon monoxide reacts with water to form carbon dioxide and more hydrogen in the water-gas shift reaction (3).

The partial oxidation of CH_4:

$$CH_4 + 0.5O_2 \rightleftharpoons 2H_2 + CO \ (\Delta H^\theta = -36 \text{ kJ/mol } CH_4) \quad (13)$$

The partial oxidation reaction is a mildly exothermic process. It can be combined with endothermic reactions, such as SR and dry reforming, to increase the efficiency of both reactions [71]. Normally, a non-catalytic, large-scale partial oxidation process can yield syngas with a H_2/CO ratio of about 2, an ideal ratio for downstream synthesis processes such as methanol production. Catalytic partial oxidation (CPO), based on short-contact time conversion of methane, hydrocarbons, or biomass on catalysts (like nickel), is suitable for small-scale applications. Early work established that nickel-based catalysts are highly active towards partial oxidation reaction. Until now, many potential alternatives have been discovered, including supported cobalt or iron catalysts, supported noble-metal catalysts [72], and transition metal carbide catalysts [73].

Autothermal reforming (ATR) is a hybrid process that combines SR (reaction (2)) and partial oxidation (reaction (13)) [74]. The heat needed for endothermic reforming is provided in situ by methane oxidization. However, it is found that nickel-based catalysts were deactivated slowly during ATR of the biogas. The deactivation of the catalysts attributed to the oxidation of metallic Ni [75, 76].

The reforming of methane with various combinations with $O_2/H_2O/CO_2$ is called mixed reforming or TR [77]. This process involves the simultaneous reaction of CH_4 with H_2O (reaction (2)), CO_2 (reaction (8)), and O_2 (reaction (13)). The process has been developed to combine the advantages of these reforming reactions. By changing the ratio of H_2O, CO_2, and O_2 in the feed stream, it's possible to control the H_2/CO ratio of the product stream. The advantages of TR include (1) producing gases with H_2/CO ratios ranging from ~1 to 3, (2) minimizing the energy requirements by incorporating partial oxidation of methane into this process, and (3) inhibiting the carbon deposition by using extra stream or O_2 in the system [45]. Currently, most studies focused on the development of new nickel-based catalysts for the TR process [36], and biogas was also proposed as a carbon neutral feedstock for this technology [78, 79].

10.3.4. *Gasification*

10.3.4.1. *Coal gasification*

Historically, coal was gasified to produce coal gas, which was subsequently piped to customers to burn for illumination, heating, and cooking [80]. Currently, large-scale instances of coal gasification are primarily for electricity generation. Coal gasification yields the syngas consisting primarily of carbon monoxide (CO) and hydrogen (H_2), carbon dioxide (CO_2), methane (CH_4), and water vapor (H_2O). The unbalanced reaction is shown as

$$CH_{0.8} + O_2 + H_2O \rightarrow CO + CO_2 + H_2 + \text{other species} \qquad (14)$$

During the coal gasification reaction, the coal is heated in the gasifier under a well-controlled oxygen and steam environment. Oxygen and water molecules partially oxidize the coal without resulting in combustion. The primary products are a desired gaseous mixture with by-products like tar and phenols.

Countries such as Russia, the United States, and China have an abundant, domestic resource of coal. The use of coal to produce hydrogen for the transportation sector can reduce the total energy consumption and create jobs through the creation of a domestic industry. The production of hydrogen from coal also offers environmental benefits to achieve extremely low SO_x, NO_x, and particulate emissions from burning coal-derived gases [81].

10.3.4.2. *Biomass gasification*

Biomass gasification converts solid biomass into a gaseous combustible gas through a sequence of thermochemical reactions. The gas is a low-heating value fuel, with a calorific value between 1,000 and 1,200 kcal/Nm^3. Nearly 2.5–3.0 Nm^3 of gas can be obtained through gasification of about 1 kg of air-dried biomass [82]. A simplified example reaction using glucose as a surrogate for cellulose is shown as

$$C_6H_{12}O_6 + O_2 + H_2O \rightarrow CO + CO_2 + H_2 + \text{other species} \qquad (15)$$

Biomass does not gasify as easily as coal. In general, drying biomass to 10–20% moisture content is considered the optimum for minimizing the size and cost of the biomass gasification plants. The gasifier is the core of this technology. Fixed-bed gasifiers are suitable for small to medium-scale applications with thermal requirements of up to a few megawatts thermal (MWt). Most fixed-bed gasifiers are air-blown, producing low-energy product gases. When applied for large-scale applications, fixed-bed gasifiers may encounter bridging problems and nonuniform bed temperatures, leading to uneven gas flow, hot spots, ash deformation, and slagging [83].

Bubbling fluid bed and circulating fluid bed designs are more suitable for large-scale applications. Both designs target a biomass size of approximately 2.0–2.5″ minus to maintain a transport velocity. A smaller biomass size might benefit some technologies, however at the expense of increased capital and operating costs.

Biomass gasification is a multiple-step process. The first step of biomass pyrolysis decomposes biomass in the absence of oxygen below 600°C to produce the gas mixture and other hydrocarbon compounds. The second stage is used to reform these hydrocarbons (tars) with a catalyst to yield a clean syngas mixture of hydrogen, carbon monoxide, and carbon dioxide. Following the tar reforming, a shift reaction step converts the carbon monoxide with steam to carbon dioxide. The product hydrogen is then separated and purified.

Two types of catalysts can be used for biomass gasification [84]. The first type of catalysts, such as dolomites and alkali metals, is added directly to the biomass prior to gasification via wet impregnation or dry mixing. The primary purpose of using these catalysts is to reduce the tar. Nickel catalysts are used in the tar reforming reactor.

Biomass is the most abundant renewable resource in the world, and its effective utilization recycles carbon dioxide and results in low net GHG emissions. However, biomass gasification is not a mature technology. Applications for heat, co-combustion in coal plants, and combined heat and power generation show limited market penetration and some are dependent on the government regulations and support [85]. Economic challenges to hydrogen production via biomass gasification involve reducing costs associated with capital equipment and biomass feedstocks. And technological barriers mainly include scaling up, tar reduction, and gas cleaning [86].

10.3.5. *Electrolysis*

Electrolysis of water powered by electricity is a promising option for hydrogen production from renewable resources [22]. Electrolysis splits water into hydrogen and oxygen as shown in reaction (16).

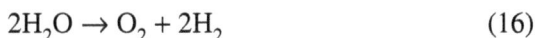

$$2H_2O \rightarrow O_2 + 2H_2 \qquad (16)$$

The electrolysis process occurs in a reactor called electrolyzer, consisting of an anode and a cathode separated by an electrolyte. The use of different electrolytes works through different mechanisms. For a polymer electrolyte membrane (PEM) electrolyzer operating at 70–90°C, water molecules react at the anode to form oxygen and protons. The protons diffuse through the electrolyte membrane to the cathode, on which protons combine with the electrons from the electrical flow to form hydrogen gas.

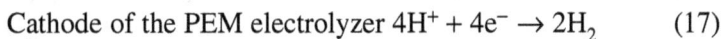

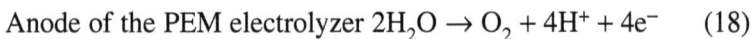

$$\text{Cathode of the PEM electrolyzer } 4H^+ + 4e^- \rightarrow 2H_2 \qquad (17)$$

$$\text{Anode of the PEM electrolyzer } 2H_2O \rightarrow O_2 + 4H^+ + 4e^- \qquad (18)$$

If a liquid alkaline solution of sodium hydroxide or potassium hydroxide is used as the electrolyte, hydroxide ions (OH⁻) migrates through the electrolyte from the cathode to the anode. On the cathode side, hydrogen is produced as shown in reactions (19) and (20). The commercial alkaline electrolyzers typically operate at 100–150°C [87].

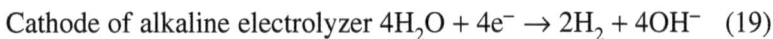

$$\text{Cathode of alkaline electrolyzer } 4H_2O + 4e^- \rightarrow 2H_2 + 4OH^- \qquad (19)$$

$$\text{Anode of alkaline electrolyzer } 4OH^- \rightarrow 2H_2O + O_2 + 4e^- \qquad (20)$$

For the solid oxide electrolyzer, a solid ceramic material is applied as the electrolyte to form hydrogen at 700–800°C. Water combines with electrons at the cathode to form hydrogen gas and negatively charged oxygen ions (O²⁻). The O²⁻ ions pass through the membrane and form oxygen gas at the anode [22].

Both PEM and alkaline technologies can deliver hydrogen with high purity on site and on demand. Commercial electrolyzers are available from small-scale distributed hydrogen production to large-scale, central production facilities that could be tied directly to electricity produced by using renewable resources such as solar energy and wind energy. Hydrogen production via electrolysis powered by renewable energy can reduce environmental footprint and possibly lead to zero-carbon emission. Current challenges still rely on improving the efficiency of electrolyzers and reducing the production cost [88].

10.3.6. *Other hydrogen production processes*

The thermochemical direct conversion for water dissociation into hydrogen and oxygen operates at a high temperature at 2,000°C. The operation temperature could be reduced by introduction of chemicals such as iron oxide into the process. The high-temperature thermochemical cycles might achieve excellent efficiencies (>40%). The thermochemical water splitting technology consuming concentrated solar power or the waste heat of nuclear power reactions is considered as a long-term technology pathway with low or no GHG emissions [89].

Photoelectrochemical and photobiological processes are subject to low efficiency of typically less than 1% efficient (solar to hydrogen) [90]. The photoelectrochemical process is equipped with semiconducting electrodes in a photoelectrochemical cell to convert light energy into chemical energy of hydrogen. The photobiological processes harvest hydrogen from biological systems such as algae. Both processes require further developments to meet the long-term energy requirements.

10.4. Closing remarks

Most hydrogen gas is produced globally via SR of the natural gas. For the near term, this production method will continue to dominate. Processes of coal gasification and alkaline electrolysis are commercially available on a large scale. Other advanced processes require further research and development efforts to economically produce hydrogen from sustainable resources.

In terms of the feedstock, technologies based on renewable resources, such as methane produced by AD, biomass gasification, and wind-generated electricity, are anticipated to begin reaching the cost targets in the midterm. In the longer term, solar-energy-based technologies are expected to become viable for near-net zero carbon emission.

Acknowledgments

Authors would like to acknowledge the support from the College of Chemistry and Environmental Engineering, Shenzhen University and the School of Chemical Engineering and Pharmacy, Wuhan Institute of Technology.

Useful resources

Haldor Topsoe, Hydrogen. https://www.topsoe.com/processes/hydrogen.
National Renewable Energy Laboratory (NREL), Hydrogen Production and Delivery. http://www.nrel.gov/hydrogen/proj_production_delivery.html.
National Renewable Energy Laboratory (NREL), Renewable Electrolysis. https://www.nrel.gov/hydrogen/renewable-electrolysis.html.
US Department of Energy's Alternative Fuels Data Center, Hydrogen Production and Distribution. https://www.afdc.energy.gov/fuels/hydrogen_production.html.
US Department of Energy's Office of Energy Efficiency and Renewable Energy (EERE), Hydrogen Production. https://energy.gov/eere/fuelcells/hydrogen-production.

References

1. Topsøe, H., Clausen, B. S., and Massoth, F. E. (1996). Hydrotreating catalysis. In: *Catalysis: Science and Technology*, J. R. Anderson and M. Boudart, eds., Springer, Berlin, pp. 1–269.
2. Boyano, A., Morosuk, T., Blanco-Marigorta, A. M., and Tsatsaronis, G. (2012). Conventional and advanced exergoenvironmental analysis of a steam methane reforming reactor for hydrogen production. *Journal of Cleaner Production*, 20(1), 152–160.

3. US Department of Energy (2016). Hydrogen Production: Natural Gas Reforming. http://energy.gov/eere/fuelcells/hydrogen-production-natural-gas-reforming.

4. Basile, A. and Iulianelli, A. (2014). *Advances in Hydrogen Production, Storage and Distribution*, Elsevier, Amsterdam.

5. US Department of Energy (2017). Hydrogen Production Pathways. https://energy.gov/eere/fuelcells/hydrogen-production-pathways.

6. US Department of Energy (2017). Hydrogen Production Processes. https://energy.gov/eere/fuelcells/hydrogen-production-processes.

7. US Department of Energy (2017). Hydrogen Resources. https://energy.gov/eere/fuelcells/hydrogen-resources.

8. Wikipedia (2017). Natural gas. https://en.wikipedia.org/wiki/Natural_gas.

9. Gebauer, R. (2004). Mesophilic anaerobic treatment of sludge from saline fish farm effluents with biogas production. *Bioresource Technology*, 93(2), 155–167.

10. US EPA (2015). Overview of Greenhouse Gases. https://www.epa.gov/ghgemissions/overview-greenhouse-gases.

11. US EPA (2015). Causes of Climate Change. https://19january2017snapshot.epa.gov/climate-change-science/causes-climate-change_.html.

12. US EPA (2015). Climate Change Indicators in the United States. https://www.epa.gov/climate-indicators.

13. Zhang, B. and Wang, L. (2014). Anaerobic digestion of organic wastes. In: *Sustainable Bioenergy Production*, L. Wang, ed., CRC Press, Taylor & Francis, Boca Raton, pp. 407–421.

14. Wikipedia (2017). Biogas. https://en.wikipedia.org/wiki/Biogas.

15. Malmberg (2009). What can biogas be used for?

16. Mézes, L., Bai, A., Nagy, D., Cinka, I., and Gabnai, Z. (2017). Optimization of raw material composition in an agricultural biogas plant. *Trends in Renewable Energy*, 3(1), 61–75.

17. ClimateTechWiki (2011). Biogas for cooking and electricity. http://www.climatetechwiki.org/technology/biogas-cook.

18. Zhang, B., Wang, L. J., and Li, R. (2015). Production of biogas from aquatic plants. In: *Aquatic Plants: Composition, Nutrient Concentration and Environmental Impact*, C. E. Rodney, ed., Nova Science Publishers, Inc., New York, pp. 33–46.

19. Wikipedia (2017). Coal. https://en.wikipedia.org/wiki/Coal.

20. US Department of Energy (2017). Hydrogen Production: Coal Gasification. https://energy.gov/eere/fuelcells/hydrogen-production-coal-gasification.

21. US Department of Energy (2017). Hydrogen Production: Biomass Gasification. https://energy.gov/eere/fuelcells/hydrogen-production-biomass-gasification.
22. US Department of Energy (2017). Hydrogen Production: Electrolysis. https://energy.gov/eere/fuelcells/hydrogen-production-electrolysis.
23. Rostrup-Nielsen, J. (2004). Steam reforming of hydrocarbons. A historical perspective. In: *Studies in Surface Science and Catalysis*, B. Xinhe and X. Yide, eds., Elsevier, Amsterdam, pp. 121–126.
24. Melo, F. and Morlanés, N. (2005). Naphtha steam reforming for hydrogen production. *Catalysis Today*, 107, 458–466.
25. Turner, J. A. (2004). Sustainable hydrogen production. *Science*, 305(5686), 972–974.
26. Mattos, L. V., Jacobs, G., Davis, B. H., and Noronha, F. B. (2012). Production of hydrogen from ethanol: review of reaction mechanism and catalyst deactivation. *Chemical Reviews*, 112(7), 4094–4123.
27. Martin, S., Kraaij, G., Ascher, T., Wails, D., and Wörner, A. (2015). An experimental investigation of biodiesel steam reforming. *International Journal of Hydrogen Energy*, 40(1), 95–105.
28. Wang, D., Czernik, S., and Chornet, E. (1998). Production of hydrogen from biomass by catalytic steam reforming of fast pyrolysis oils. *Energy & Fuels*, 12(1), 19–24.
29. Arakawa, H., Aresta, M., Armor, J. N., Barteau, M. A., Beckman, E. J., Bell, A. T., Bercaw, J. E., Creutz, C., Dinjus, E., Dixon, D. A., Domen, K., DuBois, D. L., Eckert, J., Fujita, E., Gibson, D. H., Goddard, W. A., Goodman, D. W., Keller, J., Kubas, G. J., Kung, H. H., Lyons, J. E., Manzer, L. E., Marks, T. J., Morokuma, K., Nicholas, K. M., Periana, R., Que, L., Rostrup-Nielson, J., Sachtler, W. M. H., Schmidt, L. D., Sen, A., Somorjai, G. A., Stair, P. C., Stults, B. R., and Tumas, W. (2001). Catalysis research of relevance to carbon management: progress, challenges, and opportunities. *Chemical Reviews*, 101(4), 953–996.
30. Rostrup-Nielsen, J. R., Sehested, J., and Nørskov, J. K. (2002). Hydrogen and synthesis gas by steam- and CO_2 reforming. *Advances in Catalysis*, 47, 65–139.
31. Christensen, T. S. (1996). Adiabatic prereforming of hydrocarbons — an important step in syngas production. *Applied Catalysis A: General*, 138(2), 285–309.
32. Rostrup-Nielsen, J. R., Christensen, T. S., and Dybkjaer, I. (1998). Steam reforming of liquid hydrocarbons. *Studies in Surface Science and Catalysis*, 113, 81–95.

33. Linde Engineering (2017). CO shift conversion. http://www.linde-engineering.com/en/process_plants/hydrogen_and_synthesis_gas_plants/gas_generation/co_shift_conversion/index.html.

34. Yang, L. and Ge, X. (2016). Biogas and syngas upgrading. In: *Advances in Bioenergy*, Y. Li and X. Ge, eds., Elsevier, Amsterdam, pp. 125–188.

35. van de Loosdrecht, J. and Niemantsverdriet, J. W. (2013). Synthesis gas to hydrogen, methanol and synthetic fuels. In: *Chemical Energy Storage*, R. Schloegl, ed., de Gruyter, Berlin.

36. Izquierdo, U., Barrio, V. L., Bizkarra, K., Gutierrez, A. M., Arraibi, J. R., Gartzia, L., Bañuelos, J., Lopez-Arbeloa, I., and Cambra, J. F. (2014). Ni and RhNi catalysts supported on zeolites L for hydrogen and syngas production by biogas reforming processes. *Chemical Engineering Journal*, 238, 178–188.

37. Effendi, A., Hellgardt, K., Zhang, Z. G., and Yoshida, T. (2005). Optimising H_2 production from model biogas via combined steam reforming and CO shift reactions. *Fuel*, 84(7–8), 869–874.

38. Giaconia, A., Monteleone, G., Morico, B., Salladini, A., Shabtai, K., Sheintuch, M., Boettge, D., Adler, J., Palma, V., Voutetakis, S., Lemonidou, A., Annesini, M. C., Exter, M. d., Balzer, H., and Turchetti, L. (2015). Multi-fuelled solar steam reforming for pure hydrogen production using solar salts as heat transfer fluid. *Energy Procedia*, 69, 1750–1758.

39. Angeli, S. D., Turchetti, L., Monteleone, G., and Lemonidou, A. A. (2016). Catalyst development for steam reforming of methane and model biogas at low temperature. *Applied Catalysis B: Environmental*, 181, 34–46.

40. Ross, J. R. H. (2005). Natural gas reforming and CO_2 mitigation. *Catalysis Today*, 100(1–2), 151–158.

41. Shekhawat II, D., Spivey, J. J., and Berry, D. A. (2011). *Fuel Cells: Technologies for Fuel Processing*, Elsevier, Amsterdam.

42. Dai, C., Zhang, S., Zhang, A., Song, C., Shi, C., and Guo, X. (2015). Hollow zeolite encapsulated Ni-Pt bimetals for sintering and coking resistant dry reforming of methane. *Journal of Materials Chemistry A*, 3(32), 16461–16468.

43. Rostrupnielsen, J. R. and Hansen, J. H. B. (1993). CO_2-reforming of methane over transition metals. *Journal of Catalysis*, 144(1), 38–49.

44. Pakhare, D. and Spivey, J. (2014). A review of dry (CO_2) reforming of methane over noble metal catalysts. *Chemical Society Reviews*, 43(22), 7813–7837.

45. Bradford, M. C. J. and Vannice, M. A. (1999). CO_2 reforming of CH_4. *Catalysis Reviews*, 41(1), 1–42.

46. Yang, C., Li, R., Cui, C., Liu, S., Qiu, Q., Ding, Y., and Wu, Y. (2016). Catalytic hydroprocessing of microalgae-derived biofuels: a review. *Green Chemistry*, 18(13), 3684–3699.

47. Ashcroft, A. T., Cheetham, A. K., Jones, R. H., Natarajan, S., Thomas, J. M., Waller, D., and Clark, S. M. (1993). An in situ, energy-dispersive x-ray diffraction study of natural gas conversion by carbon dioxide reforming. *The Journal of Physical Chemistry*, 97(13), 3355–3358.

48. Kapokova, L., Pavlova, S., Bunina, R., Alikina, G., Krieger, T., Ishchenko, A., Rogov, V., and Sadykov, V. (2011). Dry reforming of methane over $LnFe_{0.7}Ni_{0.3}O_{3-\delta}$ perovskites: influence of Ln nature. *Catalysis Today*, 164(1), 227–233.

49. Arbag, H., Yasyerli, S., Yasyerli, N., and Dogu, G. (2010). Activity and stability enhancement of Ni-MCM-41 catalysts by Rh incorporation for hydrogen from dry reforming of methane. *International Journal of Hydrogen Energy*, 35(6), 2296–2304.

50. Debek, R., Motak, M., Duraczyska, D., Launay, F., Galvez, M. E., Grzybek, T., and da Costa, P. (2016). Methane dry reforming over hydrotalcite-derived Ni-Mg-Al mixed oxides: the influence of Ni content on catalytic activity, selectivity and stability. *Catalysis Science & Technology*, 6(17), 6705–6715.

51. Özkara-Aydınoğlu, Ş., Özensoy, E., and Aksoylu, A. E. (2009). The effect of impregnation strategy on methane dry reforming activity of Ce promoted Pt/ZrO_2. *International Journal of Hydrogen Energy*, 34(24), 9711–9722.

52. Yan, Q., Lu, Y., To, F., Li, Y., and Yu, F. (2015). Synthesis of tungsten carbide nanoparticles in biochar matrix as a catalyst for dry reforming of methane to syngas. *Catalysis Science & Technology*, 5(6), 3270–3280.

53. Djinović, P., Črnivec, I. G. O., and Pintar, A. (2015). Biogas to syngas conversion without carbonaceous deposits via the dry reforming reaction using transition metal catalysts. *Catalysis Today*, 253, 155–162.

54. Bonura, G., Cannilla, C., and Frusteri, F. (2012). Ceria–gadolinia supported NiCu catalyst: a suitable system for dry reforming of biogas to feed a solid oxide fuel cell (SOFC). *Applied Catalysis B: Environmental*, 121–122, 135–147.

55. Kohn, M. P., Castaldi, M. J., and Farrauto, R. J. (2014). Biogas reforming for syngas production: the effect of methyl chloride. *Applied Catalysis B: Environmental*, 144, 353–361.

56. Li, X., Li, D., Tian, H., Zeng, L., Zhao, Z.-J., and Gong, J. (2017). Dry reforming of methane over Ni/La_2O_3 nanorod catalysts with stabilized Ni nanoparticles. *Applied Catalysis B: Environmental*, 202, 683–694.

57. Damyanova, S., Pawelec, B., Arishtirova, K., and Fierro, J. L. G. (2011). Biogas reforming over bimetallic PdNi catalysts supported on phosphorus-modified alumina. *International Journal of Hydrogen Energy*, 36(17), 10635–10647.

58. San-José-Alonso, D., Juan-Juan, J., Illán-Gómez, M. J., and Román-Martínez, M. C. (2009). Ni, Co and bimetallic Ni–Co catalysts for the dry reforming of methane. *Applied Catalysis A: General*, 371(1–2), 54–59.

59. Lai, M.-P., Lai, W.-H., Horng, R.-F., Chen, C.-Y., Chiu, W.-C., Su, S.-S., and Chang, Y.-M. (2012). Experimental study on the performance of oxidative dry reforming from simulated biogas. *Energy Procedia*, 29, 225–233.

60. Serrano-Lotina, A. and Daza, L. (2014). Influence of the operating parameters over dry reforming of methane to syngas. *International Journal of Hydrogen Energy*, 39(8), 4089–4094.

61. Parkhomenko, K., Tyunyaev, A., Martinez Tejada, L. M., Komissarenko, D., Dedov, A., Loktev, A., Moiseev, I., and Roger, A.-C. (2012). Mesoporous amorphous silicate catalysts for biogas reforming. *Catalysis Today*, 189(1), 129–135.

62. Gunduz Meric, G., Arbag, H., and Degirmenci, L. (2017). Coke minimization via SiC formation in dry reforming of methane conducted in the presence of Ni-based core–shell microsphere catalysts. *International Journal of Hydrogen Energy*, 42(26), 16579–16588.

63. Whang, H. S., Choi, M. S., Lim, J., Kim, C., Heo, I., Chang, T.-S., and Lee, H. (2017). Enhanced activity and durability of Ru catalyst dispersed on zirconia for dry reforming of methane. *Catalysis Today*, 293, 122–128.

64. Shang, Z., Li, S., Li, L., Liu, G., and Liang, X. (2017). Highly active and stable alumina supported nickel nanoparticle catalysts for dry reforming of methane. *Applied Catalysis B: Environmental*, 201, 302–309.

65. Barrai, F., Jackson, T., Whitmore, N., and Castaldi, M. J. (2007). The role of carbon deposition on precious metal catalyst activity during dry reforming of biogas. *Catalysis Today*, 129(3–4), 391–396.

66. Benito, M., García, S., Ferreira-Aparicio, P., Serrano, L. G., and Daza, L. (2007). Development of biogas reforming Ni-La-Al catalysts for fuel cells. *Journal of Power Sources*, 169(1), 177–183.

67. Zanganeh, R., Rezaei, M., and Zamaniyan, A. (2013). Dry reforming of methane to synthesis gas on NiO–MgO nanocrystalline solid solution catalysts. *International Journal of Hydrogen Energy*, 38(7), 3012–3018.

68. Lucrédio, A. F., Assaf, J. M., and Assaf, E. M. (2014). Reforming of a model sulfur-free biogas on Ni catalysts supported on Mg(Al)O derived from

hydrotalcite precursors: effect of La and Rh addition. *Biomass and Bioenergy*, 60, 8–17.

69. Bereketidou, O. A. and Goula, M. A. (2012). Biogas reforming for syngas production over nickel supported on ceria–alumina catalysts. *Catalysis Today*, 195(1), 93–100.

70. Tu, X. and Whitehead, J. C. (2012). Plasma-catalytic dry reforming of methane in an atmospheric dielectric barrier discharge: understanding the synergistic effect at low temperature. *Applied Catalysis B: Environmental*, 125, 439–448.

71. York, A. P. E., Xiao, T., and Green, M. L. H. (2003). Brief overview of the partial oxidation of methane to synthesis gas. *Topics in Catalysis*, 22(3), 345–358.

72. Al-Sayari, S. A. (2013). Recent developments in the partial oxidation of methane to syngas. *The Open Catalysis Journal*, 6, 17–28.

73. Christian Enger, B., Lødeng, R., and Holmen, A. (2008). A review of catalytic partial oxidation of methane to synthesis gas with emphasis on reaction mechanisms over transition metal catalysts. *Applied Catalysis A: General*, 346(1–2), 1–27.

74. Dybkjaer, I. (1995). Tubular reforming and autothermal reforming of natural gas — an overview of available processes. *Fuel Processing Technology*, 42(2), 85–107.

75. Araki, S., Hino, N., Mori, T., and Hikazudani, S. (2009). Durability of a Ni based monolithic catalyst in the autothermal reforming of biogas. *International Journal of Hydrogen Energy*, 34(11), 4727–4734.

76. Luneau, M., Gianotti, E., Meunier, F. C., Mirodatos, C., Puzenat, E., Schuurman, Y., and Guilhaume, N. (2017). Deactivation mechanism of Ni supported on Mg-Al spinel during autothermal reforming of model biogas. *Applied Catalysis B: Environmental*, 203, 289–299.

77. Song, C. and Pan, W. (2004). Tri-reforming of methane: a novel concept for catalytic production of industrially useful synthesis gas with desired H_2/CO ratios. *Catalysis Today*, 98(4), 463–484.

78. Izquierdo, U., Barrio, V. L., Requies, J., Cambra, J. F., Güemez, M. B., and Arias, P. L. (2013). Tri-reforming: a new biogas process for synthesis gas and hydrogen production. *International Journal of Hydrogen Energy*, 38(18), 7623–7631.

79. Vita, A., Pino, L., Cipitì, F., Laganà, M., and Recupero, V. (2014). Biogas as renewable raw material for syngas production by tri-reforming process over $NiCeO_2$ catalysts: optimal operative condition and effect of nickel content. *Fuel Processing Technology*, 127, 47–58.

80. Wikipedia (2017). Coal gasification. https://en.wikipedia.org/wiki/Coal_gasification.

81. US Department of Energy (2017). Gasification Systems. https://energy.gov/fe/science-innovation/clean-coal-research/gasification.

82. TERI (2017). Biomass Gasifier for Thermal and Power applications. http://www.teriin.org/technology/biomass-gasifier.

83. Worley, M. and Yale, J. (2012). Biomass Gasification Technology Assessment: Consolidated Report. National Renewable Energy Laboratory (NREL), Golden, CO., pp. 358. DOI: 10.2172/1059145.

84. Sutton, D., Kelleher, B., and Ross, J. R. H. (2001). Review of literature on catalysts for biomass gasification. *Fuel Processing Technology*, 73(3), 155–173.

85. Kirkels, A. F. and Verbong, G. P. J. (2011). Biomass gasification: still promising? A 30-year global overview. *Renewable and Sustainable Energy Reviews*, 15(1), 471–481.

86. Asadullah, M. (2014). Barriers of commercial power generation using biomass gasification gas: a review. *Renewable and Sustainable Energy Reviews*, 29, 201–215.

87. Zeng, K. and Zhang, D. (2010). Recent progress in alkaline water electrolysis for hydrogen production and applications. *Progress in Energy and Combustion Science*, 36(3), 307–326.

88. dos Santos, K. G., Eckert, C. T., de Rossi, E., Bariccatti, R. A., Frigo, E. P., Lindino, C. A., and Alves, H. J. (2017). Hydrogen production in the electrolysis of water in Brazil, a review. *Renewable and Sustainable Energy Reviews*, 68, Part 1, 563–571.

89. US Department of Energy (2017). Hydrogen Production: Thermochemical Water Splitting. https://energy.gov/eere/fuelcells/hydrogen-production-thermochemical-water-splitting.

90. Florida Solar Energy Center (FSEC) (2014). Solar Hydrogen Production. http://www.fsec.ucf.edu/en/consumer/hydrogen/basics/production-solar.htm.

Index

CATALYTIC SCIENCE SERIES

(Continued from page ii)